Microbial Diversity
Modern Trends

Microbial Diversity Modern Trends

R.C. Ram
Asha Sinha

2024

Daya Publishing House®
A Division of
Astral International Pvt. Ltd.
New Delhi – 110 002

First Published, 2007

Reprinted, 2024

ISBN: 9789359191577 (HB)

Disclaimer:

Every possible effort has been made to ensure that the information contained in this book is accurate at the time of going to press, and the publisher and author cannot accept responsibility for any errors or omissions, however caused. No responsibility for loss or damage occasioned to any person acting, or refraining from action, as a result of the material in this publication can be accepted by the editor, the publisher or the author. The Publisher is not associated with any product or vendor mentioned in the book. The contents of this work are intended to further general scientific research, understanding and discussion only. Readers should consult with a specialist where appropriate.

Every effort has been made to trace the owners of copyright material used in this book, if any. The author and the publisher will be grateful for any omission brought to their notice for acknowledgment in the future editions of the book.

Published by : **Daya Publishing House®**
A Division of
Astral International Pvt. Ltd.
– ISO 9001:2015 Certified Company –
4736/23, Ansari Road, Darya Ganj,
New Delhi-110 002
Ph. 011-43549197, 23278134
E-mail: info@astralint.com
Website: www.astralint.com

Preface

Microorganisms are widespread in nature and participate in a number of processes that ensure the continuation of life on earth. Microorganisms are relevant to all of us in multitude of says. Sometimes the influences of microorganisms in human life is beneficial where as at other time it is detrimental. India harbours a rich biodiversity. Though its area is only 2 per cent of the world landmass, it has over 5 per cent of all known species of plants and animals. Of the 30 millions or more species though to be occurring on earth, only 1.4 million have so far been named and described but information on species abundance and their functional attributes is altogether lacking. The task is more difficult for microbes which provide the basis for functional stability of ecosystems. An underlying theme of various contributors, Mycologist, Microbiologists, as well as Plant Pathologists to identify to various microorganisms from plants and soils. All the contributors have done there best to work about the microbial diversity.

This volume contains 25 reviews papers and research papers on the various areas of microbial diversity. It mainly includes fungal, bacterial and nematode diversity, mushrooms cultivation, bio-fertilizers, disease resistance, biological control, plant disease management, plant diseases and role of microbes in soil fertility. This will provide a modem trend in basic and applied research related to plants and microbial diversity. The papers of this volume are mostly invited ones from the eminent academicians from various scientists, teachers and researchers of various university and institutions. We are thankful to all the contributors for their authoritative and thought providing reviews and research papers for this volume. We will always welcome suggestions from both teachers and students for the improvement of possible subsequent edition. We are thankful to the publisher, Daya Publishing House, New Delhi.

R.C. Ram

Asha Sinha

Preface

Contents

©R.C. Ram & Asha Sinha (*Editors*)
***Microbial Diversity: Modern Trends*, 2007** **1–22**
Daya Publishing House, New Delhi

Chapter 1

Legume-Rhizobia Symbiosis in Sustainable Agriculture

A. Hemantaranjan
Department of Plant Physiology, Institute of Agricultural Sciences, Banaras Hindu University, Varanasi – 221 005, India

ABSTRACT

Legumes play a critical role in natural ecosystems, agriculture, and agroforestry, where their ability to fix N in symbiosis makes them excellent colonizers of low-N environments, and economic and environmentally friendly crop, pasture, and tree species. The legumes and their association with *Rhizobium* spp. in the broad sense have always been extremely important agronomically. Most of the research on rhizobia-legume symbioses has focused on papilionoid legumes and their symbionts, many of which have been selected for agronomic performance. Significance of certain micronutrients as well as growth regulators in efficient biological nitrogen fixation was established from our laboratory, which have great relevance in sustainable agriculture. One of the most exciting discoveries in the area of symbiotic plant-microbe interactions has been the identification of common genetic determinants underlying AM and *Rhizobium* symbioses. The legumes have given a special insight into the mycorrhizal association by enabling identification of genes required for this symbiosis. Further, nitrogen fixing bacteria that form stem- and root-nodules with legumes are all members of the Rhizobiaceae family and are currently subdivided into 6 genera and 27 species. The genera currently accepted are *Azorhizobium, Bradyrhizobium, Mesorhizobium, Allorhizobium, Rhizobium* and *Sinorhizobium*. There are some non-symbiotic species within these genera. Genetically improved strains of rhizobia that can be used in commercially available inoculums need to be developed. Constraints to N_2 fixation include drought, soil acidity, N fertilization, and nutrient limitations. Many cultivars also

E-mail: hemantaranja@satyam.net.in

show only limited ability to fix N_2 in symbiosis. There is a crucial need to increase drought tolerance in legumes; increasing salinity tolerance is a parallel requirement in many areas. Evidence is accumulating that over-expression of selected individual genes involved in N and P acquisition can improve nutrient uptake. Flavonoids and/or isoflavonoids released from the root of the legume host induce transcription of nodulation genes in compatible rhizobia, in turn; signal the host plant to begin nodule formation. The origin of the *nodA, nodB,* and *nodC* genes therefore may be crucial.

Studies on antioxidants in the rhizobia-legume symbiosis are certainly an important area of research which needs to be elaborated. The complete reduction of molecular oxygen to water requires four electrons and is catalyzed by cytochrome oxidase in aerobic bacteria and mitochondria. ROS are produced at high rates when plants are exposed to abiotic, biotic, or xenobiotic stress. Similarly, the term reactive nitrogen species (RNS) refers to nitrogen-derived molecules with variable reactivity and includes free radicals (nitric oxide) and non-free radicals (peroxynitrite). GSH is important for stress tolerance and the symbiotic process. Glutathione reductases are found in the cytosol, mitochondria, and bacteroids. The CuZnSOD may be associated with cell wall growth in the meristems, infection threads, and apoplast, and with the plant's response to bacterial infection. Last but not the least, identification of active chemical compounds establishes the existence of a potentially important molecular link between the heterotrophic soil food web and the autotrophic plant. Amino acid exudation under normal conditions is a phenomenon that probably reflects both active manipulation and passive uptake by microorganisms, as well as diffusion and adsorption to soil, all of which help overcome the innate capacity of plant roots to reabsorb amino acids.

In this review, it has been endeavoured to assemble important literature with reference to legume-rhizobium symbiosis informing critically the significance and relationship of fungi, mycorrhiza and the whole of rhizobiaceae in the nodulation followed by the biological nitrogen fixation, which well covers the topics of phosphorus acquisition, flavonoids as signals, nod factors and nodulation genes at length with subsequent coverage over the important aspects of the molecular biology and biochemistry of antioxidants relevant to the rhizobium-legume symbiosis elaborating major antioxidants of nodules including SODs and catalase in the protection of nodulation and last but not the least illustrating specifically that microbial products trigger amino acid exudation from plant roots.

Introduction

The world is on the brink of a new agriculture, one that involves the marriage of plant biology and agroecology under the umbrella of biotechnology and germplasm improvement. Although N and P fertilizers will continue to play a major role in intensive agriculture, depletion of natural resources, loss of biodiversity, and long-term unsustainability necessitate alternative strategies be investigated and implemented to buffer against food insecurity and environmental degradation. Legumes play a critical role in natural ecosystems, agriculture, and agroforestry, where their ability to fix N in symbiosis makes them excellent colonizers of low-N environments, and economic and environmentally friendly crop, pasture, and tree species. Legume yields unfortunately continue to lag behind those of cereals. A research orientation that better balances the needs of third-world or sustainability-oriented agriculture with the breakthrough technologies of genomics and bioinformatics is needed. It requires stronger and more adventurous breeding programs, better use of marker-assisted technologies, and emphasis on disease resistance, enhanced N fixation, and tolerance to edaphic soil constraints. It also requires extension of existing low-cost technologies, such as rhizobial inoculation, to the small farmer.

Research on the physiology and genetics of symbiotic nitrogen fixation has recently shown tremendous progress thanks to the application of molecular biology methods to the study of this very important process. Nitrogen fixation converts dinitrogen gas (N_2) into plant useable ammonium. The nitrogen fertilizer industry has been converting N_2 to useable nitrate or ammonium plant fertilizers for many years. Unfortunately, this process requires a tremendous amount of energy and, therefore, is quite expensive to produce, but necessary for the farmer to use in order to produce high yield per acre of crop plants. Additionally, the manufacture and application of fertilizers pollutes the environment.

Bean (*Phaseolus vulgaris*) and soybean (*Glycine max*), staple crops in the Americas and Asia, respectively, were each domesticated more than 3,000 years ago (Hymowitz and Singh, 1987; Kaplan and Lynch, 1999). Use of legumes in pastures and for soil improvement dates back to the Romans, with Varro (37 BC; cited by Fred *et al.* 1932) noting "Legumes should be planted in light soils, not so much for their own crops as for the good they do to subsequent crops."

Under subsistence conditions, the percentage of legume protein N in the diet can reach twice this figure. In rank order, bean, pea (*Pisum sativum*), chickpea (*Cicer arietinum*), broad bean (*Vicia faba*), pigeon pea (*Cajanus cajan*), cowpea (*Vigna unguiculata*), and lentil constitute the primary dietary legumes (National Academy of Science, 1994). Legumes [predominantly soybean and peanut (*Arachis hypogeae*)] provide more than 35 per cent of the world's processed vegetable oil, and soybean and peanut are also rich sources of dietary protein for the chicken and pork industries. Unfortunately, improvement in legume crop yields has not kept pace with those of cereals.

Drought problems for legumes are likely to worsen with the projected rapid expansion of water-stressed areas of the world from 28 to 30 countries today to 50 countries encompassing 3 billion people by 2030 (Postel, 2000). There is a crucial need to increase drought tolerance in legumes; increasing salinity tolerance is a parallel requirement in many areas. The more drought-tolerant legumes, such as cowpea, are deeply rooted and may have reduced leaf size with thickened cuticles to reduce water loss. Less tolerant legumes such as beans can be selected for early maturity, efficiency in the partitioning of nutrients toward reproductive structures, and phenotypic plasticity (Beaver *et al.*, 2003).

For the past 20 years, rhizobial and legume biologists have pursued a scientific investigation based on this biodiversity for the purposes of understanding the complexities of the agriculturally and environmentally important nitrogen-fixing symbiosis epitomized by nodulation. By recognizing that the evolution of the rhizobia-legume symbiosis is more akin to an interwoven tapestry than to a continuous thread, we may have a better chance of understanding the uniqueness of this association. Thus, genome projects and scientific pursuits that include a diversity of legumes and rhizobial species will better inform us as to which genes/proteins are conserved among all hosts and symbionts and help us determine whether the ability to fix N_2 into ammonia can be transferred to crops other than legumes. The "Holy Grail" awaits.

On the reverse end of the spectrum, not all members of the legume family nodulate. Of the three different subfamilies of legumes–Caesalpinoideae, Mimosoideae, and Papilionoideae–members of the basal subfamily, Caesalpinoideae, are mostly non-nodulating (Nod^-). Thus, nodulation and presumably nitrogen-fixing ability are not 100 per cent correlated even within the legume family. Nodulation may have originated multiple times in the Fabaceae: once in the only caesalpinoid that is confirmed to be nodulated, *Chamaecrista* genus; once in the mimosoid line; and lastly, at the base of the papilionoid line (Doyle, 1998). Alternatively, there may have been a single origin of nodulation with multiple losses. In any case, more than 90 per cent of the Papilionoideae and Mimosoideae are nodulated, whereas less than 25 per cent of the Caesalpinoideae form nodules. Moreover, other plant families can establish interactions with nitrogen-fixing bacteria exclusive of rhizobia.

The legumes and their association with *Rhizobium* spp. in the broad sense have always been extremely important agronomically. The use of crop rotations to enhance the productivity of non-legume crops was vividly described by the Romans, who were probably aware of an even older tradition in Greece. In this review, it has been endeavoured to assemble important literature with reference to legume-rhizobium symbiosis informing critically the significance and relationship of fungi, mycorrhiza and the whole of rhizobiaceae in the nodulation followed by the biological nitrogen fixation, which will covers the topics of phosphorus acquisition, flavonoids as signals, nod factors and nodulation genes at length with subsequent coverage over the important aspects of the molecular biology and biochemistry of antioxidants relevant to the rhizobium-legume symbiosis elaborating major antioxidants of nodules including SODs and catalase in the protection of nodulation and last but not the least illustrating specifically that microbial products trigger amino acid exudation from plant roots.

Fungal Footprints in the Nodulation

Members of eight different families, known as actinorhizal plants, are nodulated by *Frankia* spp. nitrogen-fixing actinomycetes. Various grasses, including such agronomically important ones as sugarcane (*Saccharum officinarum*), maize (*Zea mays*), and rice (*Oryza sativa*), associate with different nitrogen-fixing bacteria, among them species of *Glucoacetobacter, Azospirillum, Herbaspirillum,* and *Azoarcus;* these associations do not, however, result in nodule formation. Some of the positive responses on plant growth exhibited by these so-called "associative" nitrogen-fixing interactions are due to the production of phytohormones, but nitrogen fixation has also been demonstrated (Sevilla *et al.*, 2001). Nevertheless, what makes rhizobia and *Frankia* spp. different from the associative nitrogen-fixing microbes is that most of the rhizobia or *Frankia* spp.-fixed nitrogen is transferred to and assimilated by the plant for the plant's growth.

One of the most exciting discoveries in the area of symbiotic plant-microbe interactions has been the identification of common genetic determinants underlying AM and *Rhizobium* symbioses (Duc *et al.*, 1989). The initial molecular and genetic observations support a hypothesis that the mechanism underlying the compatible response of legume roots to rhizobial infections may have evolved in part from a pre-existing pathway that regulates the more ancient AM symbiosis (Albrecht *et al.*, 1999).

It is well known that the rhizobia-legume interaction falls into cross inoculation groups, whereby certain rhizobial strains nodulate only certain legumes. For example, *Sinorhizobium meliloti* effectively nodulates species of *Medicago, Melilotus,* and *Trigonella,* whereas *Rhizobium leguminosarum* bv *viciae* induces nitrogen-fixing nodules on *Pisum, Vicia, Lens,* and *Lathyrus* spp. Closely related to the pea (*Pisum sativum*) strain is *R. leguminosarum* bv *trifolii,* which initiates nodules only on species of clover (*Trifolium*). New molecular data indicate the presence of common plant genetic elements underlying parasitic nematodes and rhizobial endosymbioses (Koltai *et al.*, 2001). Parallel insight into symbiont-specific plant functions may provide the long-awaited answer to the question regarding the unique evolutionary event(s) that predisposed legume plants and *Parasponia* to form nitrogen-fixing nodular symbiosis with rhizobia.

The orthologous genes *LjSYMRK* and *NORK,* recently characterized from *Lotus japonicus* (Stracke *et al.*, 2002) and alfalfa (*Medicago sativa*) (Endre *et al.*, 2002), respectively, may represent footprints of such pathway(s). These genes were identified in genetic screens for plant mutants resistant to both rhizobial (Nod$^-$ or non-nodulation phenotype) and AM (Myc$^-$ or non-mycorrhizal phenotype) infections and were subsequently cloned and used to characterize the orthologous *sym19* and *dmi2* loci from pea (*Pisum sativum*) and barrel medic (*Medicago truncatula*), respectively. SYMRK/NORK function is required early on during symbiotic interaction for root hair curling, and *nork* mutants were found to be incapable of performing calcium spiking, one of the earliest root hair responses to Nod factors and rhizobial infection. In contrast to *legume-Rhizobium* symbiosis, signaling events during AM symbiosis are poorly understood and the nature of a hypothetical AM factor(s) that could activate SYMRK/NORK-dependent signaling is unknown. By analogy to Nod factors, chitin-like molecules derived from the fungal cell wall represent good candidates for such factors. However, whether SYMRK/NORK receptor kinase represents a component of a perception apparatus or works downstream from a symbiont-specific perception mechanism remains an open question. Considering that pharmacological studies implicate heterotrimeric G-proteins and phosphoinositide (PI)-mediated signaling as downstream elements of the Nod factor transduction pathway (Pingret *et al.*, 1998), it will be interesting to see if SYMRK/NORK work as G-protein-coupled receptors. Characterization of the remaining loci required for both symbioses will undoubtedly expose the signaling mechanisms involved and define to what extent the species-specific nodular symbiosis has "learned" from the evolutionarily much older AM symbiosis of plants and fungi.

The Mycorrhizal Relation

Since the original coupling of Nod^- and Myc^- (inability to establish a mycorrhizal association) in pea and *Vicia faba* mutants by Duc *et al.* (1989), a number of papilionoid legumes that normally nodulate have been shown to be both Nod^- and Myc^-. A Myc^- mutant allele has also been described for bean. The various mutants can be described as those that are blocked very early in the signaling cascade (before Ca^{2+} spiking) and those blocked later (after Ca^{2+} spiking). Those Nod^- Myc^- legume mutants are blocked before Ca^{2+} spiking presumed to be altered in a receptor that is common to the mycorrhizal and nodulation pathways (*Pssym8, Pssyml9, Mtdmil, Mtdmi2,* and MN1008). A group of mutants, *Masym3* and the *L. japonicus sym* mutants, have not been tested yet for Ca^{2+} spiking, or if they have, not all mutant alleles have been examined. Therefore, it is not known whether they are blocked before or after Ca^{2+} spiking. In addition, the results on the *L. japonicus sym* mutants have been generated from several different laboratories, so it is not yet known whether some of the mutations are allelic. *Pssym10, Ljsym1, Ljsym5,* and *Ljsym70* mutants are Nod^-Myc^+ and presumably blocked before Ca^{2+} spiking (to date, the *Ljsym* mutants have not been tested) and upstream of the NORK-type receptor. This upstream gene(s) could encode a Nod factor-binding protein and/or a nodulation specific signaling protein. Also, many legume mutants that are very likely blocked after Ca^{2+} spiking, but before infection thread formation, have been found. They include those mutated in *NIN,* which has been characterized as a transcription factor (Schauser *et al.*, 1999). Others blocked after infection thread development may have mutations in genes that encode elements of the signal transduction pathway leading to nodule morphogenesis.

The connection between mycorrhizal development and nodulation as well as the fact that the early nodulin (*ENOD*) genes are expressed in both symbioses suggest that nodulation may have evolved from the more ancient mycorrhizal condition. The legumes have given a special insight into the mycorrhizal association by enabling identification of genes required for this symbiosis. Of course, it is predicted that other mutations are likely to affect the mycorrhizal symbiosis, but not nodulation. However, such mutants have yet to be described. The mycorrhizal association is believed to have originated more than 400 MYA based on fossil evidence, and so it is possible that the nodulation symbiosis may have adapted some components of a much older symbiotic pathway.

The Rhizobiaceae

Nitrogen fixing bacteria that form stem- and root-nodules with legumes are all members of the Rhizobiaceae family and are currently subdivided into 6 genera and 27 species. The genera currently accepted are *Azorhizobium, Bradyrhizobium, Mesorhizobium, Allorhizobium, Rhizobium* and *Sinorhizobium.* There are some non-symbiotic species within these genera. When the term rhizobia is used, it refers to those bacterial species within these six genera known to be symbiotic bacteria capable of eliciting and invading root and stem nodules on legume plants and differentiating into nitrogen fixing bacteroids (van Rhijin and Vanderleyden, 1995). Rhizobia show tremendous ability to adapt to very different environments found in the rhizosphere area of different legumes. For example, it has been shown by various researchers that soils around legumes contain a rather large population of rhizobia not capable of establishing a symbiotic relationship. These rhizobia can occupy the soil in the rhizospheres of legumes and other plants (Schloter *et al.*, 1997 and Saito *et al.*, 1998). These non-symbiotic rhizobia can become symbiotic by the horizontal acquisition of a symbiotic plasmid or a chromosomal symbiotic island (Sullivan and Roson, 1998), which allows them to synthesize and secrete lipo-chitin Nod signals that are host-specific and required for the establishment of an effective nitrogen-fixing symbiotic relationship (Spaink, 2000). Interestingly, viable rhizobial cells have been found in water where they can establish an effective symbiotic relationship with aquatic legumes of the genera *Aeschynomene* and *Sesbania* (Chaintreuil *et al.*, 2000; Wang and Martinez-Romero, 2000). Some of these legumes present a special case where nodules are formed on the stem, as well as the roots.

The classical nitrogenases have been extensively studied in *Azotobacter vinelandii* (an obligate aerobe), *Clostridium pasteurianum* (an obligate anaerobe), *Klebsiella pneumonia* (a facultative anaerobe), *Rhodospirillum rubrum* (a photosynthetic bacterium), *Anabaena* sp. 7120 (a heterocyst-forming cyanobacterium), and *Bradyrhizobium japonicum* (a symbiotic bacteria). Two alternative nitrogenases were flrst reported in *Azotobacter* and subsequently found in a variety of diazotrophs (Loveless and Bishop, 1999). One of these nitrogenase contains vanadium, in place of molybdenum, while the other contains only iron. Most recently, a unique nitrogen fixing system was reported in *Streptomyces thermoautotrophicus,* a thermophilic, aerobic, chemolithoautotrophic bacterium. This system couples carbon monoxide oxidation to superoxide reduction and subsequent N_2 reduction by a molybdenum nitrogenase (Ribbe *et al.*, 1997).

Tremendous progress has been made in recent years in understanding the molecular basis for this fascinating symbiosis. However, there are still gaps in our knowledge (*e.g.*, host recognition and preferential rhizobial species selection, ultrastructure and function of the nodule and symbiosome, plant cell receptors, Nod-signal action, etc.), which need to be filled before we can enjoy the full benefits of legume-rhizobia symbiosis. Nonetheless, there are many genes in rhizobia involved in this symbiotic process that have yet to be identified and studied. The recent application of genomics to this field promises to accelerate the pace of discovery. In 1997, Frieberg *et al.* reported the full sequence of the symbiotic plasmid of *Rhizobium* sp. NGR234 and recent reports have revealed the entire genomic sequence of *Mesorhizobium loti* (Kaneko *et al.*, 2000) and *Sinorhizobium meliloti* (Galibert *et al.*, 2001). Currently, on-going sequencing projects promise the full sequence of *B. japonicum, R. leguminosarum* and *R. etli.* These developments will usher in the age of functional genomics (DNA microarray analysis, proteomics, etc.), which will ultimately identify all the bacterial genes essential for the nodulation process. In addition to the genes directly involved in the nodulation process, other genes involved in symbiotic effectiveness and/or competitiveness of rhizobial strains need to be studied. Genetically improved strains of rhizobia that can be used in commercially available inoculums need to be developed. According to Maier and Tripplet (1996), three obstacles remain before this can be a reality.

First, genes that specifically enhance nitrogen fixation need to be isolated. Second, genes that enhance the ability of the innoculum strain of rhizobia to successfully compete against indigenous strains of rhizobia need to be found. Third, a stable plasmid, which stays in the rhizobial cell and can carry/ express these genes needs to be developed. Progress has been made in this area although many obstacles still remain.

Biological Nitrogen Fixation

Most of the research on rhizobia-legume symbioses has focused on papilionoid legumes and their symbionts, many of which have been selected for agronomic performance. Significance of certain micronutrients as well as growth regulators in efficient biological nitrogen fixation in French bean (Hemantaranjan and Garg, 1986; Hemantaranjan, 1988; Hemantaranjan, 1991; Hemantaranjan, 1994a) and soybean (Hemantaranjan, 1994b; Hemantaranjan, 1996; Hemantaranjan, 1999) were established from our laboratory, which have great relevance in sustainable agriculture. Similar studies with growth regulators and micronutrients are needed to be investigated for efficient biological nitrogen fixation that could maintain a long nodule life avoiding an early senescence in all pulse crops for supplementing optimum amount of fixed nitrogen to crop plants. That is also the prime need of time to maintain soil integrity on the one hand and economizing the use of synthetic nitrogenous fertilizers manufactured at the cost of fossil fuels. Nevertheless, there is also the task that grain legume cultivars must be duly screened regarding their performance either under natural abiotic stress conditions or through inducing stresses like salinity, drought and heavy metal pollution with a view to select susceptible and tolerant genotypes for making extensive physiological and biochemical studies to find out the limiting factors in terms of biomarkers for legume-rhizobium symbiosis and plant productivity resulting from specific stresses, since pulses are highly sensitive to nodulation under water and salinity stresses. The stepwise investigations concerning applied aspects–as far as correcting the field soils, economic use of N-fertilizers, nutrient availability etc. are concerned and scientific aspects, as has been seriously going on extensively at global scale to critically understand the intricacies in legume-rhizobium symbiosis, will certainly bring about revolution in sustainability of agriculture. That would be, no doubt, in addition to many other areas of research struggling hard to achieve success in the same theme.

The establishment of the nitrogen-fixing symbiosis between rhizobia and legumes is a multistep process that involves the differentiation of root cortical cells to form a new organ (called the root nodule), the adhesion of the rhizobia to root hairs, and the subsequent internalization and transport of the rhizobia to the root nodule along infection threads produced by the plant Formation of symbiotically effective root nodules involves signaling between host and microsymbiont. Additional P is needed for signal transduction and membrane biosynthesis. Phosphorus concentrations in the nodule are often significantly higher than those in shoot or root tissue (Israel, 1987). Flavonoids and/ or isoflavonoids released from the root of the legume host induce transcription of nodulation genes in compatible rhizobia, leading to the formation of lipochitooligosaccharide molecules that, in turn, signal the host plant to begin nodule formation (Long, 1996). Numerous changes occur in host and bacterial gene expression during infection, nodule development, and function (Vance, 2002), with approximately 100 host legume and rhizobial genes involved.

A hallmark trait of legumes is their ability to develop root nodules and to fix N_2 in symbiosis with compatible rhizobia. This is often a critical factor in their suitability for the uses. Legume and actinorhizal nodules can be indeterminate, growing by means of an apical meristem, but determinate nodules, those lacking a persistent apical meristem, are only found in legumes. Moreover, some legumes such as lupins or *Sesbania rostrata* develop nodules that fall into an intermediate category. Unlike the lateral root that is initiated from cell divisions in the pericycle, the legume nodule originates from cell divisions in the outer or inner cortex, depending on whether a determinate or indeterminate nodule is formed. One of the driving forces behind agricultural sustainability is effective management of N in the environment (Graham and Vance, 2000). Application of fertilizer N increased approximately 10-fold to 90 million Mt between 1950 and 1995 (Frink *et al.*, 1999) with significant energy consumption for N fertilizer synthesis and application. Further increases in N needs for agriculture are projected for the period to 2030 (Tilman, 1999), and these needs will contribute to environmental pollution. To the extent that farming practices can make use of the more economically viable and environmentally prudent N_2 fixation (Peoples *et al.*, 1995; Vance, 2001), agriculture and the environment will benefit. The ability of legumes to sequester C has also been seen as a means to offset increases in atmospheric CO_2 levels while enhancing soil quality and tilth.

Giller (2001) suggests that rates of N_2 fixation of 1 to 2 kg N ha^{-1} growing season day^{-1} should be possible in all legumes. Rates reported by Unkovich and Pate (2000) and van Kessel and Hartley (2000) are clearly less than this, with the latter authors reporting a decline in average N_2 fixation rate for both soybean and beans over the period since 1985. Constraints to N_2 fixation include drought (Sinclair *et al.*, 1987), soil acidity, N fertilization, and nutrient limitations. Many cultivars also show only limited ability to fix N_2 in symbiosis.

Management of soil acidity for temperate and tropical regions has often differed but increasingly depends on acid tolerant legume cultivars and rhizobia (Howieson *et al.*, 2000), with soil liming only to a pH at which Al and Mn are no longer toxic. Identification of additional acid-tolerant host and rhizobial germplasm and the deployment of acid tolerance genes such as occur in *R. tropici* CIAT899 (Graham *et al.*, 1982) are priority areas.

Phosphorus Acquisition in Nitrogen Fixation

Plants dependent on symbiotic N_2 fixation have ATP requirements for nodule development and function (Ribet and Drevon, 1996) and need additional P for signal transduction and membrane biosynthesis. Phosphorus concentrations in the nodule are often significantly higher than those in shoot or root tissue (Israel, 1987). Al-Niemi *et al.* (1997) suggest that bacteroids can be P limited even when plants have received otherwise adequate P levels. Given this requirement for symbiosis, approaches leading to improved P acquisition and use in legumes (rhizosphere acidification, acid phosphatase secretion, root architectural changes at low P, enhanced P transport and use-efficiency, and functional differences in mycorrhizal symbioses) all need additional study. White lupine (*Lupinus albus*) and common bean are excellent model legumes for such studies. Each undergoes change in root architecture and rhizosphere chemistry at low P (Johnson *et al.*, 1996; Nielsen *et al.*, 1998; Miller *et al.*, 2001), improving soil exploration and phosphate scavenging. Transgenic alfalfa has proved extremely useful in understanding the genetic and molecular basis of low soil P, acid, and aluminium stress responses (Tesfaye *et al.*, 2001). Results in this study highlight the need for more effective transformation and regeneration protocols in the more recalcitrant legumes, including bean and cowpea. Progress in the study of nodulation and N_2 fixation under drought or salinity stress has been minimal, largely because the legume and the process of nodulation are more susceptible to these constraints than is the microorganism.

P is second only to N as the most limiting element for plant growth (Bieleski, 1973; Vance *et al.*, 2000). The amount of Pin plants ranges from 0.05 per cent to 0.30 per cent of total dry weight. The concentration gradient from the soil solution to the plant cell exceeds 2,000-fold, with an average free P of 1 μM in the soil solution (Bieleski, 1973; Ragothama, 1999). This concentration is well below the

Km for plant uptake. Thus, although bound P is quite abundant in many soils, it is largely unavailable for uptake (Bieleski, 1973; Schactman *et al.*, 1998). Crop yield on 40 per cent of the world's arable land is limited by P availability. P is unavailable because it rapidly forms insoluble complexes with cations and is incorporated into organic matter by microbes. The acid-weathered soils of the tropics and subtropics are particularly prone to P deficiency and aluminum (Al) toxicity (von Uexküll and Mutert, 1995). In intensive agriculture, a grain crop yield of 7 metric tons Ha^{-1} requires the addition of 90 to 120 kg P ha^{-1} (Bieleski, 1913). However, even under adequate P fertilization, only 20 per cent or less of that applied is removed by the 1st year's growth. This results in P loading of prime agricultural land. Runoff from P-loaded soils is a primary factor in eutrophication and hypoxia of lakes and marine estuaries in the developed world.

Sustainable management of P in agriculture requires that plant biologists discover mechanisms in plants that enhance P acquisition and exploit these adaptations to make plants more efficient at acquiring P, develop P-efficient germplasm, and advance crop management schemes that increase soil P availability. In extensive agriculture of the developing world, von Uexküll and Mutert (1995) recommend that the first phase of reclaiming acid soils low in P and N is the use of a legume cover crop supported by liming and conservative P application. Incorporation of the legume residue results in higher soil organic matter content and increased P and N availability. Snapp (1998) identified long duration legumes having indeterminate growth and low harvest index as critical to improving P and N sustainability in small holders and subsistence agriculture in Africa. On the Indian subcontinent and in other semiarid regions having weathered soils, intercropping of pigeon pea (*Cajanus cajan*) with cereals is a proven management scheme for increasing available N and P. In India, more than 90 per cent of pigeon pea production is in intercropping. Ae *et al.* (1990) demonstrated that this success is due, at least in part, to exudation of piscidic acid from roots of pigeon pea, which enhances the availability of phosphate from iron-phosphate (Fe-PO_4). Mycorrhizal inoculation of pigeon pea further improved P uptake in this study. Thus, in addition to providing an immediate source of dietary N, incorporation of pigeon pea residues after seed harvest make P and N more available to a subsequent crop.,

Evidence is accumulating that over-expression of selected individual genes involved in N and P acquisition can improve nutrient uptake. As the arsenal of plant genes involved in these processes expands, we will undoubtedly see more successful examples. Because of the urgent need for plant germplasm having improved N and P use efficiency, research programs that combine traditional plant breeding and transgenic technology will be imperative. The following recommendations deserve attention:

1. Re-emphasize the use of legumes and symbiotic N_2 fixation to improve soil N and P fertility while reducing fossil fuel consumption and providing a source of dietary N;
2. Develop intercropping schemes that foster efficient N and P use;
3. Continue to isolate, characterize, and develop fundamental understanding of individual genes holding promise of application to improving N and P use;
4. Enhance the expression of genes and increase the synthesis of gene products, such as those involved in transport of nutrients and exudation of organic acids, through both traditional plant breeding and transgenic technology and incorporate these traits into adapted germplasm; and
5. Assess the factors limiting rhizobial and mycorrhizal interactions with plants with the goal of site- (region) specific inoculation.

Flavonoids as Signals of Nodule Development

Flavonoids and/or isoflavonoids released from the root of the legume host induce transcription of nodulation genes in compatible rhizobia, leading to the formation of lipochitooligosaccharide molecules that, in turn, signal the host plant to begin nodule formation (Long, 1996). More than 4,000 different flavonoids have been identified in vascular plants, and a particular subset of them is involved in mediating host specificity in the legumes (Perret *et al.*, 2000). All flavonoids consist of two benzene rings linked through a heterocyclic pyran or pyrone ring. Specific substitutions on the ring produce flavonols, flavones, flavanones, as well as isoflavonoids, which are derived from a migration of the B ring from the 2 to the 3 positions.

Isoflavonoids are restricted to the legume family. Isoflavonoids, daidzein and genistein, produced by soybean (*Glycine max*), are effective inducers of *Bradyrhizobium japonicum nod* genes, but inhibit *S. meliloti nod* gene expression. *S. meliloti nod* genes can be induced by luteolin. This specificity enables rhizobia to distinguish their hosts from other legumes. The specific flavonoid not only induces *nod* gene expression, but also rhizobial chemotaxis. However, other than the isoflavones, most flavonoids are not unique to legumes. It appears that, it is the next stage, once the flavonoids are perceived, where another level of specificity comes into play. Flavonoids are perceived as aglycones, which induce rhizobial *nod* genes by interacting with the gene product of *nodD*, a LysR-type regulator. This interaction results in a conformational change in the NodD protein such that it binds to *nod* box elements in the promoters of the *nod* genes (Perret *et al.*, 2000). The intensive expression of these genes leads to the synthesis of Nod factor molecules–lipochitooligosaccharides that usually consist of four or five *N*-acetylglucosamines, β-1-4 linked, with the terminal nonreducing sugar N-acylated with a fatty acid of 16 to 18 carbon residues. Nod factors can be chemically modified with acetate, sulfate, or carbamoyl groups, or can have different sugars, such as Ara, Man, Fuc, or substituted Fuc. The degree of saturation of the acyl tail can also vary (Perret *et al.*, 2000). The assemblage of these substitutions results in a specific Nod factor that is recognized by a particular legume.

Sensitivity of Legumes: Nod Factor

Legumes are unique in their response to Nod factors in that they actively promote entry of bacteria into the root. However, as yet we do not know how legumes evolved the ability to recognize such signals or how entry is actually accomplished. Attention must be given to the broader aspects of the legume-rhizobia association. Darwin's revolution of biology could not have occurred without the unrestrained view he had of the organisms around him. Variation among individuals gave Darwin the insight to understand the origin of species.

One of the key traits that differentiate the nodulating legumes from other plant species is their sensitivity to Nod factor. Early responses to Nod factor include ion flow across the plasma membrane and an associated depolarization of the membrane, followed by periodic oscillations in intracellular calcium, referred to as "calcium spiking"; these are followed by deformation of root hairs and initiation of cortical cell division (Downie and Walker, 1999). Root hair curling (which involves entrapment of the bacteria) and infection thread growth require the presence of the bacteria. Nodulation appears to have an absolute requirement for Nod factor because rhizobia that do not synthesize Nod factor do not nodulate, and legume mutants that are incapable of perceiving Nod factor or transducing it along a signal transduction pathway are Nod$^-$ (Downie and Walker, 1999). The identity of a Nod factor receptor(s) in legumes is unknown, but research is well under way (Cullimore *et al.*, 2001).

A biochemical approach has led to the characterization of high-affinity binding sites for Nod factors. One of these, NFBS2, is located in the plasma membrane and exhibits different selectivities for

Nod factors in alfalfa (*Medicago sativa*) and bean (*Phaseolus vulgaris;* Gressent *et al.,* 1999). In *Dolichos biflorus,* an unusual lectin with Nod factor-binding activity has been characterized (Etzler *et al.,* 1999). This protein has an apyrase activity and has been named lectin nucleotide phosphohydrolase. It is not clear, however, whether Nod factor-binding proteins are unique to legumes. Furthermore, to date, there is no evidence for linking a Nod factor-binding protein to a Nod$^-$ mutation.

The study of Nod$^-$ plant mutants has also yielded leads for identifying the proteins involved in Nod factor perception and signal transduction. Many Nod$^-$ mutants have been identified in commercially and agronomically important legumes such as pea, bean, alfalfa, sweet clover, and others, and more recently, model legumes such as *Medicago truncatula* and *Lotus japonicus* have been used for genetic studies. Transposon tagging in *L. japonicus* led to identification of *NIN* (nodule inception), which encodes a transcription factor, and the first cloned gene that is directly involved in nodule development (Schauser *et al.,* 1999). *nin* mutants showed root hair curling in response to rhizobia, but did not develop infection threads or nodules. *NORK* encodes a Leu-rich repeat kinase that could be a receptor, but so far it has not been shown whether or not it directly interacts with Nod factor. It is possible that NORK could interact with a Nod factor-binding protein via its Leu-rich repeats.

Nodulation Genes

The ability to fix atmospheric N_2 is very widespread among bacteria and Archaea, although interestingly, this capacity is restricted to prokaryotes. Therefore, there are many different diazotrophs that, if equipped with the ability to invade plants, could theoretically evolve to establish a nitrogen-fixing symbiosis. One change, however, would be to uncouple the regulation of nitrogen fixation in plants from the microbial requirement for fixed nitrogen, something rhizobia have done very efficiently.

As mentioned previously, there are many bacteria that grow endophytically within plants, but what distinguishes the rhizobia is their ability to make "Nod factors," molecules required to program the specialized infection process and nodule morphogenesis. The biosynthesis of Nod factors has been thoroughly reviewed (Perret *et al.,* 2000). Although Nod factors can carry many substituents, which are important for nodulating specific legumes, their basic structure requires the action of only three gene products, NodA, NodB, and NodC. NodC is an N-acetyl glucosaminyl transferase that produces the chitin backbone from UDP-N-acetyl glucosarnine. NodB removes an acetyl group from the terminal residue of the chitin oligomer, and NodA catalyzes the transfer of a fatty acyl chain onto the resulting free amino group using acyl-ACP from fatty acid biosynthesis.

The origin of the *nodA, nodB,* and *nodC* genes therefore may be crucial. It is likely that they came from outside, the Rhizobiaceae because, like most of the nodulation and nitrogen fixation genes, they have a G + C content that is significantly lower than the average G + C content of rhizobia; they also have a different codon usage from most chromosomal genes (Galibert *et al.,* 200 I). NodC is one of a large class of bacterial β-glucosyl transferases, many of which can incorporate N-acetyl glucosamine into cell wall polysaccharides. For example, *Streptococcus pyogenes* produces a polymer of alternating β-inked GlcNAc and GlcUA. Furthermore, the peptidoglycan of many bacteria is composed of a backbone of alternating β-14-linked GlcNAc and N-acetyl muramic acid (which is the lactic acid ether of GIcNAc). It is possible that a NodC-like protein could have evolved from such a bacterial enzyme. There are several NodB-Iike proteins in databases, and it is easy to imagine how a simple glucosamine-deacetylase like NodB could have been recruited.

Several discoveries relevant to papilionoid legumes may not apply to all symbiotic interactions, particularly for understanding the evolution of nodulation. For example, extended infection threads are required for pea and alfalfa nodulation, but there are examples where infection threads are almost nonexistent and bacteria spread interstitially as in peanut (*Arachis hypogaea;* Chandler, 1978). In some tropical legumes and also in *Parasponia* sp., rhizobia are not released into membrane-bound symbiosomes; rather, they fix nitrogen within specialized fixation threads (Hirsch and LaRue, 1998). There are reports of nitrogen-fixing bacteria, based on acetylene reduction assays, in more basal legumes (Bryan *et al.*, 1996), but these studies remain foundation.

Quite a lot of the mutations affecting host-specific modifications of Nod factors delay nodulation, but in many cases, the process continues normally (Ardourel *et al.*, 1994). However, an interesting phenotype was described for a mutant of *R. leguminosarum* bv *viciae* lacking all of the host-specific nodulation genes but retaining the *nodABC* genes and their regulator (Walker and Downie, 2000). On vetch (*Vicia sativa*), many hundreds of root hairs were heavily infected, but infection threads and nodules were not formed. Extra studies are evidently needed.

The potential to induce cell division and create a nodule greatly enhances the efficiency of the symbiosis. It may be significant that minimalist Nod factor structures can induce early signaling events, whereas more highly substituted Nod factors are required to initiate cell division, nodule primordia, and infection thread structures (van Brussel *et al.*, 1992; Ardourel *et al.*, 1994; Walker and Downie, 2000). This has led to the idea that there may be different levels of recognition of Nod factors. Some plant genes involved in processes related to cell division, such as cell cycle control and nuclear endo-reduplication, have been identified (Cebolla *et al.*, 1999; Charon *et al.*, 1999; Roudier *et al.*, 2000). Pingret *et al.* (1998) suggested a role for a G protein-mediated signaling pathway for induction of legume early nodulin genes, based on inhibitor studies and the induction of gene expression by mastoparan, a G protein agonist. However, mastoparan did not induce calcium spiking in root hairs (Walker *et al.*, 2000), and taken at face value, this would imply that a role for G protein-mediated signaling could be downstream of calcium spiking.

Molecular Biology and Biochemistry of Antioxidants in the Rhizobia-legume Symbiosis

Studies on antioxidants in the rhizobia-legume symbiosis are certainly an important area of research which needs to be elaborated. The complete reduction of molecular oxygen to water requires four electrons and is catalyzed by cytochrome oxidase in aerobic bacteria and mitochondria. However, 1 per cent to 3 per cent of all oxygen consumed by respiration is inevitably reduced to superoxide radicals and hydrogen peroxide (H_2O_2). These and other oxygen-derived molecules with moderate to very high reactivity are known as reactive oxygen species (ROS). The term includes free radicals (molecules with one or more unpaired electrons, such as the superoxide and hydroxyl radicals) and non-free radicals (molecules with no unpaired electrons, such as H_2O_2 and singlet oxygen). The main sources of ROS in plants under physiological conditions are respiration, photosynthesis, and N_2 fixation. In addition, ROS are produced at high rates when plants are exposed to abiotic, biotic, or xenobiotic stress. Similarly, the term reactive nitrogen species (RNS) refers to nitrogen-derived molecules with variable reactivity and includes free radicals (nitric oxide) and non-free radicals (peroxynitrite).

The superoxide radical, H_2O_2, and nitric oxide have moderate reactivity toward biomolecules and, thus, may have some direct detrimental effects in plants. The superoxide radical inactivates dihydroxy-acid dehydratase (required for the synthesis of branched chain amino acids) and aconitase (required for the operation of the Krebs cycle) by oxidizing the iron-sulfur clusters at the active site, and ribonucleotide reductase (required for DNA synthesis) by oxidizing an essential Tyr radical. Also, H_2O_2 can inactivate Calvin cycle enzymes, metalloproteins such as superoxide dismutases

(SODs), and hemoproteins such as nodule leghemoglobin (Dalton, 1995; Scandalios *et al.*, 1997). However, the real threat of the superoxide radical and H_2O_2 is their potential to act as precursors of the hydroxyl radical. The hydroyxl radical can readily oxidize amino acid residues of proteins, fatty acids of phospholipids, and deoxy-Rib and bases in DNA (Halliwell and Gutteridge, 1999). Nitric oxide can directly inhibit iron-containing proteins (Neill *et al.*, 2002), but its toxicity stems mainly from its ability to react with the superoxide radical to form peroxynitrite. This compound can induce lipid peroxidation, nitration of Tyr residues of proteins, oxidation of thiols, and nitration or deamination of DNA bases (Halliwell and Gutteridge, 1999). The superoxide radical and H_2O_2 are involved in lignification of cell walls, defense against pathogen attack, and sensing of, and subsequent adaptation to, stressful conditions. H_2O_2 can also induce programmed cell death during the plant's hypersensitive response to infection by modulating gene expression (Neill *et al.*, 2002). Nitric oxide also acts as a signal molecule and is involved in the control of gene expression, hypersensitive response, antioxidant defense, organogenesis, and stomatal closure (Neill *et al.*, 2002; Lamattina *et al.*, 2003).

Major Antioxidants of Nodules

Ascorbate (vitamin C) is a water-soluble reductant that can be found in nodules at concentrations of I to 2 mM. Ascorbate is required for the progression of the cell cycle and for cell elongation. The latter effect has been attributed to its participation as cofactor of prolyl hydroxylase (required for the synthesis of Hyp-rich proteins of the cell wall) and to the ability of apoplastic ascorbate to alter the properties of the plasma membrane or to inhibit the cross-linking of Hyp-rich proteins by phenols (Horemans *et al.*, 2000). Ascorbate regenerates the α-tocopherol oxidized by ROS at the membrane-cytosol interface, is a direct scavenger of most ROS, and is the substrate of ascorbate peroxidase (APX).

GSH is an important antioxidant in its own right but also as a substrate for glutathione reductase and glutathione peroxidase (GSH-PX). However, in some legumes, homoglutathione (γ-Glu-Cys-β-Ala) may partially or completely replace GSH. Homoglutathione is the major tripeptide in nodules of soybean (*Glycine max*), common bean (*Phaseolus vulgaris*), and mung bean (*Vigna radiata*), whereas GSH is predominant in nodules of pea (*Pisum sativum*), alfalfa (*Medicago sativa*), and cowpea (*Vigna unguiculata*). In each case, the major thiol is present at concentrations of 0.5 to 1 mM. The synthesis of GSH and homoglutathione proceeds through two ATP-dependent steps catalyzed, respectively, by γ-glutamylcysteine synthetase and a specific glutathione or homoglutathione synthetase. The enzymes from pea, mung bean, and tobacco (*Nicotiana tabacum*) leaves have been partially purified and localized to the cytosol and plastids (Rennenberg, 1997). The GSH may be internally consumed by bacteroids in metabolic reactions and maintenance of cellular redox status rather than being exported to the plant (Iturbe-Ormaetxe *et al.*, 2001). Recently, a mutant of *Rhizobium tropici* has been isolated which is deficient in glutathione synthetase and contains only 3 per cent of the GSH present in the wild-type strain (Riccillo *et al.*, 2000). This mutant is sensitive to weak organic acids and to osmotic and oxidative stress and the addition of GSH restores the responses to these stresses to wild-type levels. Interestingly, the mutant can form effective nodules on bean, but it is out competed by the wild-type strain, indicating that GSH is important for stress tolerance and the symbiotic process (Riccillo *et al.*, 2000).

Glutathione reductases are found in the cytosol, mitochondria, and bacteroids, but are probably present also in nodule plastids because the enzyme is abundant in chloroplasts and root plastids (Bielawski and Joy, 1986) and because cDNA clones encoding a putative plastid glutathione reductase have been isolated (Tang and Webb, 1994). It is likely that the enzymes of nodule mitochondria and plastids are coded for by a single gene, as occurs with the enzymes of pea leaves (Creissen *et al.*, 1995). Some nodules, such as those of bean and soybean, synthesize homoglutathione rather than GSH; therefore, the enzyme is functionally a homoglutathione reductase. Monodehydroascorbate reductases are ubiquitous flavoproteins of plants that occur in soluble and membrane-bound isoforms. In nodules, two isoforms have been found, and at least one of them is associated with the cell wall. Only low enzyme levels are detected in the cytosol (Dalton *et al.*, 1993). Thus, it appears that recycling of ascorbate through the ascorbate-GSH pathway in the cytosol is mainly accomplished by dehydroascorbate reductase, whereas monodehydroascorbate reductase may be involved in regeneration of apoplastic ascorbate, synthesis of Hyp-rich proteins, and lignification of cell walls.

There are further indications of the importance of the ascorbate-GSH pathway for N_2 fixation. The activities of all four enzymes are much higher (2- to 36-fold) in nodules than in uninfected roots. The enzyme activities and thiol contents are also substantially higher (1.5- to 5.5-fold) in effective than in ineffective nodules. Also, treatment of plants with fixed nitrogen (urea) inhibits N_2 fixation concomitantly with three enzyme activities of the pathway, indicating that there is a link between N_2 fixation and antioxidant defenses. The most compelling evidence for the connection between antioxidants and N_2 fixation comes from observations that direct infusion of ascorbate into stems of soybean plants leads to an increase in leghemoglobin content, a 4-fold increase in rates of N_2 fixation, and a substantial delay in nodule senescence (Bashor and Dalton, 1999).

SODs and Catalases for Protection of Nitrogen Fixation

SODs are a family of metalloenzymes that catalyze the dismutation of superoxide radicals into molecular oxygen and H_2O_2. Three classes of SODs, differing in their metals at the active site, may coexist in plants, and all of them have been found in the nodule plant fraction. The CuZnSOD is predominant in the nodule apex especially in the infection threads, cytosol adjacent to cell walls, and apoplast; the MnSOD is abundant in the infected zone, especially in the infected cells. An additional CuZnSOD isozyme, the plastid CuZnSOD, is localized to the amyloplasts, whereas MnSOD is also found in the bacteroids and bacteria within infection threads. The distinct tissue localizations of "cytosolic" CuZnSOD and MnSOD suggest specific functions for the two enzymes. The CuZnSOD may be associated with cell wall growth in the meristems, infection threads, and apoplast, and with the plant's response to bacterial infection. The MnSOD would play a role related to the protection and functioning of symbiotic tissue in mature nodules.

The structures of the genes encoding cytosolic CuZnSOD (*sodCc*) and mitochondrial MnSOD (*sodA*) of *L. japonicus* have been determined. The *sodCc* gene consists of eight exons; interestingly, the flrst intron is in the 5′-UTR, as occurs for the pea *apxI* gene. The *sodA* gene has six exons with no apparent special features. The FeSODs are the most enigmatic class of SODs; in fact, the corresponding gene (*sodS*) was once thought to be present, or expressed, only in a few families of higher plants. The FeSODs, when present, appear to be localized exclusively in the chloroplast stroma. Two types of FeSOD were clearly recognized: the typical FeSOD localized in the plastids of alfalfa and pea nodules and an unusual FeSOD localized in the cytosol of cowpea nodules (Moran *et al.*, 2003). Bacteroids possess a MnSOD in the cytoplasm and a CuZnSOD in the periplasmic space. These enzymes are encoded by the respective bacterial *sodA* and *sodC* genes. Catalases decompose H_2O_2 to water and molecular oxygen without consuming reductants and, thus, may provide plant cells with an energy-efficient mechanism to remove H_2O_2 (Scandalios *et al.*, 1997). However, catalases have a much lower affinity for H_2O_2 than APXs and are expected to be active only at subcellular sites where H_2O_2 or catalase concentrations are very high, such as the peroxisomes. The production of oxygen by catalase would appear to be unfavorable in the case of bacteroids because nitrogenase and other proteins with high redox potential are readily inactivated by excess oxygen. Therefore, it is interesting that bacteroids do not contain peroxidases but have catalases instead.

Further, plant membranes are mainly protected against lipid peroxidation and other types of oxidative damage by small lipophilic molecules such as tocopherols, ubiquinol, lipoic acid, and flavonoids. α-Tocopherol is found at concentrations of 15 μg g^{-1} in both young and old soybean nodules (Evans *et al.*, 1999). Ubiquinol (the reduced form of ubiquinone) and lipoic acid are abundant in membranes of mitochondria and other organelles, where they act as potent inhibitors of lipid peroxidation. However, they have not been quantified in nodules. Flavonoids and other polyphenols are found in nodules at concentrations of 0.4 to 4 mM. Some of these compounds have important antioxidant properties, protecting membranes by neutralizing lipid radicals (Moran *et al.*, 1997). Polyamines are also abundant in nodules and inhibit lipid peroxidation in vitro, probably by their ability to associate with phospholipids and stabilize membranes (Fujihara *et al.*, 1994).

Plant cells also have an adequate protection against iron-mediated toxicity. Iron in the free form or bound to small chelators is potentially toxic because it can catalyze formation of hydroxyl radicals. The exceptions seem to be phytic acid and certain phenolic compounds that are able to chelate iron in a catalytically inactive form and may inhibit oxidative damage of lipids and proteins in vitro (Moran *et al.*, 1997). The supply of free iron must be tightly regulated because plants require a steady, low iron supply for the synthesis of iron-proteins, DNA, and some hormones (Briat and Lobreaux, 1997). This need must be carefully balanced against the potential toxicity of excess iron. The protein, ferritin stores up to 4,500 atoms of iron in a form that avoids the deleterious effects of iron while controlling its availability for metabolic purposes. Plant ferritins are composed of a central iron-filled cavity surrounded by a shell of 24 identical subunits. Nodules have an active iron metabolism and abundant ferritin, which is localized in plastids and amyloplasts, much like the ferritin of leaves that is localized exclusively in the chloroplasts (Lucas *et al.*, 1998; Matamoros *et al.*, 1999). The ferritin protein and transcript increase very early in nodulation. Later in nodule development, the ferritin protein (but not its transcript) declines concomitantly with the increase in nitrogenase and leghemoglobin proteins. This suggests that ferritin is a reservoir of iron and supplies it for nitrogenase and leghemoglobin synthesis. Because changes in ferritin protein and transcript are not coordinated, ferritin expression may be posttranscriptionally regulated (Ragland and Theil, 1993). The ferritin content increases during natural and stress-induced senescence of nodules, although variations can be found depending on legume species and nodule tissue. This is most probably due to induction of ferritin expression by the iron released during degradation of leghemoglobin and other iron proteins (Lucas *et al.*, 1998; Matarnoros *et al.*, 1999).

Microbial Products Trigger Amino Acid Exudation from Plant Roots

Plants naturally cycle amino acids across root cell plasma membranes, and any net efflux is termed exudation. The dominant ecological view is that microorganisms and roots passively compete for amino acids in the soil solution, yet the innate capacity of roots to recover amino acids present in ecologically relevant concentrations is unknown. Plant roots normally are surrounded by a collection of functionally and nutritionally interconnected species known as the soil food web (Brussaard *et al.*, 1997). Root-colonizing bacteria and fungi, which form a resource base for this heterotrophic complex, use root exudates as a carbon source while simultaneously supplying mineralized nitrogen to the plant as they are consumed by nematodes and protozoa (Griffiths, 1994; Ferris *et al.*, 1998). Root exudation of soluble compounds, specifically amino acids, is actually a net release composed of both influx and efflux components (Jones and Darrah, 1994). The passive efflux of amino acids is driven primarily by large differences in concentration between the inside (*e.g.* 10 *μM*) and outside *e.g.* 0.1-10 *μM*) of root cells, while influx is fueled by outward pumping proton-ATPases that create and maintain an electrochemical potential difference across the plasma membrane to support uptake into roots by proton-coupled amino acid transporters (Farrar *et al.*, 2003). The extent to which plants compete effectively for amino acid exudates with root-colonizing microorganisms, therefore, depends on the innate capacity of root cells to recover released compounds. However, the realistic capacity is as yet open-ended because quantitative relationships between amino acid influx and effiux rates in axenic roots have not been measured at ecologically relevant concentrations.

Phillips *et al.* (2004) found that, in the absence of culturable microorganisms, the influx rates of 16 amino acids (each supplied at 2.5 μM) exceed efflux rates by 5 per cent to 545 per cent in roots of alfalfa (*Medicago sativa*), *Medicago truncatula*, maize (*Zea mays*), and wheat (*Triticum aestivum*). Several microbial products, which are produced by common soil microorganisms such as *Pseudomonas* bacteria and *Fusarium* fungi, significantly enhanced the net effiux (*i.e.* exudation) of amino acids from roots of these four plant species. In alfalfa, treating roots with 200 μM phenazine, 2,4-diacetylphloroglucinol, or zearalenone increased total net effiux of 16 amino acids 200 per cent to 2,600 per cent in 3 h. Data from ^{15}N tests suggest that 2,4-diacetylphloroglucinol blocks amino acid uptake, whereas zearalenone enhances efflux. The importance of identifying potential enhancers of root exudation lies in understanding that such compounds may represent regulatory linkages between the larger soil food web and the internal carbon metabolism of the plant. Z is an antifungal compound produced by various species of the soil fungus *Fusarium* (Jimenez *et al.*, 1996). If these compounds increase the availability of amino acids in root exudates in addition to impairing the growth of competing microorganisms, then clearly this combination of benefits could promote growth of the organisms producing them. By extension, anything that micropredators, such as protozoa, nematodes, or collembola, do physically or chemically to simulate the synthesis or release of these compounds in microorganisms could enhance foundation resources for the entire food web. Therefore, amino acid exudation under normal conditions is a phenomenon that probably reflects both active manipulation and passive uptake by microorganisms, as well as diffusion and adsorption to soil, all of which help overcome the innate capacity of plant roots to reabsorb amino acids. Thus, this identification of active chemical compounds establishes the existence of a potentially important molecular link between the heterotrophic soil food web and the autotrophic plant.

Future Scope

Optimistically, the discussion above has given the reader a real sense of the complexity of the legume nodulation process. Obviously, incredible advancement has been made in recent years in understanding the molecular basis for this captivating symbiosis. Along the way, one can expect discoveries that will have ultimate application to the development of legume agriculture. Likewise exciting will be the basic discoveries that not only reveal the mysteries of nodulation but aid in our understanding of the basis of plant growth and development. It is indeed an exciting time to work on symbiotic biological nitrogen to really sustain our agriculture. Nevertheless, there are still gaps in our understanding (*e.g.*, host recognition and preferential rhizobial species selection, ultrastructure and function of the nodule and symbiosome, plant cell receptors, Nod-signal action, etc.) which need to be filled before we can get pleasure from the full benefits of legume-rhizobia symbiosis. Knowledge of rhizobia and its role in this symbiotic relationship is far ahead of our knowledge of the legume's role. Nonetheless, there are many genes in rhizobia involved in this symbiotic process that have yet to be recognized and studied. The recent application of genomics to this field promises to accelerate the pace of discovery. It is predicted that other mutations are likely to affect the mycorrhizal symbiosis, but not nodulation. Moreover, this nonpathogenic association between prokaryote and eukaryote is a fascinating phenomenon for investigation of basic biological principles.

References

Ae, N., Arihara, J., Okada, K., Yoshihara, T. and Johansen, C. (1990). Phosphorus uptake by pigeon pea and its role in cropping systems of the Indian subcontinent. *Science*, 248: 477–480.

Al-Niemi, T.S., Kahn, M.L. and McDermott, T.R. (1997). P metabolism in the bean *Rhizobium tropici* symbiosis. *Plant Physiol.*, 113: 1233–1242.

Ardourel, M., Demont, N., Debelle, F., Maillet, F., de Billy, F., Promé, J-C., Denarié, J. and Truchet, G. (1994). *Rhizobiwn meliloti* lipooligosaccharide nodulation factors: different structural requirements for bacterial entry into target root hair cells and induction of plant symbiotic developmental responses. *Plant Cell*, 6: 1357–1374.

Bashor, C.J. and Dalton, D.A. (1999). Effects of exogenous application and stem infusion of ascorbate on soybean (*Glycine max*) root nodules. *New Phytol.*, 1142: 19–26.

Beaver, J.S., Rosas, J.C., Myers, J., Acosta, J., Kelly, J.D., Nchimbi-Msolla, S., Misangu, R., Bokosi, J., Temple, S. and Amaud-Santana, E. *et al.* (2003). Contributions of the bean/cowpea CRSP program to cultivar and germplasm development in common bean. *Field Crops Res.*

Bielawski, W. and Joy, K.W. (1986). Reduced and oxidised glutathione and glutathione-reductase activity in tissues of *Pisum sativum*. *Planta*, 169: 267–272.

Bieleski, R.L. (1973). Phosphate pools, phosphate transport, and phosphate availability. *Ann. Rev. Plant Physiol.*, 24: 225– 252.

Briat, J.F. and Lobreaux, S. (1997). Iron transport and storage in plants. *Trends Plant Sci.*, 2: 187–193.

Brussaard, L., Behan-Pelletier, V.M., Bignell, D.E., Brown, V.K., Didden, W., Folgarait, P., Fragoso, C., Freckman, D.W., Gupta, V. and Hattori, T., *et al.* (1997). Biodiversity and ecosystem functioning in soil. *Ambio.*, 26: 563–570.

Bryan, J.A., Berlyn, G.P. and Gordon, J.C. (1996). Towards a new concept of the evolution of symbiotic nitrogen fixation in the Leguminosae. *Plant Soil*, 186: 151–159.

Cebolla, A., Vinardell, J.M., Kiss, E., Olah, B., Roudier, F., Kondorosi, A. and Kondorosi, E. (1999). The mitotic inhibitor *ccs52* is required for endoreduplication and ploidy–dependent cell enlargement in plants. *EMBO J.*, 18: 4476–4484.

Chaintreuil, C., Girand, E., Prin, Y., Lorquin, J., Ba, A., Gillis, M., de Lajudie, P. and Dreyfus, B. (2000). Photosynthetic bradyrhizobia are natural endophytes of the African wild rice *Oryza breviligualata*. *Appl. Environ. Microbial.*, 66: 5437–5447.

Chandler, M.R. (1978). Some observations on the infection of *Arachis hypogaea* L. by *Rhizobium*. *J. Exp. Bot.*, 29: 749–755.

Charon, C., Sousa, C., Crespi, M. and Kondorosi, A. (1999). Alteration of *enod40* expression modifies *Medicago truncatula* root nodule development induced by *Sinorhizobium meliloti*. *Plant Cell*, 11: 1953–1966.

Creissen, G., Reynolds, H., Xue, Y. and Mullineaux, P. (1995). Simultaneous targeting of pea glutathione reductase and of a bacterial fusion protein to chloroplasts and mitochondria in transgenic tobacco. *Plant J.*, 8: 167–175.

Cullimore, J.V., Ranjeva, R. and Bono, J.J. (2001). Perception of lipo-chitooligosaccharidic Nod factors in legumes. *Trends Plant Sci.*, 6: 24–30.

Dalton, D.A., Baird, L.M., Langeberg, L., Taugher, C.Y., Anyan, W.R., Vance, C.P. and Sarath, G. (1993). Subcellular localization of oxygen defense enzymes in soybean [*Glycine max* (L.) Merr.] root nodules. *Plant Physiol.*, 102: 481–489.

Dalton, D.A. (1995). Antioxidant defenses of plants and fungi. In: *Oxidative Stress and Antioxidant: Defenses in Biology*, (Ed.) S. Ahmad. Chapman and Hall, New York, pp. 298–355.

Downie, J.A. and Walker, S.A. (1999). Plant responses to nodulation factors. *Curr. Opin. Plant Biol.*, 2: 483–489.

Doyle, J.J. (1998). Phylogenetic perspectives on nodulation: evolving views of plants and symbiotic bacteria. *Trends Plant Sci.*, 3: 473–478.

Duc, G., Trouvelot, A., Gianinazzi-Pearson, V. and Gianinazzi, S. (1989). First report of non-mycorrhizal plant mutants (Myc$^-$) obtained in pea (*Pisum sativum* L.). and fababean (*Viciafaba* L.). *Plant Sci.*, 60: 215–222.

Etzler, M.E., Kalsi, G., Ewing, N.N., Roberts, N.J., Day, R.B. and Murphy, J.B. (1999). A Nod factor binding lectin with apyrase activity from legume roots. *Proc. Natl. Acad. Sci.*, USA, 96: 5856–5861.

Evans, P.J., Gallesi, D., Mathieu, C., Hernandez, M.J., de Felipe, M., Halliwell, B. and Puppo, A. (1999). Oxidative stress occurs during soybean nodule senescence. *Planta*, 208: 73–79.

Farrar, J., Hawes, M., Jones, D. and Lindow, S. (2003). How roots control the flux of carbon to the rhizosphere. *Ecology*, 84: 827–837.

Ferris, H., Venette, R.C., vander Meulen, H.R. and Lau, S.S. (1998). Nitrogen mineralization by bacterial-feeding nematodes: Verification and measurement. *Plant and Soil*, 203: 159–171.

Fred, E.W., Baldwin, I.L. and McCoy, E. (1932). *Root Nodule Bacteria and Leguminous Plants*. University of Wisconsin Studies, Madison, WI.

Freiberg, C., Fellay, R., Bairoch, A., Broughton, W.J., Rosenthal, A. and Perret, X. 1997. Molecular basis of symbiosis between *Rhizobium* and legumes. *Nature*, 387: 394–401.

Frink, C.R., Waggoner, P.E. and Ausubel, S.H. (1999). Nitrogen fertilizer: retrospect and prospect. *Proc. Natl. Acad. Sci.*, USA, 96: 1175–1180.

Fujihara, S., Abe, H., Minakawa, Y., Akao, S. and Yoneyama, T. (1994). Polyamines in nodules from various plant-microbe interactions. *Plant Cell Physiol.*, 35: 1127–1134.

Galibert, F., Finan, T.M., Long, S.R., Pühler, A., Abola, P., Ampe, F., Barloy-Hubler, F., Barnett, M.J., Becker, A. and Boistard, P. (2001). The composite genome of the legume symbiont *Sinorhizobium meliloti*. *Science*, 293: 668–672.

Giller, K.E. (2001). *Nitrogen Fixation in Tropical Cropping Systems*. CABI Publishing, Wallingford, UK.

Graham, P.H. and Vance, C.P. (2000). Nitrogen fixation in perspective: An overview of research and extension needs. *Field Crop Res.*, 65: 93–106.

Graham, P.H., Viteri, S.E., Mackie, F., Vargas, A.T. and Palacios, A. (1982). Variation in acid soil tolerance among strains of *Rhizobium phaseoli*. *Field Crops Res.*, 5: 121–128.

Gressent, F., Drouillard, S., Mantegazza, N., Samain, E., Geremia, R.A., Canut, H., Niebel, A., Driguez, H., Ranjeva, R. and Cullimore, J. (1999). Ligand specificity of a high-affinity binding site for lipo-chitooligosaccharidic Nod factors in *Medicago*-cell suspension cultures. *Proc. Natl. Acad. Sci.*, USA, 96: 4704–4709.

Griffiths, B.S. (1994). Soil nutrient flow. In: *Soil Protozoa,* (Ed.) J.F. Darbyshire. CAB International, Wallingford, UK, pp. 65–91.

Halliwell, B. and Gutteridge, J.M.C. (1999). *Free Radicals in Biology and Medicine,* 3rd edn. Oxford University Press, Oxford.

Hemantaranjan, A. and Garg, O.K. (1986). Introduction of nitrogen fixing nodules through iron and zinc fertilization in non-nodule forming French beans (*Phaseolus vulgaris* L.). *J. Plant Nutrition,* 9: 281–288.

Hemantaranjan, A. (1988). Iron fertilization in relation to nodulation and nitrogen fixation in *Phaseolus vulgaris* L. *J. Plant Nutrition,* 11: 829–842.

Hemantaranjan, A. (1991). Yield potentiality of *Phaseolus vulgaris* L. with special reference to vanadium, zinc and iron. In: *Physiological Strategies for Crop Improvement,* 1: 49–55.

Hemantaranjan, A. (1994a). Physiological significance of aluminium, vanadium and zinc on the sensitivity of nitrogen fixation in *Phaseolus vulgaris* L. In: *New Perspectives in the Research of Hardly Known Trace Elements* (Ed.) Prof. Istvan Pais. Budapest, Hungary, 6: 45–62.

Hemantaranjan, A. (1994b). Foliar application of iron and certain growth regulators on nodule senescence of soybean (*Glycine Max* L.). In: *Plant Productivity Under Environmental Stresses,* 1: 413–418.

Hemantaranjan, A. (1996). Iron activity, nitrogen fixation and yield potentiality of *Glycine max* L. in relation to certain trace elements. In: *Proc. New Pers. Res. Trace El Life Proc.,* (Ed.) Istvan Pais. Budapest, Hungary, 7: 313–318.

Hemantaranjan, A. (1999). Physiological significance of boron and cobalt on iron activity, nitrogen fixation and yield potentiality of *Glycine max* L. In: *Plant Physiol. Sustainable Agril.,* (Ed.) G.C. Srivastava *et al.,* Pointer Publishers, Jodhpur, 27: 163–169.

Hirsch, A.M. and LaRue, T.A. (1998). Is the legume nodule a modified root or stem or an organ *sui generis? Crit. Rev. Plant Sci.,* 16: 361–392.

Horemans, N., Foyer, C.H., Potters, G. and Asard, H. (2000). Ascorbate function and associated transport systems in plants. *Plant Physiol. Biochem.,* 38: 531–540.

Howieson, J.G., O'Hara, G.W. and Carr, S.J. (2000). Changing roles for legumes in Mediterranean agriculture: Developments from an Australian perspective. *Field Crops Res.,* 65: 107–122.

Hymowitz, T. and Singh, R.J. (1987). Taxonomy and speciation. In: *Soybeans: Improvement, Production and Uses,* (Ed.) J.R. Wilcox. Agronomy Monograph 16. American Society of Agronomy, Madison, WI, pp. 23–48.

Israel, D.W. (1987). Investigation of the role of phosphorus in symbiotic dinitrogen fixation. *Plant Physiol.,* 84: 835–840.

Iturbe-Ormaetxe, I., Matamoros, M.A., Rubio, M.C., Dalton, D.A. and Becana, M. (2001). The antioxidants of legume nodule mitochondria. *Mol. Plant-Microbe Interact.,* 14: 1189–1196.

Jimenez, A., Hernandez, J., del Rio, L.A. and Sevilla, F. (1997). Evidence for the presence of the ascorbate-glutathione cycle in mitochondria and peroxisomes of pea leaves. *Plant Physiol.,* 114: 275–284.

Johnson, J.F., Vance, C.P. and Allan, D.L. (1996). Phosphorus deficiency in *Lupinus albus*-altered lateral root development and enhanced expression of phosphoenolpyruvate carboxylase. *Plant Physiol.,* 112: 31–41.

Jones, D.L. and Darrah, P.R. (1994). Amino-acid influx at the soil-root interface of *Zea mays* L. and its implications in the rhizosphere. *Plant Soil,* 163: 1–12.

Kaneko, T., Nakamura, Y., Sato, S., Asamizu, E., Kato, T., Sasamoto, S., Watanabe, A., Idesawa, K., Ishikawa, A. and Kawashima, K. (2000). Complete genome structure of the nitrogen-fixing symbiotic bacterium *Mesorhizobium loti. DNA Res.,* 7: 331–338.

Kaplan, L. and Lynch, T.F. (1999). *Phaseoius* (Fabaceae) in archeology: AMS radiocarbon dates and their significance in pre-Colombian agriculture. *Econ. Bot.,* 53: 261–272.

Koltai, H., Dhandaydham, M., Opperman, C., Thomas, J. and Bird, D. (2001). Overlapping plant signal transduction pathways induced by a parasitic nematode and a rhizobial endosymbiont. *Mol. Plant Microbe Interact.,* 14: 1168–1177.

Lamattina, L., Garcia-Mata, C., Graziano, M. and Pagnussat, G. (2003). Nitric oxide: The versatility of an extensive signal molecule. *Ann. Rev. Plant Biol.,* 154: 109–136.

Long, S.R. (1996). *Rhizobium* symbiosis: Nod factors in perspective. *Plant Cell,* 8: 1885–1898.

Loveless, T.M. and Bishop, P.E. (1999). Identification of genes unique to Mo-independent nitrogenase systems in diverse diazotrophs. *Can. J. Microbiol.,* 45(4): 312–317.

Lucas, M.M., Van de Sype, G., Herouart, D., Hernandez, M.J., Puppo, A. and de Felipe, M.R. (1998). Immunolocalization of ferritin in determinate and indeterminate legume root nodules. *Protoplasma,* 204: 61–70.

Maier, R.J. and Triplett, E.W. (1996). Toward more productive, efficient, and competitive nitrogen fixing symbiotic bacteria. *Crit. Rev. Plant Sci.,* 15: 191–234.

Matamoros, M.A., Baird, L.M., Escuredo, P.R., Dalton, D.A., Minchin, F.R., Iturbe-Ormaetxe, I., Rubio, M.C., Moran, J.F., Gordon A.J. and Becana, M. (1999). Stress-induced legume root nodule senescence: Physiological, biochemical, and structural alterations. *Plant Physiol.,* 121: 97–111.

Miller, S.S., Liu, J.Q., Allan, D.L., Menzhuber, C.J., Fedorova, M. and Vance, C.P. (2001). Molecular control of acid phosphatase secretion into the rhizosphere of proteoid roots from phosphorus-stressed white lupin. *Plant Physiol.,* 127: 594–606.

Moran, J.F., Klucas, R.V., Grayer, R.J., Abian, J. and Becana, M. (1997). Complexes of iron with phenolic compounds from soybean nodules and other legume tissues: Prooxidant and antioxidant properties. *Free Radic. Biol. Med.,* 22: 861– 870.

Neill, S.J., Desikan, R., Clarke, A., Hurst, R.D. and Hancock, J.T. (2002). Hydrogen peroxide and nitric oxide as signalling molecules in plants. *J. Exp. Bot.,* 53: 1237–1247.

Peoples, M.B., Herridge, D.F. and Ladha, J.K. (1995). Biological nitrogen fixation: An efficient source of nitrogen for sustainable agricultural production. *Plant Soil,* 174: 3–28.

Perret, X., Stahelin, C. and Broughton, W.J. (2000). Molecular basis of symbiotic promiscuity. *Microbiol. Mol. Biol. Rev.,* 64: 180–201.

Phillips, D.A., Joseph, C.M., Yang, G.P., Martinez-Romero, E., Sanborn, J.R. and Volpin, H. (1999). Identification of lumichrome as a *Sinorhizobium* enhancer of alfalfa root respiration and shoot growth. *Proc. Natl. Acad. Sci.,* USA, 96: 12275–12280.

Pingret, J-L., Journet, E-P. and Barker, D.G. (1998). *Rhizobium* Nod factor signaling: evidence for a G protein-mediated transduction mechanism. *Plant Cell,* 10: 659–671.

Postel, S. (2001). Growing more food with less water. *Sci. Am.*, 284: 46–51.

Ragland, M. and Theil, E.C. (1993). Ferritin (mRNA, protein) and iron concentrations during soybean nodule development. Plant Mol. Biol., 21: 555–560.

Ragothama, K.G. (1999). Phosphate acquisition. *Ann. Rev. Plant Physiol. & Plant Mol. Biol.*, 50: 665–693.

Rennenberg, H. (1997). Molecular approaches to glutathione biosynthesis. In: *Sulphur Metabolism in Higher Plants*, (Eds.) W.J. Cram, L.J. DeKok, C. Brunold, H. Rennenberg. Backhuys Publishers, Leiden, The Netherlands, pp. 59–70.

Ribbe, M., Gudkari, D. and Meyer, D. (1997). N_2 fixation by *Streptomyces thermoautotrophicus* involves a molybdenum-dinitrogenese and a manganese-superoxide reductase that couple N_2 reduction to the oxidation of superoxide produced from O_2 by molybdenum-CO dehydrogenase. *J. Biol. Chem.*, 272(42): 26627–26633.

Ribet, J., and Drevon, J.J. (1996). The phosphorus requirement of N_2 fixing and urea-fed *Acacia mangium. New Phytol.*, 132: 383–390.

Riccillo, P.M., Muglia, C.I., de Bruijn, F.J., Roe, A.J., Booth, I.R. and Aguilar, O.M. (2000). Glutathione is involved in environmental stress responses in *Rhizobium tropici*, including acid tolerance. *J Bacteriol.*, 182: 1748–1753.

Roudier, F., Fedorova, E., Gyorgyey, J., Feher, A., Brown, S., Kondorosi, A., Kondorosi, E. (2000). Cell cycle function of a *Medicago sativa* A2–type cyclin interacting with a PSTAIRE-type cyclin-dependent kinase and a retinoblastoma protein. *Plant J.*, 23: 73–83.

Saito, A., Mitsui, H., Hattori, R., Minamisawa, K., Tsutomu Hattori, T. (1998). Slow-growing and Oligotrophic soil bacteria phylogenetically close to *Bradyrhizobiumjaponicum. FEMS Microbiol. Ecol.*, 25: 277–286.

Scandalios, J.G., Guan, L. and Polidoros, A.N. (1997). Catalases in plants: Gene structure, properties, regulation, and expression. In: *Oxidative Stress and the Molecular Biology of Antioxidant Defenses*, (Ed.) J.G. Scandalios. Cold Spring Harbor Laboratory Press, Cold Spring Harbor, N.Y., pp. 343–406.

Schactman, D.P., Reid, R.J. and Ayling, S.M. (1998). Phosphorus uptake by plants: from soil to cell. *Plant Physiol.*, 116: 447–453.

Schauser, L., Roussis, A., Stiller, J. and Stougaard, J. (1999). A plant regulator controlling development of symbiotic root nodules. *Nature*, 402: 191–195.

Schloter, M., Whiehe, W., Asimus, B., Steindl, H., Becke, H., Hoflilch, G. and Hartmann, A. (1997). Root Colonization of different plants by plant-growth-promoting *Rhizobium teguminosarum* bv. *Trifotii* studied with Monospecific Polyclonal antisera. *Appl. Environ. Microbiol.*, 63: 2039–2046.

Sevilla, M., Burris, R.H., Gunapala, N. and Kennedy, C. (2001). Comparison of benefit to sugarcane plant growth and $^{15}N_2$ incorporation following inoculation of sterile plants with *Acetobacter diazotrophicus* wild-type and *Nif* mutant strains. *Mol. Plant-Microbe Interact.*, 14: 358–366.

Sinclair, T.R., Muchow, R.C., Bennet, J.M. and Hammond, L.C. (1987). Relative sensitivity of nitrogen and biomass accumulation to drought in field-grown soybean. *Agron. J.*, 79: 986–991.

Snapp, S.S. (1998). Phosphorus and sustainability of sub-Saharan African smallholder farms. In: *Phosphorus in Plant Biology: Regulatory Roles in Molecular, Cellular, Organismic and Ecosystem Processes*, (Eds.) J.P. Lynch, J. Deikman. American Society Plant Physiology, Rockville, MD, pp. 71–83.

Spaink, H.P. (2000). Root nodulation and infection factors produced by rhizobial bacteria. *Ann. Rev. Microbiol.*, 54: 257–288.

Stougaard, J. (2000). Regulators and regulation of legume root nodule development. *Plant Physiol.*, 124: 531–540.

Stracke, S. *et al.*, (2002). A plant receptor-like kinase required for both bacterial and fungal symbiosis. *Nature*, 417(6892): 959–962.

Sullivan, J.T. and Ronson, C.W. (1998). Evolution of rhizobia by acquisition of a 500-kb symbiosis island that integrates into a phe-tRNA gene. *Proc. Natl. Acad. Sci.*, USA, 95: 5145–5149.

Tang, X. and Webb, M.A. (1994). Soybean root nodule cDNA encoding glutathione reductase. *Plant Physiol.*, 104: 1081–1082

Tilman, D., Fargione, J., Wolf, B., D'Antonio, C., Dobson, A., Howarth, R., Schindler, D., Schlesinger, W.H., Simberlof, D. and Swackhamer, D. (2001). Forecasting agriculturally driven global environmental change. *Science*, 292: 281–284.

Unkovich, M.J. and Pate, J.S. (2000). An appraisal of recent field measurements of symbiotic N_2 fixation by annual legumes. *Field Crops Res.*, 65: 211–222.

van Brussel, A.A.N., Bakuizen, R., van Spronsen, P.C., Spaink, H.P., Tak, T., Lugtenberg, B.J.J. and Kijne, J.W. (1992). Induction of preinfection thread structures in the leguminous host plant by mitogenic lipooligosaccharides of *Rhizobium*. *Science*, 257: 70–72.

van Kessel, C. and Hartley, C. (2000). Agricultural management of grain legumes: Has it led to an increase in nitrogen fixation? *Field Crops Res.*, 65: 165–181.

van Rhijin, P. and Vanderleyden, J. (1995). The *Rhizobium*-plant symbiosis. *Microbiol. Rev.*, 59: 124–142.

Vance, C.P., Graham, P.H. and Allan, D.L. (2000). Biological nitrogen fixation. Phosphorus: A critical future need. In: *Nitrogen Fixation: From Molecules to Crop Productivity*, (Eds.) F.O. Pedrosa, M. Hungria, M.G. Yates and W.E. Newton. Kluwer Academic Publishers, Dordrecht, The Netherlands, pp. 506–514.

Vance, C.P. (2001). Symbiotic nitrogen fixation and phosphorus acquisition: Plant nutrition in a world of declining renewable resources. *Plant Physiol.*, 127: 390–397.

Vance, C.P. (2002). Root-bacteria interactions: Symbiotic nitrogen fixation. In: *Plant Roots: The Hidden Half*, 3rd edn., (Eds) Y. Waisel, A. Eshel, U. Kafkati. Marcel Dekker Publishers, New York, pp. 839–867.

von Uexküll, H.R. and Mutert, E. (1995). Global extent, development and economic impact of acid soils. *Plant Soil*, 171: 1–15.

Walker, S.A. and Downie, J.A. (2000). Entry of *Rhizobium leguminosarum* bv. *viciae* into root hairs requires minimal Nod factor specificity, but subsequent infection thread growth requires *nodO* or *nodE*. *Mol. Plant-Microbe Interact.*, 13: 754–762.

Walker, S.A., Viprey, V. and Downie, J.A. (2000). Dissection of nodulation signaling using pea mutants defective for calcium spiking induced by nod factors and chitin oligomers. *Proc. Natl. Acad. Sci.*, USA, 97: 13413–13418.

Wang, E.T. and Martinez-Romero, E. (2000). *Sesbania herbaceae-Rhizobium huautlense* nodulation in flooded soils and comparative characterization of *S. herbacea*-nodulating rhizobia in different environments. *Microbiol. Ecol.*, 40: 25–32.

©R.C. Ram & Asha Sinha (*Editors*)
***Microbial Diversity: Modern Trends*, 2007**
Daya Publishing House, New Delhi

23–34

Chapter 2

Diversity in Fungi Associated with Seeds

M.N. Khare and M.S. Bhale

Department of Plant Pathology, Jawaharlal Nehru Krishi Vishwa Vidyalaya, Jabalpur – 482 004, M.P.

Introduction

Diversity in the association of microorganisms is the rule of nature. It depends upon the ecological situations including the food available, competition and the substrate. Certain times only one species is found associated and several are also involved. Generally a sequence is followed. Pathogens and non-pathogens both are involved exhibiting antagonistic, synergistic and normal relationship.

In seeds the pathogens are associated externally, internally, extraembryal, intraembryal, concomitant contaminant, as inert matter and associated with inert matter. With them non-pathogens and weak pathogens are also found whose survival depends upon moisture requirement. They may be hydrophilic or xerophilic. The seed borne fungi may be present in form of hyphae, conidia, oospores, chlamydospores, sclerotia, microsclerotis, spores, hyalospores and phaeospores. Seeds provide natural substrate for the growth of associated fungi.

In seed health testing standard blotter method is followed with various modifications. After incubation seeds are examined under stereoscopic wild type microscope. Most of the fungi are identified based on habit characters. Spore structures are also observed under a compound microscope for confirmation of the species. A great diversity in mycoflora is observed with seeds. A number of fungal genera and species are found associated with seeds of a sample and even with e single seed (Neergaard, 1977). Some such diversified fungi are described in this article as observed associated with seeds.

Hyphomycetes

This group of fungi have septate hyphae and grow externally or internally in the seed producing microsclerotia, stromata, hyphopodia, appressoria, setae, conidiophores and conidia. These are of various types. For identification Ellis (1971; 1976) and Subramanian (1971) are excellent monographs. Detailed keys have been developed for specific seed borne important genera whose several species are associated with seeds. Some such important genera and their species are described herewith giving important distinguishing characters as they appear on seed and the, conidia/spores under both stereoscopic wild type microscope and compound microscope.

Alternaria

On seed the mycelium is greyish or olivaceous, scanty or profuse, conidiophores grow singly or in groups, conidia are usually in chains, ovoid, cylindrical or obclavate, smooth or echinulate, beaked with short or long beak, muriform. The species can be identified based on the variability in these characters (Neergaard, 1945).

A. alternata (*A. tenuis*)

It forms long chains of conidia on the seed which vary in shape. The perfect state is *Pleospora infectoria*. It is observed associated with seeds of various crops. *A. brassicicola* has conidia without beak and is seed borne in crucifers like Brassicas. On the contrary conidia of *A. brassicicae* have a long filiform beak. *A. lini* has small conidia and is seed borne in linseed causing-blight. *A. linicola* has cylindrical conidia with long beak and causes blight in linseed. *A. carthami* conidia usually have a long beak, it is responsible for causing blight in Safflower. *A. sesami* conidia also have a long hyaline beak. It causes severe blight in sesame. *A. solani* has conidia beaked, obclavate, formed singly and attacks-members of solanaceae family like tomato, chili, brinjal etc. resulting in severe blight *A. porri* is seed borne in onion causing purple blotch. Conidia are formed singly. They are smooth, obclavate to elongate oval, tapering and ending into a long, filiform beak which may be bi or tri-furcate (Khare and Khare, 2002). *A. longipes* produces conidia singly or in chains, obclavate, tapering into a beak 1/3 to 1/2 of the total length of the conidium. It causes brown spot in tobacco and is seed borne. *A. padwickii* is responsible for stack burn disease of paddy and is highly seed borne. Conidia are straight or curved, fusiform to obclavate, mostly four transverse septa, hyaline later turning light brown. Many more species of *Alternaria* are seed borne in crops which need critical observation and identification (Figure 2.1).

Fusarium

Several species of *Fusarium* cause diseases like wilt and rots in crop plants and a number of them are seed borne. They are identified on seed itself on blotters after incubation based on colony characters (Ram Nath *et al.*, 1970; Booth, 1971). They produce macro and micro conidia as well as chlamydospores which can be confirmed by examining on a slide under compound microscope.

F. moniliforme develops dusty white mycelium on the seed which forms microconidia in chains. The chains can be observed by touching the colony with a small piece of cellotape and mountine it on a slide and examining under a microscope. Macroconidla are produced in sporodochia which are dirty white. The fungus is found associated with seeds of many crops like, paddy, wheat, maize, sorghum, pearl millet, pulses, vegetables etc. The perfect stage is *Gibberella fujikuroi* which forms brownish perithecia. *F. equiseti* forms white fluffy mycelium on the seed on which orange coloured pionnotes are formed. The macroconidia are falcate, tapering at both ends with a foot cell at the base. It is commonly observed on seeds of maize, sorghum, paddy, okra, capsicum, green gram etc. *F. solani* is recognised on the seed by lemon shaped, milky white droplets containing microconidia. They are one celled, small, oval, hyaline. Macroconidia are formed in slimy, dirty white pionnotes on the mycelium on the seed. It has been observed associated with seeds of capsicum, tomato, paddy, sorghum,

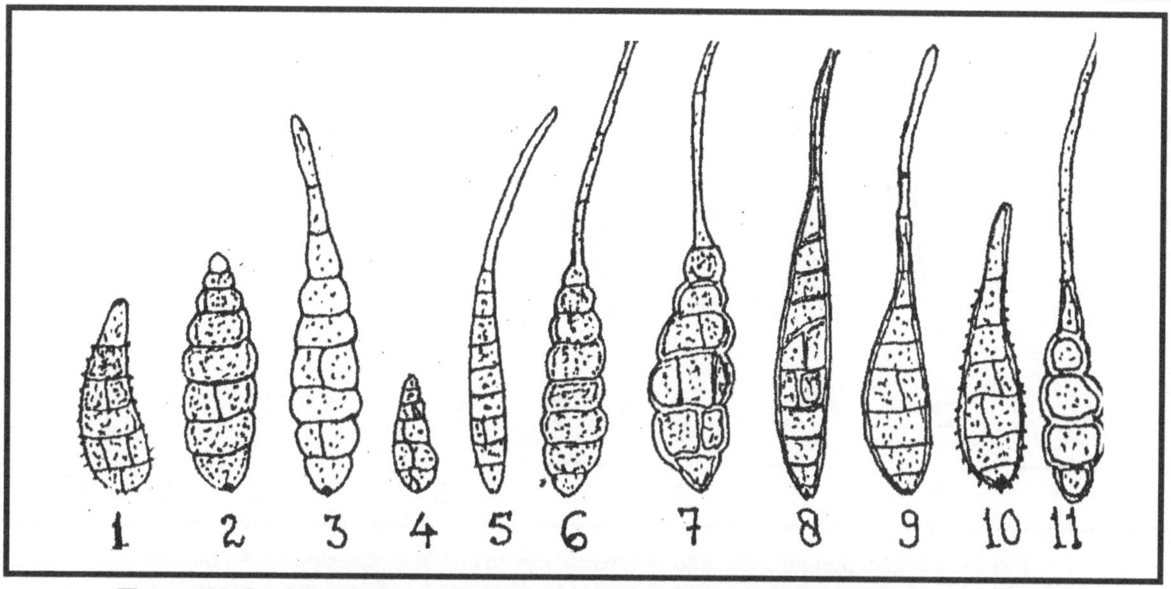

Figure 2.1: Conidia of Eleven Species of *Alternaria* with Distinguishing Characters
(1) *A. alternata*, (2) *A. brassicicola*, (3) *A. brassicae*, (4) *A. lini*, (5) *A. linicola*, (6). *A. carthami*, (7) *A. sesami*, (8) *A. solani*, (9) *A. porri*, (10) *A. longipes*, (11) *A. padwickii*

coriander, cumin etc. *F. semitectum* is seed borne on crops like paddy, green gram, okra, pearl-millet, sorghum, maize etc. Highly branched, shiny conidiophores are formed on the mycelium and at the tip macroconidia are formed. *F. poae* is seed borne usually in monocots like wheat, barley, oat etc. Hyphae are dull white, mlcroconidial masses are irregularly arranged along the hyphae. Macroconidoa develop on pionnotes. The colony emits fruity odour. *F. graminearum* is seed borne on seeds of wheat, paddy, etc. A good growth of white mycelium develops on seed surface and brownish pionnotes are formed. The macroconidia are pointed at both ends and possess a curved foot cell. *F. longipes* forms a rich growth of whitish hyphae giving pinkish stain to the seed and blotter on which the hyphae grow at the base of the seed. Sporodochia and pionnotes are formed. Macroconidia are much elongate with whip like apical part and a long foot cell. It is found associated with paddy, sorghum, cumin, coriander etc. *F. avenaceum* has been observed with seeds, of wheat, barley, paddy, oat, onion etc. It forms white mycelium on the seed on which orange coloured, slimy sporodochia are formed. The macroconidia are long, filiform, curved, pointed at both ends. *F. chlamydosporum* develops cottony mycelial growth on seed surface on which yellowish to brown sporodochia or pionnotes are formed. Chlamydospores are produced singly or in chains. Conidia are small, one to three septate, fusiform to ellipsoldal. It is seed borne in pearl millet, sorghum, paddy etc. In *F. dimerum* white to pinkish mycelium grows on the seed surface on which orange coloured circular pionnotes and sporodochia are formed. Conidia are small, falcate, hyaline, pointed at two ends, usually with one septa. It has been observed on seeds of paddy, brinjal etc. (Figure 2.2).

Drechslera

Effuse grey or blackish mycelium develops on the seed surface and even on the blotter close to the seed from which erect, simple, dark olivaceous, often geniculate conidiophores come out. The conidia are brownish black, usually straight, cylindrical, rounded at the tip. Variations are observed in *Drechslera*

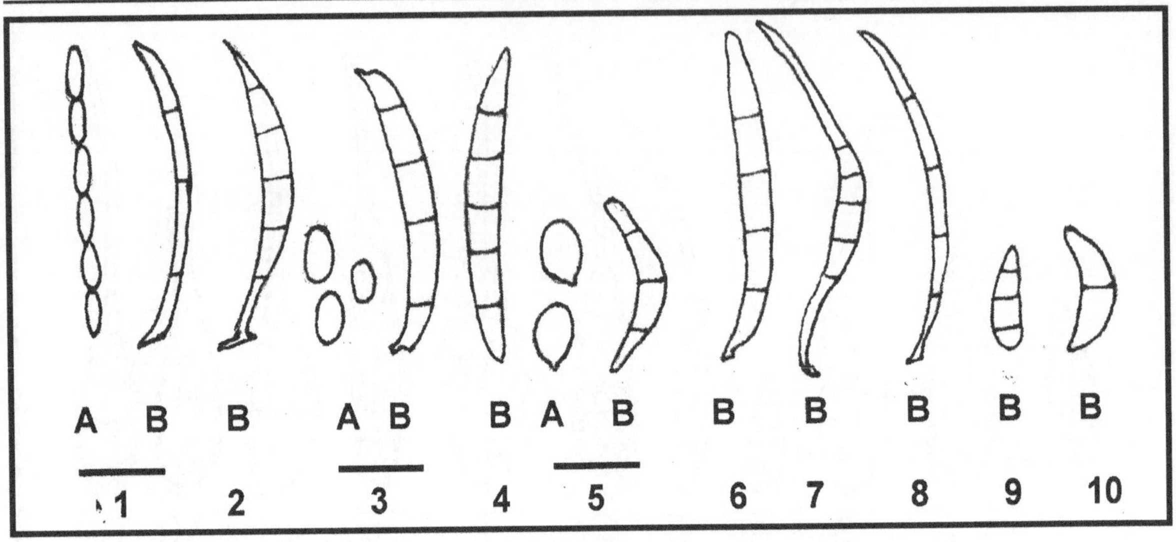

Figure 2.2: Microconidia (A) and Macroconidia (B) of Ten Species of *Fusarium*
(1) *F. moniliforme*, (2) *F. equiseti* (3) *F. solani*, (4) *F. semitectum*, (5) *F. poae*, (6) *F. graminearum*, (7) *F. longipes*, (8) *F. avenaceum*, (9) *F. chlamydosporum*, (10) *F. dimerum*.

species on seeds in the blotter test and their conidial morphology under stereoscopic wild microscope and compound microscope.

D. bicolor produces grey to light brown mycelium on the seed, conidiophores brown, straight or flexuous with scars. Conidia mostly straight, broader in the middle, tappering at the ends and rounded at the tips, divided by 3–14 pseudosepta, central cells dark brown, the end cells pale cut off by a dark septum. It has been found associated with seeds of Brassicas, coriander, lesser millets, tomato, paddy, pearl millet, sorghum, wheat etc. *D. graminea* forms dark mycelium on the seed from which dark brown straight, short conidiophores emerge. Conidia are straight, broad at the base, taper towards tip, several pseudosepta, dark hilum, secondary conidiophores and conidia are formed usually from the terminal cell of the conidium. It is seed borne in barley. *D. halodes* produces conidiophores singly or in pairs, straight or flexuous, dark brown, upper part geniculate. Conidia acropleurogenus, dark brown, ellipsoid, broader in the middle, dark septa at two ends of the middle cell visible, hilum protruberant, pseudosepta 3–18. It has been reported seed borne in paddy, wheat sorghum, lesser millets, barley, sesame, cowpea etc. *D. hawaiiensis* forms blackish hyphae on seed producing shorter conidiophores. They are geniculate. Conidia are formed in clusters towards the tip. They are straight, oblong, rounded at the ends, 4 to 7 septate, brownish. The fungus is seed borne in several crops like okra, oat, brassicas, capsicum, coriander, cucurbits, barley, paddy, blackgram, greengram, cowpea, wheat etc. *D. rostrata* grows greyish mycelium on the seed. Conidiophores are-straight or flexuous, brownish bearing conidia which are straight or slightly curved, rostrate, broader at the middle with 6 to 16 pseudosepta, end cells hyaline, cut off by thick dark septa, hilum protruberant. The fungus has been recorded on seeds of several crops like wheat, sugarbeet, linseed, capsicum, linseed, cucurbits, paddy, maize, pearl millet, sorghum etc. *D. longirostrata* is different from *D. rostrata* in having very long slender, acropleurogenous, rostrate conidia, broadest near the basal septum, narrowing towards the tip, basal and apical septum dark and thick, hilum protrubelrant. The fungus is seed borne in wheat, paddy, sorghum, pearl millet, cowpea, greengram, cumin etc. *D. maydis* forms dark brown to blackish mycelium on seed surface. Conidiophores are straight or flexuous forming distinctly curved conidia, light brown, 5 to 11 pseudosepta, taper towards rounded end. It is found associated with seeds of paddy, pearl millet, lesser millets, sorghum, maize, capsicum, coriander etc. *D. oryzae* is responsible for brown spot in paddy. The fungus grows on seeds as fluffy mycelium covering the entire seed or a part of it. Conidiophores are flexuous, brownish bearing fuliginous to olivaceous brown, curved conidia which are widest in the middle, tapering to rounded ends, 7 to 11 septate. Besides paddy it has been found associated with wheat, sorghum etc. *D. sorokiniana* produces grey to dark grey mycelium on seed. Conidiophores are short, brown erect, forming dark brown, straight or slightly curved conidia which are widest in the middle with rounded ends and 6 to 9 pseudosepta. It is seed borne on grasses, oat, paddy, sorghum, pearl millet, wheat, barley, capsicum, soybean, coriander, sunflower etc. *D. teres* forms dark brown to black, straight conidiophores with swollen base. The conidia are light brown, straight, cylindrical, rounded at the ends, basal cell inflated, 1 to 6 pseudosepta. It causes net blotch in barley and is highly seed borne. *D. tetramera* produces brownish, short conidiophores bearing conidia in clusters which are brown, straight with broadly rounded ends, lighter towards end cells, three septate, hence divided into four cells. It is seed borne in oat, coriander, cucurbits, cumin, carrot, radish, sesame, cowpea, wheat, sorghum, greerigram etc. *D. turcica* forms long, straight, flexuous, geniculate, dark brown conidiophores. Conidia are straight or slightly curved, lighter coloured, ellipsoid or obclavate, widest in the middle with tapering ends and rounded apex, hilum distinctly protruberant. It is seed borne in maize, sorghum and grasses. *D. victoriae* develops straight or flexuous light brown conidiophores with slightly curved broadly fusiform, 5 to 12 septate conidia. It is seed borne in grasses, wheat, barley, sorghum, soybean etc.

Curvularia

Several species of *Curvularia* are associated with seeds of different crops. They cause leaf spots, leaf blights, kernel rot, root rot, seedling blights, grain discolourations and grain deformation. *Curvularia* species are more important in monocots than dicots. Detailed information about taxonomy is dealt by Ellis (1971) and Subramanjan (1971). Benoit and Mathur (1970) published a guideline for easy

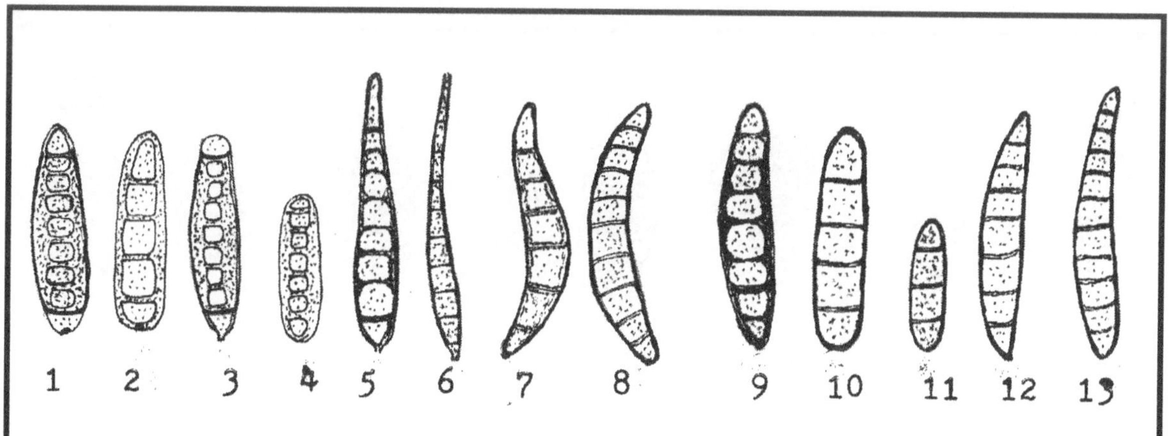

Figure 2.3: Conidia of 13 Species of *Drechslera* Seed Borne in Nature

(1) *D. bicolor*, (2) *D. graminea*, (3) *D. halodes*, (4) *D. hawaliensis*, (5) *D. rostrata*, (6) *D. longirostrata*, (7) *D. maydis*, (8) *D. oryzae*, (9) *D. sorokiniana*, (10) *D. teres*, (11) *D. tetramera*, (12) *D. turcica*, (13) *D. victoriae*

identification of 15 *Curvularia* species based on growth characters on seeds in standard blotter method after incubation. The species can be confirmed by examining conidia under compound microscope. *Curvularia* species in general develop grey to black mycelium on seeds which grow even on blotter around the seed. The identification of species is based on the characters of conidiophores and conidia.

In *C. pallescens* the conidiophores are simple, erect or flexuous, geniculate, small, slender. Conidia are three septate, straight or slightly curved, third cell from the base larger, pointed apical end cells. The central cells are darker. *C. lunata* forms conidiophores dark brown, unbranched, twisted on seeds of some crops. Conidia three septate, third cell from the base broader and darker, subhyaline rounded apical cell. *C. uncinata* has brown, simple conidiophores with strongly curved hookshaped, four septate, middle cell largest, broadest and darkest, bent at 90° or more, two apical cells shorter with rounded ends, blackish scar is present at the base. *C. inaequalis* has conidiophores brown, simple, geniculate towards the tip. Conidia are borne acropleurogenously or acrogenously, four septate, unequally ventricose–fusiform in shape or straight, the third' cell from the base broader and darker, the apical cells with rounded tip, the basal cell has a scar. *C. geniculata* forms brown, simple, erect, straight conidiophores with conidia four setate, geniculate or straight and fusiform, the third cell from the base slightly bigger, end cells slightly hyaline. *C. eragrostidis* produces conidiophores straight, brown, forming barrer-shaped or ellipsoidal conidia with a thick dark median band. *C. intermedia* produces conidia acropleurogenously, curved or ellipsoid with unequal sides, thick median band at the widest part of the conidium. In *C. cymbopogonis* conidia are formed acropleurogenously. They are straight, obovate or elliptical, black, four septate, broad in the middle, dark septa on both the sides of the bigger cell (Figure 2.4).

Cercospora

Several species of *Cercospora* causing diseases or crops are seed borne, they are differentiated through the characters of conidia. Monographs on *Cercospora* by Chupp (1953) and Vasudeva (1963) can be used for identification. Besides good information is available in the books by Ellis (1971; 1976) and Subramanian (1971). The species cause leaf spots, leaf blights, seed discolourations etc.

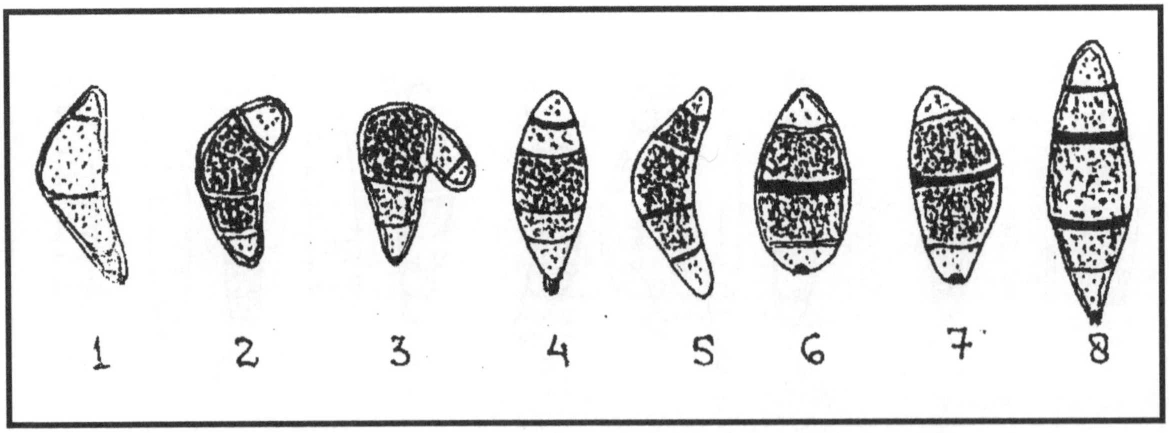

Figure 2.4: Spore Characters of Eight *Curvularia* Species on Seed

(1): *C. pallescens*, (2) *C. lunata*, (3) *C. uncinata*, (4) *C. inaeqalis*, (5) *C. geniculata*, (6) *C. eragrostidis*, (7) *C. intermedia*, (8) *C. cymbopogonis*

C. arachidicola forms conidia subhyaline, obclavate, mildly curved, 3 to 12 septate, 3–5 × 15– 45 µ, rounded at the base. The main host is groundnut. *C. personata* forms cylindric conidia, medium dark olivaceous, slightly spindle shaped, straight or slightly curved, 1–8 septate, bluntly rounded ends, 5–7.5 × 20–70 µ. It is also seed borne in groundnut. *C. beticola* is seed borne in beet root crop. On seed hyaline, acicular-conidia are formed which are slightly curved, 2.5–4 × 50–200 µ. In *C. capsici* the conidia are hyaline, acicular, 2.5–4 × 50–200 µ. It is seed borne in capsicum. *C. carotae* has conidia cylindro-obclavate, hyaline to sub hyaline, 1 to 6 septa, 2–3.5 × 30–115 µ. The main host is carrot. In *C. carthami* conidia are hyaline, acicular, straight or slightly curved, 2.5–5 × 50–300 µ. The main host is safflower. *C. kikuchii* is seed borne in soybean and is responsible for causing purple stain of seeds. The conidiophores grow from cravices in seed coat. The conidia are long, hyaline, acicular, 2.5–5 × 50–375 µ. *C. nicotianae* is seed borne in tobacco. The conidia are hyaline, acicular, base truncate, tipe acute to subacute, 3–4 × 35–150 µ. *C. oryzae* is seed borne in paddy. Conidia are hyaline, cylindric to-cylindro-obclavate, 1 to 4 septate, tip blunt, 3–6 × 40–60 µ. *C. sesami* on sesame seed produces acicular, hyaline, straight to curved conidia with truncate base and acute tip, 2–4.5 × 30–150 µ. *C. sojina* produces conidia hyaline, cylindrical to obclavate, not acicular, base subtruncate, tip obtuse, 4–8 × 20–80 µ. It is seed borne in soybean. *C. sorghi* seed borne in sorghum, maize, grasses etc. It produces hyaline, acicular to cylindric conidia, 2–4 × 30–70 µ.

Myrothecium

Three species of *Myrothecium* have been observed associated with seeds of various crops. They have become more important due to their pathogenic potential. They can be distinguished on the structural formation of sporodochia, hyphal growth and conidial mass (Nguyen *et al.*, 1973).

In case of *M. roridum* on seed the sporodochia are sessile, confluent, viscid, greenish black later turn black with whitish hyphae around the sporodochium forming a white margin. Setae are absent. Conidia are one celled, small, cylindrical with rounded ends, hyaline but on maturity turn black. It is

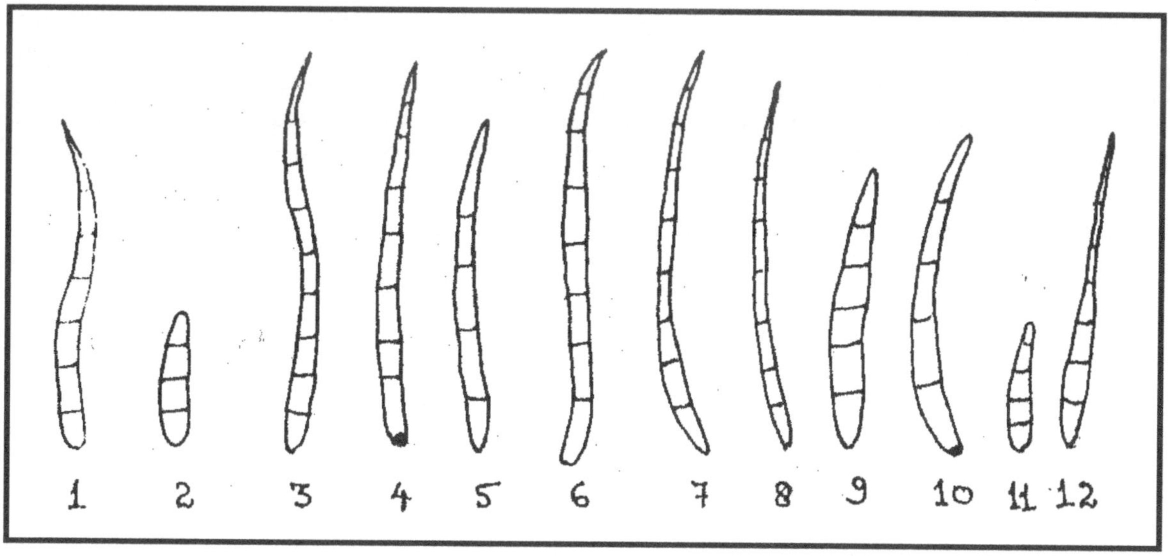

Figure 2.5: Conidia of 12 Seed Borne Species of *Cercospora*

(1) *C. arachidicola*, (2) *C. personata*, (3) *C. beticola*, (4) *C. capsici*, (5) *C. carotae*, (6) *C. carthami*, (7) *C. kikuchii*, (8) *C. nicotianae*, (9) *C. oryzae*, (10) *C. sesami*, (11) *C. sojina*, (12) *C. sorghii*

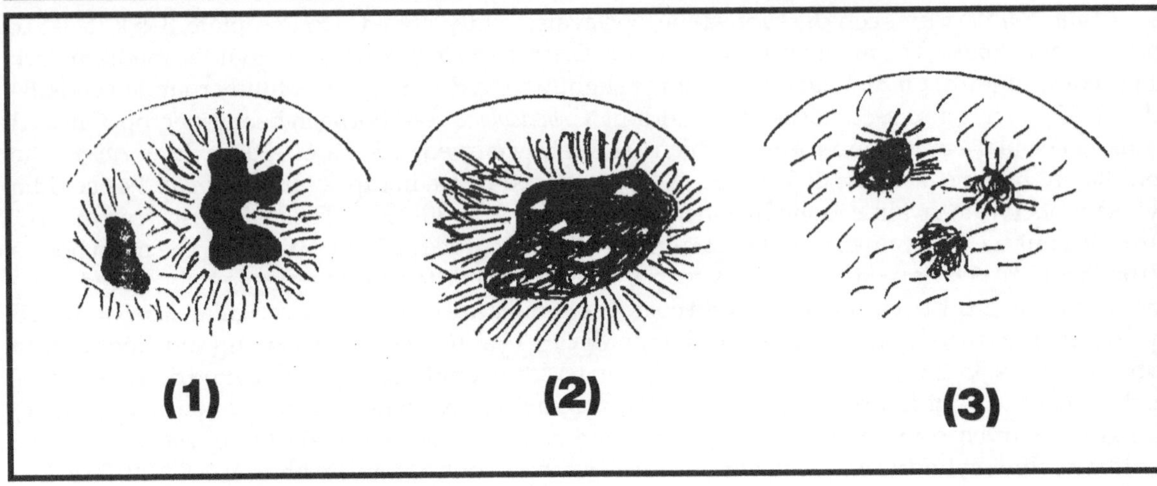

Figure 2.6: Sporodochia of Three Species of *Myrothecium* on Seed
(1) *M. roridum*, (2) *M. verrucaria*, (3) *M. leucotrichum*

seed borne in cowpea, soybean, holyhock, other legumes etc. *M. verrucaria* develops sporodoch1a similar to *M. roridum* but-short contorted white hyphae emerge from the spore mass, hence the black mass of conidia is not continuous. *M. leucotrichum* forms circular, discoid sporodochia surrounded by-white long-straight setae. Conidia are one celled, small, cylindrical with rounded ends, black on maturity.

Colletotrichum

Species of *Colletotrichum* are responsible for causing anthracnose, wither tip, stem rot, fruit rot, pre and post emergence rots in many crops. They are seed borne both extraembryal and intraembryal. Seed to plant infection is local as well as systemic. They form acervuli where the mycelium produces setae, conidiophores and conidia. Their identification on seed is based on setae size in relation to the conidial mass in acervuli under stereobinocular wild type microscope and shape of conidia under compound microscope (Kulshrestha *et al.*, 1976).

When in acervuli the setae are longer than the spore mass and the setae are numerous with fusoid conidia, it is *C. dematium* but in *C. graminicola* the conidia are falcate. In *C. lindemuthianum* the setae are few but more than two and the conidia are dumble shaped but in *C. lini* the setae are two or less and conidia are allantoid. In *C. gloeosporioides* the setae are shorter than the spore mass and the conidia are straight, cylindrical with both ends rounded, but in *C. acutatum* both ends of the conidia are pointed.

C. dematium is seed borne in paddy, capsicum, cowpea, okra, soybean etc. *C. graminicola* in maize, sorghum, cowpea, blackgram etc. *C. lindemuthianum* in beans etc., *C. lini* in linseed, *C. gloeosporioides* in sorghum, Tropaeolum etc., *C. acutatum* in capsicum, Zinnia etc.

Aspergillus

Several species of *Aspergillus* are associated with seed. They produce long, erect conidiophores with a bulbous head called vesicle. A number of sterigmata in one or two layers are formed on which conidial chains are produced, in basipetal order. For identification of the species Raper and Fennell

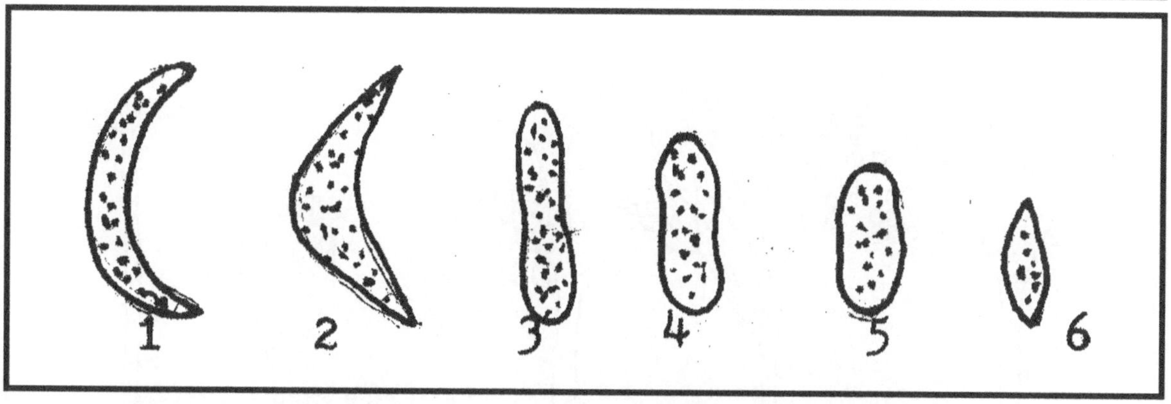

Figure 2.7: Spore Forms of Six Species of *Colletotrichum* Seed Borne on Crops
(1) *C. dematium*, (2) *C. graminicola*, (3) *C. lindemuthianum*, (4) *C. lini*, (5) *C. gloeosporioides*, (6) *C. acutatum*

(1965) and Thom and Raper (1945) can be consulted. *A. restrictus, A. repens, A. ruber, A. amsteldami, A. candidus, A. ochraceus, A. flavus, A. niger, A. fumigatus* are associated with seeds under storage. They affect the seed quality in respect of nutritive value as well as germination. Some species produce toxins which are poisonous to consumers like *A. flavus* which forms aflatoxins. *A. parasiticus, A. fumigatus, A. ochraceus* also produce harmful toxins. *A. nidulans* and *A. glaucus* are also harmful resulting in *Aspergillosis*. Some carcinogenous like aflatoxin.

Penicillium

It produces long conidiophores bearing broom like whorls-called penicillus which are verticillately branched several times forming phialides on which conidia are formed in chains in basipetal manner. Several species of *Penicillium* are seed borne mostly resulting in storage problem. Examples are *P. chrysogenum, P. purpurogenum, P. cyclopium, P. oxalicum, P. patulum*. Some species like *P. citrinum, P. islandicum, P. rubrum* and *P. viricatum* are toxin producers.

Botrytis

It forms conidiophores which are tall, straight or flexus smooth, brown, branched dichotomously or trichotomously which are swollen in the form of head at the end. Conidia are generally single celled, pale brown, spherical or obovoid or ellipsoid. Several species of *Botrytis* are seed borne, *B. allii* (onion), *B. anthrophila* (red clover), *B. fabae* (Faba bean), *B. ricini* (castor), *B. cinerea* (chickpea) etc.

Stemphylium

The mycelial growth on the seed is similar to *Alternaria* being brown to blackish, cottony. Conidiophores are unbranched and straight. Conidia are single, oblong, rounded at the ends, verrucose or echinulate, muriform. Important seed borne species are *S. botryosum* (clover, onion, bean), *S. callistephi* (china-aster), *S. loti* (bird's foot trefoil), *S. sarciniforme* (clover), *S. solani* (tomato, brinjal), *S. trifoli* (*Trifolium repensa*), *S. lycopersici* (tomato) etc.

Cladosporium

Mycelial growth on the seed is velvety, blackish forming conidiophores straight or flexuous, unbranched but branched at the uppermost part only. Conidia are single or in short chains, simple,

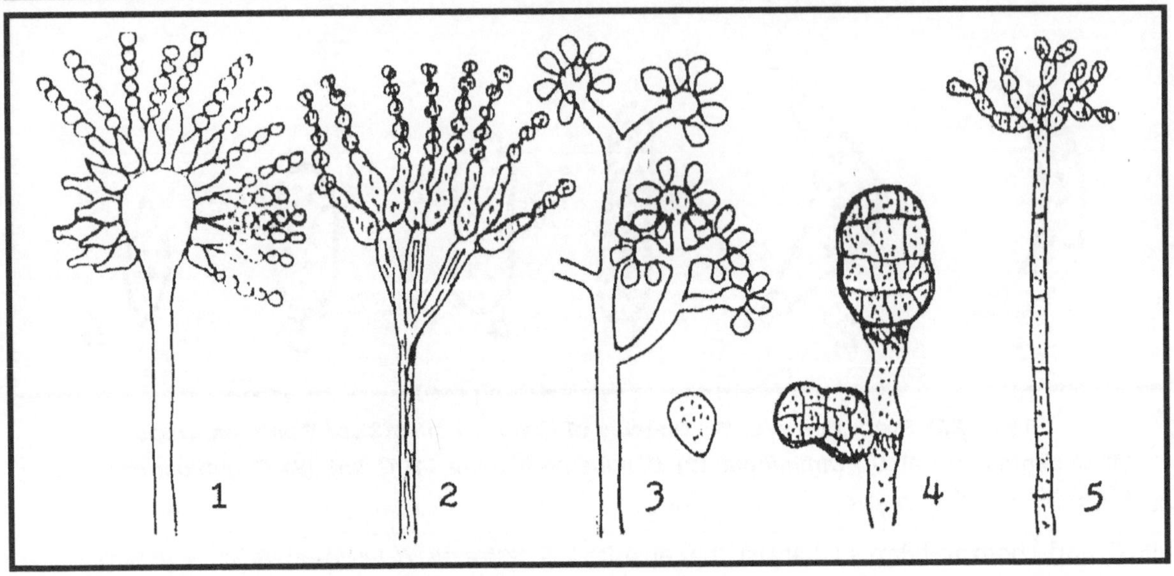

Figure 2.8: Conidiophores and Conidia of (1) *Aspergillus*, (2) *Penicillium*, (3) *Botrytis*, (4) *Stemphylium*, (5) *Cladosporium*

ellipsoid or ovoid with a scar at each, end, verrucose or echnulate, 0 to 3 septa. The seed borne species are *C. cladosporioides* (pea), *C. herbarum* (greengram, sorghum), *C. cucumerinum* (cucurbits) etc.

Sphaeropsodaceae (*Phomaceae*)

Most of the pycnidium forming seed borne fungi belong to this family. They produce dark, globose or pyriform pycnidia witt an ostiole to release the pycnidiospores formed inside. There is great diversity in different genera and species in the structure and shape of pyondium and pycnidiospores (Sutton, 1964 and Neergaard, 1977).

Macrophomina

M. phaseolina has mycelial and sclerotial stage named *Rhizoctonia bataticola*. The fungus is seed borne in many crops like soybean, blackgram, greengram, cowpea, groundnut, sesame, sorghum, millets etc. It is externally as well as internally seed borne. Blackish hyphae are formed and pear like carbonaceous pycnidia are formed with hyphal growth near the ostiole. Spores are elongated, cylindrical to ellipsoid, one celled, hyaline. They come out in a cirrhus.

Phoma

Pycnidia are globular with protruding ostiole. Spores are one celled, small, hyaline. Important seed borne species are *P. apiicola* (celery), *P. betae* (beet root seed), *P. destructiva* (capsicum, okra), *P. exigua* (bean, cowpea, linseed), *P. glumarum* (paddy), *P. lingam* (crucifers), *P. medicaginis* (soybean, pea).

Phomopsis

It forms dark pycnidia with ostiole, erumpent, nearly globose with two types of hyaline spores, one celled ovoid or fusoid alpha spores and the other filiform curved beta spores. Important seed borne species are *P. sojae* (soybean, lima bean), *P. vexans* (brinjal) and *P. heveae* (rubber seeds).

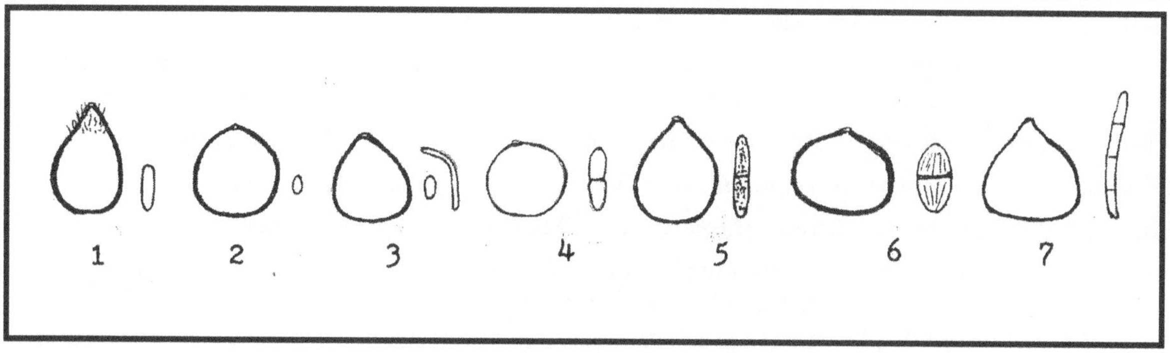

Figure 2.9: Pycnidial and Spore Characters of (1) *Macrophomina*, (2) *Phoma*, (3) *Phomopsis*, (4) *Ascochyta*, (5) *Diplodia*, (6) *Botryodiplodia*, (7) *Septoria*

Ascochyta

In *Ascochyta* the pycnidia are separate, globose, ostiolate with two celled hyaline, ovoid or oblong spores. Important seed borne species are *A. fabae* (broad bean), *A. gossypii* (cotton), *A. pinodes* (pea), *A. pisi* (pea) and *A. rabiei* (chickpea).

Diplodia

In *Diplodia* the pycnidia ere black, globose, erumpent with ostiole. The spores are dark, two celled, ellipsoid or ovoid with notch near the septa, pigmented. Some important seed borne species are *D. frumenti* (maize), *D. maydis* (maize), *D. gossypina* (cotton) etc.

Botryodiplodia

It also produces black pycnidia with ostiole, thick walled, erumpent with two celled, dark, ovoid to elongate pycnidiospores. Striations are observed in spores on maturity. Most important seed borne species is *B. theobromae* (Cacao, cotton, groundnut, greengram).

Septoria

Pycnidia are globose, thin walled, papillate, ostiolate, dark and form hyaline, filiform, several celled spores. Important seed borne species are *S. apiicola* (celery), *S. avenae* (oats), *S. glycines* (soybean), *S. lactucae* (lettuce), *S. linicola* (linseed), *S. lycopersici* (tomato), *S. pisi* (pea), *S. secalis* (Rye), *S. tritici* (wheat) etc.

Future Strategies

Crops are being grown in different agro-ecological situations under varied atmospheric conditions which favour or disfavour association of fungi with seeds. The time of plant to seed transmission of fungal pathogens decides the seed-pathogen relationship. There are a number of crops grown in different crop zones. Great diversity exists in the mycoflora which is seed borne. It is very necessary to examine them critically as they exist with seed as mycelial structure, conidiophores, conidia, pycnidia and perfect stages after incubation. Proper care is needed to comply with the Sanitary and Phyosanitary requirements under WTO.

References

Benoit, M.A. and Mathur, S.B. (1970). Identification of species of *Curvularia* on rice seed. *Proc. Int. seed Test. Ass.*, 35: 99–119.

Booth, C. (1971). *The Genus Fusarium*. CMI, Kew Surrey, England.

Chidambaram, P., Mathur, S.B. and Neergaard, P. (1973). Identification of seed borne *Drechslera* species. *Friesia*, 10: 165–207.

Ghupp, C. (1953). A Monograph of the fungus genus *Cercospora*. *Ithaca*, New York, pp. 667.

Ellis, M.B. (1971). *Dematiaceous Hyphomycetes*. CMI, Kew, Surrey, England, pp. 608.

Ellis, M.B. (1976). *More Dematiaceous Hyphomycetes*. CMI, Kew, Surrey, England, pp. 507.

Khare, U.K. and Khare, M.N. (2002). Purple blotch of onion present status and future strategies for sustainable production. In: *Plant Pest Management*, (Ed.) P.C. Trivedi. Aavishkar Publishers, Distributors, Jaipur, pp. 262–277.

Kulshrestha, D.D., Mathur, S.B. and Neergoard, P. (1976). Identification of seed borne species of *Colletotrichum*. *Friesia*, 11: 116–125.

Neergaard, P. (1945). Danish species of *Alternaria* and *Stemphylium*. *DGISP*, Denmark, pp. 559.

Neergaard, P. (1977). *Seed Pathology*. The McMillan Press Ltd., London, pp. 1187.

Nguyen, T.H., Neergaard, P. and Mathur, S.B. (1973). Seed borne species of *Myrothecium* and their pathogenic potential. *Trans. Br. Mycol. Soc.*, 61: 347–354.

Ram Nath, Neergaard, P. and Mathur, S.B. (1970). Identification of *Fusarium* species on seeds as they occur in blotter test. *Proc. Int. Seed. Test. Ass.*, 35: 121–144.

Raper, K.E. and Fennell, D.I. (1965). *The Genus Aspergillus*. The Williams and Watkins Co., Baltimore.

Subramanian, G.V. (1971). *Hyphomycetes*. ICAR, New Delhi, pp. 930.

Sutton, B.G. (1964). Phoma and related genera. *Trans. Brti. Mycol. Soc.*, 47: 497–509.

Thom, G. and Raper, K.B. (1945). *A Manual of the Aspergilli*. The Williams and Wilkins Co., Baltimore.

Vasudeva, R.S. (1963). *Indian Cercosporae*. ICAR, New Delhi, pp. 245.

©R.C. Ram & Asha Sinha (*Editors*)
***Microbial Diversity: Modern Trends*, 2007**
Daya Publishing House, New Delhi

Chapter 3

Use of Various Agricultural Residues and By-products for Mushroom Cultivation: A Review

R.C. Ram

Department of Mycology and Plant Pathology, Institute of Agricultural Sciences, Banaras Hindu University, Varanasi – 221 005 (U.P.)

ABSTRACT

Mushrooms species grow and yield on a spectrum of agricultural residues and by-products. Chemically these agricultural residues are lignocellulosics, composed of various levels of lignin, cellulose and hemicellulose. Biodegradation and bio- transformation of lignocellulosics biomass by edible fungi represent a successful solid state fermentation for effective utilization of agricultural residues and by-products for mushroom production on large scale. Large quantities of lingo-cellulosic agricultural residues and other organic wastes are generated annually through the activities of agricultural, forest and food processing industries. Mushroom growers all over the world are facing several problems such as hunger, poverty, unemployed, multifaceted health hazards and environmental pollutions. By this way, the valuable use of agricultural residues and by-products for mushrooms cultivation can provide more food, more jobs, better family income, improved living standard, safe to human health and environmental pollutions, curb global warming and CO_2 level, clear up crop residues, road sides and forest margins.

Introduction

Mushrooms cultivation presents an economically important biotechnological industry that has distinguishly expanded all over the world in the last few years. Mushroom cultivation mainly depends on the agricultural crop residues and by-products. These crop residues and by-products are abundantly available in our country. It is primarily used as cattle feed and remaining part is burnt or spread in the fields. Agro-wastes contain about 80 per cent cellulose, hemi-cellulose and lignin which are not easily degradable, but most of the edible fungi possess enzyme system which can degrade these components. Thus their mushrooms cultivation can avoid environmental pollution by recycling agricultural wastes and can also convert straw into easily digestible animal feed.

Huge quantities of lingo-cellulosic crop residues and other organic wastes are generated annually through the activities of agricultural, forest and food processing industries. According to the estimates of Food and Agricultural Organization (FAO, 1989) about 2946 million tones of cereals straws are produced worldwide of which Asia's share is 1135 million tones and that of India is 240 million tones. In India, during 1993–94 production of paddy straw and wheat straw was approximated at 110.6 million tones and 153.6 million tones respectively (Mathew 1996). Recently about 385 million tones of agricultural wastes are available annually in India and about half of this residue remains unused. The major parts of these residues are burnt after harvesting, resulting multifaceted hazards including oxygen deficient environment, respiratory diseases (such as allergic, bronchitis, asthma, tuberculosis) and poor visibility at night. If even one per cent of this crop residue is used to produce mushrooms, India will become a major mushroom producing country in the world (Tewari and Pandey, 2002).

Edible mushroom production represented an attractive method of improving the nutritional quality of ligno-cellulosic wastes for use as an animal feed stock. Among the various physical, chemical and biological methods used for upgrading the digestibility and nutritive value of agricultural wastes, biodegradation by using white rot fungi including mushrooms have been found promising. The species of *Pleurotus* has ability to excrete hydrolyzing and oxidizing enzymes (Toyama and Ogawa, 1974; Daugulis and Bano, 1977 and Rajarathnam *et al.*, 1979) which enables them to grow and flourish over wide range of natural ligno-cellulosic waste materials.

Around 800 million people living in 46 countries are malnourished, 40,000 die everyday of hunger and hunger related diseases (Swaminathan 1995). In this context the mushroom cultivation represent one of the economically viable processes for the bioconversion of agricultural and agro-industrial wastes in to protein rich food making it a potent weapon against malnutrition in developing countries like India which has lowest per capita consumption of protein in the world (Sohi, 1982, Wood, 1989, Chang and Miles 1989, Buswell and Chang, 1993). It is also consistent with the emerging view that an ecologically oriented society must use its wastes as resources rather than discarding them as useless materials.

Even in case that the world population would not increase any more, there is an enormous amount of wastes from field, agro-industry, and wood industry. Only using 25 per cent of the yearly burned cereal straw in the world could result in a mushroom yield of 317 million metric tones (317 milliard kg) of fresh mushroom per year (Chang and Miles 1989). But on this moment, the yearly mushroom production is only 6 milliard persons or 1 kg per year or 3 gram per day (Courvoisier 1999).

In fact, counting the early available world wastes in agriculture (500 milliard kg) and forestry (100 milliard kg), we can easily grow 360 milliard kg of fresh mushroom on the total of 600 milliard kg dry wastes! This would bring us a yearly mushroom food of 60 kg per head per year, containing 4 per cent protein in fresh mushroom and we know that 30 per cent of the world population is protein deficient. On the other hand, we all know the high risk of further population growth with again more need for food (field crops) and wood (forest) swelling up the already so high mountains of wastes!

In this regard, Pope (2000) have to considered the multilateral message of agricultural wastes conversion into mushrooms in the following categories :

1. Provision of food
2. Creation of jobs
3. Enhancement of family income
4. Control burning of waste and curb global warming
5. Lowering air pollution and CO_2 level
6. Cleaning the field, road sides and forest margins
7. Protection of natural mushrooms flora
8. Forest fire prevention
9. Spent mushroom substrates for garden compost bioremediation purposes.

Why Use of Agricultural Residues and By-products for Mushroom Production

Numerous reasons for growing more mushrooms by agricultural residues and by-products as approved by Chang and Miles 1989; Stamets 1993 and Pope 1995.

1. Decreasing the mountain range of wastes.
2. Most mushrooms are fast productive and can be grown all over the year in temperate zones with heating in winter and in tropical zones with protection against dry hot seasons.
3. No fight for field is necessary because mushroom cultivation needs only small spaces and can even grow in vertical layers or on the floor in forests where they don't need one centimeter extra place.
4. Converting the wastes in to mushroom protein and vitamins, while also mushrooms represent one of the world's greatest untapped sources of tasteful food in the future. And let us not forget the medicinal value of mushrooms species.
5. Mushroom cultivation is labour intensive but labours are readily available in most tropical regions.
6. Mushrooms are type of vegetable and like other vegetables they should be available in every vegetable garden.
7. The domesticated mushrooms are possess such a rich gamma of enzymes such as cellulose, hemicellulose, ligninases, proteases etc. making it possible to biodegrade and assimilate a multitude of organic wastes substrates.
8. Finally mushrooms on wastes is a duty for all of us for the restoration of the deep damaged ecosystem and it is the only way to close the gap between endless monoculture and the original balanced growth-planning of the ecosystem on planet earth.

More than 1500 macro fungi are already widely used for human food although only a few are grown large scale, but because all these macro-fungi are active in bio-transformation of lignocellulose waste, we should ten times more stimulate the domestication of at least hundred desired wild edible species (Somasundaram *et al.*, 1992). Different types of Agricultural residues and by-products can be used for spawn production, substrates and compost preparation, substrate amendments and casing materials of all cultivated mushrooms.

Use of Agricultural Residues and By-products for Spawn Production

Mostly cereal grains or non-waste granules are used for the preparation of mushroom spawn but in some cases other substrates like straw, sawdust or composts are an excellent alternative because of their longer life, their selectivity, availability, cheapest, comfortably and antifeedant effect on mice and other grubbing animals.

Duggar (1905) for the first time introduced pure tissue culture spawn preparation technique which is no more in practice today. Sinden (1932) introduced grain spawn, with the addition of calcium salts to rye grains. Terrier (1945) used wheat and corn grain spawns for cultivation of *Psalliata compestris.*

Singer (1961) has suggested various wastes can be used for spawn production of different species of mushrooms. Apple peels and other apple residues mixed with paper waste for *Morchella.* Chopped hay and fennented horse manure for *Agaricus.* Rice bran, as additive with sawdust for shitake mushroom (*Lentinus* species). Spent straw viable *Volvariella* bed straw inoculation of new *Volvariella* beds, mostly when production becomes scanty. Chopped tobacco stem used for spawn production of *Agaricus.*

Stoller (1962) preferred rye grain to sorghum for avoiding sectoring and recommended the use of 6 g of gypsum and 1.5 g of chalk per pound of grains to avoid clumping of grains.

Kumar *et al.* (1975) found that jowar grain gives better growth of mushroom mycelium after mixing 2 per cent gypsum and 6 per cent calcium carbonate by weight, with the boiled grains.

Sivaprakasam and Kandaswamy (1981) recorded maximum yield of *Pleurotus sajor-caju* from jowar and bajra grain spawn followed by maize grain spawn.

Bisht and Harsh (1984) recommended the use of waste tea leaves (10 : 1) ratio of wheat grain and tea leaves) in place of calcium carbonate and gypsum for grain spawn production of *Agaricus bisporus* and *Volvariella volyacea.*

Chang and Miles (1989) tea leaves used in Asia for spawn of shitake mushroom and *Volvariella* mostly mixed with grains or cotton waste. Sawdust and rice bran was a good spawn substrate for *Hercium.* Sugarcane bagasse used for *Tremella.* Corncobs, hammaennilled, cotton seed hulls, peanut hulls used for shitake. Paddy straw special spawn substrate for *Volvariella.* Lotus seed husks, lotus is a kind of lilly (Nymphara) for *Volvariella* spawn. Wheat bran, additive with sawdust spawn for shitake, *Pleurotus* or *Stropharia.* Grass chaff single or mixed with wheat straw for *Stropharia.*

Poppe (1995) mostly cereal straw is chopped in parts of 1 to 3 cm, it can be mixed with sawdust or bran and is suitable for many kinds of spawn, especially for *Stropharia* and *Volvariella.* Sunflower peels used for *Stropharia.* Sawdust used as single spawn substrate or in combination with all kinds of grains for *Pleurotus, Stropharia, Pholiota, Flamulina* and *Auricularia.*

Agricultural Residues to now Available as Mushroom Substrates

Different types of lignocellulosic agricultural residues such as paddy, wheat straw, barley straw, maize, jowar and bajra stalk, ground haulms and pod shell, cotton waste, mango stiff, sunflower straw, jute coir pith, pulse straw, sugarcane bagasse, water hyacinth, rubber wood dust, some weed straws and leaf fall of tree can be used as a basal substrates. These crops residues should be chopped, dried and free from disease, rotting, contamination and other inert materials.

The species of *Pleurotus* was cultivated experimentally for the first time on tree stumps and sterilized straw by Flack (1917) and on mixture of saw dust and cereals by Kaufert (1935).

Constantin (1923) reported that occurrence of *Pleurotus eryngii* on living roots of *Erynguim alpium.* Lutz (1925) cultivated *P. eryngii* and *P. hedamardii* on stem of various species of *Erynguim, Laserpitium, Apium* and *Sium* and roots of *Celeriae.* First innovative techniques achieved by Block *et al.* (1958 and 1959) for cultivation of *Pleurotus* species.

Bano and Srivastava (1962) first carried out mass production of *Pleurotus substratum* on straw basis. Jandaik (1974) was first described successful cultivation of *Pleurotus sajor-caju* at IARI, New Delhi and found that banana pseudo stems to be the best basal substrate for its cultivation followed by paddy straw.

The prospects of *Pleurotus* cultivation in India were discussed by Kaul and Kachroo (1975). They conducted experiment in order to compare the yields of different strains of *P. ostreatus* at the Regional Research Laboratory, Jammu. Maximum production of fruit bodies (1200 g/sq. ft.) was obtained from a particular Japanese strain under identical conditions.

Bhaskaran *et al.* (1978) introduced a new method 'compact bag method' for cultivation of *Pleurotus sajor-caju* and obtained 87.5 per cent higher yield (1500 g/kg straw) than that of bamboo basket method (800 g/kg straw).

Garcha *et al.* (1981) reported the utilization of agricultural wastes for mushroom cultivation in India, and recommended the preparation of different type of composts by long method from wheat straw, mentha stalks and spent compost of wheat and paddy, along with chemical fertilizers using molasses, ginning mills wastes and poultry litters as activators. The performance of these composts were evaluated under natural conditions of North India. Maximum yield per sq. m was obtained from the compost with molasses.

Chang *et al.* (1981) observed 177.41 per cent biological efficiency of *Pleurotus sajor-caju* on paddy straw, 104.9 per cent on cotton waste and 73.22 per cent on used tea leaves.

Platt *et al.* (1982) successfully cultivated *Pleurotus ostreatus* on cotton straw following 'compact bag method' and obtained 600–700 g/kg dry straw in two flushes. In comparative trials on cotton straw and wheat straw, growth on the former was rapid and fruiting 23 days earlier.

Sathe and Nagarkar (1983) modified Zadrazil's method with the help of perforated bamboo stick inserted vertically in the centre of polythene bag, for the cultivation of oyster mushroom. They obtained fresh yield with (*P. sajor-caju*) of 900 gms, 350, and 400 gms per Kg dry straw during the first, second and third flushes respectively.

Garcha *et al.* (1984) cultivated *Pleurotus florida* and *Pleurotus sajor-caju* on wheat straw, maize, bajra, groundnut stalks, barseem, sugarcane bagasse, chilli, bhindi and potato wastes. They obtained maximum yield of *P. florida* and *P. sajor-caju* (48–60 per cent BE) from wheat and paddy straw.

Garcha *et al.* (1985) used different containers like trays, baskets and perforated polythene bags for the cultivation of different *Pleurotus* species. Polythene bags out yielded the baskets and trays, where a biological efficiency to the tune of 52–55 per cent and 57–60 per cent was obtained for *Pleurotus florida* and *Pleurotus sajor-caju* respectively.

Bhatti *et al.* (1987) obtained higher yield of *Pleurotus ostreatus* on wheat straw and cotton waste in respect of maize cobs and rice straw.

Bhattacharjee *et al.* (1989) showed that wheat grain spawns of *Pleurotus sajor-caju* was inoculated on rice straw beds, using (a) different amount of spawns, (b) different methods of spawning and (c) grain spawn pretreated with nutrients. The maximum yield of sporophores was obtained with

250 g spawn/tray, although good yields were possible with as little as 30 g/tray. The best result was obtained, when spawn was spread only on the middle layer of mushroom bed rather than on the top, bottom or 2–3 layers. Normally, the growth of spawn inoculum enhanced when wheat grain spawn was pretreated with $CaCO_3$ + $CaSO_4$ + trace elements (Zn, Fe, Mg, Mo, Mn, B and Cu), rather than with $CaCO_3$ + $CaSO_4$ alone.

Upadhyay and Vijay (1991) recorded 65–34 per cent biological efficiency (BE) of *Pleurotus florida* and *Pleurotus ostreatus* respectively from wheat straw.

Badshah *et al.* (1992) screened various substrates to choose the better one for *Pleurotus florida* and *Pleurotus ostreatus.* Wheat straw gave better yield (86.47 per cent BE) than other substrates.

Dwivedi and Singh (1994) studied the cultivation of *Pleurotus sajor-caju* on different agricultural wastes. The results showed that paddy and wheat straws as the best substrates.

Patrabansh and Madan (1995) conducted an experiment on the effect of different agricultural residues *viz., Saccharum munja, Oryza sativa, Vigna unguiculata, Abolmoschus esculentus, Zea mays* and *Cyamopsis tetragonaloba* on cultivation of *Pleurotus sajor-caju. Saccharum munja* and *Oryza sativa* were found to be best substrates for the growth of *Pleurotus sajor-caju,* followed by other substrates.

Singh (1995) screened six species of *Pleurotus* for their yield response on six different substrates. *Pleurotus florida* gave maximum yield of sporophores (63.9 per cent BE) on wheat straw followed by *P. sajor-caju, P. flabellatus, P. citrinopileatus* and *P. abalones.*

Gogoi and Adhikary (2002) studied on alternative substrates for the cultivation of *Pleurotus citrinopileatus* (Fr) Singer. They used various substrates such as, dried turmeric leaves, sugarcane baggase, tea waste and deoiled Java citronella grass. Paddy straw was kept as control. Highest yield was obtained from paddy straw (95 per cent BE) followed by dried turmeric leaves (93.7 per cent BE).

Jain and Vyas (2002) used different combinations of wheat straw with other substrates for enhancing the yield of *Pleurotus florida.* Out of nine combinations of wheat straw tested. Only two combinations, wheat straw + soyabean straw and wheat straw + tea leaves were found to show biological efficiency of *Pleurotus florida* was either slightly superior or equal to 93 and 89 per cent respectively in comparison to wheat straw alone (biological efficiency 87 per cent).

Ram and Sinha (2004) have studied on mango (*Mangifera indica* L.) three substrates *viz.,* wood saw dust, tip branch and leaf fall were used individual and in combination with paddy straw for cultivation of oyster mushroom (*Pleurotus florida*) under glass house condition. Observations with spawn run period, days taken for initiation of fruit bodies, harvesting of 1st flush and total yield were recorded.

Use of Agricultural Residues and By-products for Supplementation

It is not easy to separate all possible agricultural residues and by-products into two perfect categories like substrates and supplements (additives) because some substrates used in small amount less than 5 per cent can be called "supplement" while the same substrate can be used more than 5 per cent in which case we can consider it as a bulk ingredient in the "substrate".

Hayes (1969) recommended molasses, wet brewers" grain, malt sprouts, potato wastes, apple and grape pomace to be employed as carbohydrate sources as they correct C/N ratio and also establish the bacterial flora in compost.

Stoller (1979) Shredded paper supplemented with limestone and was a suitable casing for *Agaricus* mushroom.

Garcha *et al.* (1981) reported the utilization of agricultural wastes for mushroom cultivation in India, and recommended the preparation of different type of composts by long method from wheat straw, mentha stalks and spent compost of wheat and paddy, along with chemical fertilizers using molasses, ginning mills wastes and poultry litters as activators.

Purkayasth *et al.* (1981) banana epicarp, suitable supplement for the edible mushroom *Calocybe indica* when added to basic maize substrate.

Suman and Shamla (1990) screened different varieties of wheat (*Triticum aestivum* Linn.) for spawn production of *Agaricus bisporus* (Lange) Sing in Himachal Pradesh.

Rinker (1991) canola meal, added to *Agaricus* compost and it significantly increased yield.

Garcha and Khanna (1993) advocated biogas slurry and burnt rice husk as the other possible materials for casing soil preparation.

Marimuthu *et al.* (1994) tested *Pleurotus citrinopileatus, P. sajor-caju* and *P. platypus* for their response to amendment of substrates with neem cake, rice and wheat bran and tapioca tip.

Khanna *et al.* (1995) observed that burnt rice husk and biogas slurry can also be used as the base material for preparing casing mixtures by mixing them with either clay or loam soil or FYM.

Noble and Gaze (1995) reported sugar bet lime found to be a suitable supplement for different casing materials.

Ram (2004) studied the neem leaf powder at 2 per cent and 3 per cent level with wheat straw for cultivation of *Pleurotus sajor-caju* and *P. flabellatus.* It reduces the number of days required for completion of spawn run and increases the mushroom yield due to presence of antifungal properties in the neem leaf powder.

Agricultural Wastes used as an Alternative Casing Materials

In regions where the basic soil and peat for *Agaricus, Calocybe indica* and. *Stropharia* casing is not so evident or becoming more and more expensive, the use of some organic wastes is more than justified.

Mantel (1973) recommended the use of spent compost with slaked lime and sand (4 : 1 : 1).

Hayes *et al.* (1976) reported a number of deficiencies during crop production with the use of one year old spent compost as casing soil. Later on Hayes and Shandilya (1977) recommended farm yard manure (FYM) and loam soil (1 : 1 v/v) as a standard casing material.

Chang and Hayes (1978) Cardboard = carton, shredded as casing component and it also combination with other paper waste and several supplements, it is called gray peat.

Stoller (1979) spent compost of *Agaricus,* after certain period of bio-transformation and ion leaching, one to three years in open field, the old compost is useful as casing for *Agaricus* mushroom.

Garcha (1980) suggested FYM and two years old spent compost (1 : 1) as a good material under the conditions of Punjab state.

Garcha *et al.* (1981) paper mulch mixed with two years old spent compost gave the best result for *Agaricus* casing.

Nair *et al.* (1981 and 1989) sawdust of Eucalyptus mixed with clay loam soil an efficient for *Agaricus* casing layer. The spent casing peat stripped from the mushroom beds after cropping was an efficient component of new casing layer for *Agaricus* mushroom. Filter mud, by-product of cane sugar in Australian refineries is mixed with peat moss for *Agaricus* casing.

Garcha and Khanna (1993) advocated biogas slurry and burnt rice husk as the other possible materials for casing soil preparation.

Noble and Gaze (1995) Recycled granulated rock-wood, containing plant root parts, a waste product of the glasshouse crops, could be 50 per cent mixed with peat for a suitable *Agaricus* casing.

Khanna *et al.* (1995) observed that burnt rice husk and biogas slurry can also be used as the base material for preparing casing mixtures by mixing them with either clay or loam soil or FYM.

Singh *et al.* (2000) studied the six casing mixtures *viz.*, farm yard manure + spent compost (1 : 1, 2 : 1 and 1 : 3), farm yard manure + garden soil, farm yard manure alone and spent compost alone. Yield recorded in all the six treatments were significantly different from each other. Casing mixture prepared from farm yard manure + spent compost (2 : 1) performs best by giving highest yield.

Shandilya (2002) reported that farm yard manure with its exceptional structural water properties and presence of varied bacteria flora remains the preferred casing material for mushroom crop. A mixture of farm yard manure + vermi compost-leached + 3 years old spent compost provides the ideal and standard casing for Indian mushroom farms, till better, inexpensive, easily available and more productive substrate or mixture of materials will developed as an alternative.

Conclusions

Various agricultural residues and by-products are the most used basic substrates for *Agaricus, Pleurotus, Volvariella, Stropharia, Lentinus, Morchella* etc. Oyster mushroom (*Pleurotus* species) are largest gamma of waste substrates by its fast mycelium growth via its multilateral enzymes system biodegrading nearly all the available wastes.

The success of any nation depends upon the ability of their people to sustainably convert the natural resources into economic wealth with judicious application of schemes and technologies without endangering the environment. Huge quantities of lingo-cellulosic crop residues and other organic wastes are generated annually through the activities of agricultural, forest and food processing Industries. The major parts of these residues are burnt after harvesting, resulting multifaceted hazards including oxygen deficient environment, respiratory diseases (such as allergic, bronchitis, asthma, tuberculosis) and poor visibility at night. If even one per cent, of this crop residue is used to produce mushrooms, India will become a major mushroom producing country in the world (Tewari and Pandey, 2002).

An immense mass of agricultural residues and by-products are available on this earthly globe! Long series of estimations could lead us to the conception that more than 600 milliard kg of field and forest residues can be used every year for mushroom cultivation as a weapon against hunger. But as we read in Chang and Miles (1989): "Lack of interest of academician and scientists in fundamental studies of edible mushrooms are delaying the hugeness mushroom production". Indeed, the mycologists, biologists and plant pathologists all over the world know that our hidden fungi or "children of darkness" need more light in the future, more attention and love for more spawn more training groups and more mushroom production! We appreciate so much all "doctors without frontiers" but where are our "mycologist without frontiers"? By this way, the valuable use of agricultural residues and by-products for mushrooms cultivation can provide more food, more jobs, better family income, improved living standard, safe to human health and environmental pollutions, curb global warming and CO_2 level, clear up crop residues, road sides and forest margins. A daily availability of cultivated mushrooms will prevent over picking and loss of desired wild species. A general fire ceased will provide via mushroom substrate more organic materials for garden and floristry. Finally, our options will lead to a general worldwide biological balance and a common welfare in peaceful villages, where as the guns are replaced by mushrooms.

References

Badshah, N., Naeem-ur Rehrnan and Wahid, M. (1992). Yield and quality of mushroom grown on different substrates. *Sarhad. J. Agric.*, 8: 631–635.

Bano, Z. and Srivastava, H.C. (1962). Studies on the cultivation of *Pleurotus* species on paddy straw. *Food Sci.*, 12: 363–365.

Bhaskaran, T.L., Sivaprakasam, K. and Kandaswamy, T.K. (1978). Compact bag method: A new method of increasing the yield of *Pleurotus sajor-caju. Indian J. Mushroom*, 4(2): 10–12.

Bhattacharjee, M.K., Samajpati, N. and Chilvers, G.A. (1989). Studies on the effect of different amounts and methods of applying spawn on the yield of *Pleurotus sajor-caju* (Fr.) Singer. *Indian Journal of Mycological Res.*, 27: 15–23.

Bhatti, M.A., Mir, F.A. and Siddiq, M. (1987). Effect of different bedding materials on relative yield of oyster mushroom in successive flushes. *Pakistan J. Agric. Res.*, 8: 256–259.

Bisht, N.S. and Harsh, N.S.K. (1984). Utilization of waste tea leaves for spawn preparation of *Agaricus bisporus* and *Volvariella volvacea. Mushroom J.*, 139: 231–233.

Block, S.S., Tsao, G. and Han, L. (1958). Production of mushroom from saw dust. *J. Agric. Food Chem.*, 6: 923–927.

Block, S.S., Tsao, G. and Han, L. (1959). Experiments in the cultivation of *Pleurotus ostreatus. Mush. Sci.*, 4: 309–325.

Buswell, J.A. and Chang, S.T. (1993). Edible mushrooms: attributes and applications. In : *Genetics and Breeding of Edible Mushrooms*, (Eds.) S.T. Chang, J.A. Buswell and P.G. Miles. Gordon and Breach Science Publishers, USA, pp. 297–324.

Chang, S.T. and Hayes, W.A. 1978. *The Biology and Cultivation of Edible Mushrooms.* Academic Press, New York.

Chang, S.T., Lau, O.W. and Cho, K.Y. (1981). The cultivation and nutritional value of *Pleurotus sajor-caju. European. J. Appl. Microbiol Biotechnol.*, 12: 58–62.

Chang, S.T. and Miles, P.G. (1989). *Edible Mushrooms and their Cultivation.* CRC Press, Florida.

Constantin, J. (1923). Sur le *Pleurote* du chardon blew de la vanoi-se-compleuteus. *Comptes Rendus Acad. des Science*, Cl XXVI 19: 849–852.

Courvoisier, M. (1999). Less champignons comestible dans le monde. *Bul. Fed. Syn. Champ.*, 82: 829–837.

Dwivedi, R.R. and Singh, R.P. (1994). Studies on different *Pleurotus* species for their yield performance in North-West *"Tarai"* of U.P, *Mush Res.*, 3: 45.

Daugulis, A.J. and Bone, D.H. (1977). Submerged cultivation of edible white rot fungi on tree bark. *Eur. J. Appl. Microbiol.*, 4: 159–168.

Duggar, B.M. (1905). Some principles in mushroom growing and spawn making. *U.S. Deptt. of Agric. and Tech., Bull.*, 85: 1–60.

F.A.O. (1989). Food and Agricultural Organisation of the United Nation. *Production Year Book*, Vol. 41, Food and Agricultural Organization, Rome.

Flack, R. (1917). Uber dre Waldkultur des Austempilzes (*Agarcius ostreatus*) auf Labubholzstubbem. *Z. Forst. Jaqd. Wes.* 49: 159–165.

Garcha, H.S. and Kiran, U. (1981). Studies on mushroom composts under tropical conditions. *Mush. Sci.,* 11: 219–233.

Garcha, H.S. and Sekhon, A. (1981). Evaluating casing materials for mushroom culture in Punjab (India). *Mush. Sci.,* 11(1): 411–417.

Garcha, H.S., Dhanda, S. and Khanna, P.K. (1984). Evaluation of various organic residues for the cultivation of *Pleurotus* species (dhingri). *Mush. News Lett. Tropics,* 5: 13–16.

Garcha, H.S., Dhanda, S. and Khanna, P.K. (1985). Efficacy of container system for the production of *Pleurotus. Mush. News. Lett. Tropics,* 6: 16–20.

Garch, H., Amarjit, S. and Phutela, R. (1981). Utilization of agri-waste for mushroom cultivation in India. *Mushroom Science,* 1(1): 245–256.

Gogoi, G. and Adhikary, R.K. (2002). Suitability of certain newer plant wastes for production of oyster mushroom (*Pleurotus citrinopileatus*). *Mush.*

Hayes, W.A. (1969). Microbiological changes in composting wheat straw/horse manure mixtures. *Mush. Sci.,* 7: 173–186.

Hayes, W.A. and Shandilya, T.R. (1977). Casing soil and compost substrates used in the artificial culture of *Agaricus bisporus* the cultivated mushroom. *Indian J. Mycol. and Pl. Pathol.,* 7: 5–10.

Jain, A.K. and Vyas, D. (2002). Yields response of *Pleurotus florida* on wheat straw in combination with other substrates. *Mush. Res.,* 11(1): 19–20.

Jandaik, C.L. (1974). Artificial cultivation of *Pleurotus sajor-caju. Mushroom J.,* 22: 405.

Kachroo, J.L., Ahmad, N., Amanullah, N. and Kaul, T.N. (1979). Experiments in the improvement of compost formulae and procedure in *Agaricus bisporus* cultivation. *Mushroom J.,* 82: 434–440.

Kaufert, F. (1935). The production of asexual spores by *Pleurotus corticatus. Mycologia,* 27: 330–340.

Kaul, T.N. and Kachroo, J.L. (1975). Prospects of *Pleurotus* cultivation in India (Abst.). *1st Workshop on Mushroom Res.,* Solan, H.P.

Khan, S.M., Javed, N., Khan, S.M. and Chohan, M.A. (2002). Effect of different sterilization methods on the production of oyster mushroom (*Pleurotus ostreatus*) on different substrates. In *Proceedings of 3rd National Conference of Plant Pathology,* NARC, Islamabad, 1–3 Oct. 2001, pp. 159–160.

Kumar, S.K., Seth, P.K. and Munjal, R.L. (1975). Studies on quantities of gypsum and calcium carbonate singly and in combination on spawn production of *Agaricus bisporus. Indian J. Mush.,* 1: 27.

Lutz, L. (1925). Sur la specificite vis-vis de leurs supports des champignon du groupe du *Pleurotus eryngii. Comptes Rendus Acad. des Science,* CIXXX 13: 1044–1047.

Mathew, M. (ed.) (1996). *Manorama Year Book.* Mhalaya Manorama Company Limited, Kottayam, Kerala, pp. 800.

Mantel, E.F.K. (1973). Casing soil made from spent compost. *Indian J. Mush.,* 1: 15–16.

Mantel, E.F.K., Agarwala, R.K. and Seth, P.K. (1972). *A Guide to Mushroom Cultivation.* Ministry of Agriculture, Farm Information Unit, Directorate of Extension, New Delhi, Farm Bull. No. 2.

Nair, S.G. and Bradly, J. K. (1981). Recycling waste plant products as casing materials in mushroom cultivation. *Mushroom Science,* 11(2): 147–152.

Noble, R. and Gaze, R. (1995). Properties of casing peat types and additives and their influence on mushroom yield and quality. *Mushroom Science,* 14(1) : 305–312.

Patrabansh, S. and Madan, M (1995). The microbial conversion of different agricultural residues and its, biological efficiency. *Acta. Biotechnol.,* 15: 131–135.

Platt, M.W., Chet, I. and Henis, Y. (1982). Growth of *Pleurotus ostreatus* on cotton straw. *Mushroom J.,* 120: 425–427.

Pope, J. (1995). Cultivation of edible mushrooms on tropical agricultural wastes. Bienial Training Course, ABOS and VLIR, University, Gent.

Pope, J. (2000). Use of agricultural waste materials in the cultivation of mushrooms. In: *Science and Cultivation of Edible Fungi,* (Ed.) Van Griensven, 1: 3–23.

Purkayastha, R. and Nayak, D. (1981). Development of cultivation and analysis of proteins of a promising edible Calocybe indica. *Mushroom Sciences,* 11(2): 697–713.

Rajarathnam, S., Singh, N.S. and Bano, Z. (1979). Efficiency of carboxin and heat treatments for controlling the growth of *Sclerotium rolfsii* during culture of mushroom *Pleurotus flabellatus. Ann. Appl. Biol.,* 92: 323–328.

Ram, R. C. (2004). Bio–control of substrate fungi for better growth behaviour and yield potential of oyster mushroom. " *Biological Diversity* " *Current Trends* (Prof. R.C. Rajak Fests Crifts Volume) Page: 7–12.

Ram, R.C. and Sinha, A. (2004). Use of mango stiff with paddy straw for oyster mushroom cultivation. *Indian Phytopath.,* 57(4) (In Press).

Sathe, A.V. and Nagarkar, M.B. (1983). A modified Zadrazil's methods for the cultivation of oyster mushroom. *Current Science,* 52: 1066–1067.

Rinker, D.L. (1991). Alternative additives, supplements and casing amendments for *Agaricus bisporus. Mushroom Science,* 13(2): 781–789.

Shandilya, T.R. (2002). Present status and future prospects of casing research in India. In: *Current Vistas in Mushroom Biology and Production,* (Eds.) Upadhyay, R.C., Singh, S.K. and Rai, R.D. Mushroom Society of India, pp. 166–183.

Singh, A.K. (1995). Growth and development of edible mushrooms on sugarcane-substrates under stress conditions. *Ph.D. Thesis,* Banaras Hindu Uinversity, Varanasi, U.P., India.

Singer, R. 1961. *Mushrooms and Truffles.* Interscience Publishers, New York.

Sivaprakasam, K. and Kandaswamy, T.K. (1981). Effect of different types of spawn on sporophore production of *P. sajor-caju* (Fr.). singer on cellulose content of the substrates. *Madras Agric. J.,* 68: 628–633.

Somasundaram, R., Shashirekha, M., Bano, Z. and Rajarathanam, S. (1992). Bio-potentialities of Basidiomycetes. *Food and Vegetable Technology,* pp. 233–361.

Sohi, H. S. (1982). Role of edible mushroom in recycling of agricultural wastes and as an alternative protein source: present status of mushroom cultivation in India. In: *Frontier of Research in Agriculture,* (Ed.) S.K. Roy. ISI Calcutta, pp. 565–579.

Stamets, P. (1993). *Growing Gourmet and Medicinal Mushroom*. Ten Speed Press, Berkeley, Hong Kong.

Stoller, B.B. (1962). Some practical aspects of making mushroom spawn. *Mushroom Sci.*, 5: 170–184.

Stoller, B.B. (1979). Synthetic casing for mushroom beds. *Mushroom Science*, 10(2): 187–191.

Suman, B.C. and Sharma, V.P. (1990). Screening of different varieties of wheat (*Triticum aestivum* Linn.) for spawn production of *Agaricus bisporus* (Lange) Sing in Himachal Pradesh. *Indian J. Mycol. and Pl. Pathol.*, 20: 44–46.

Swaminathan, M.S. 1995. Agriculture, food security and employment: changing time, uncommon opportunities. *Nature and Resources*, 31: 2–15.

Terrier, L. (1945). Mode of preparation of the host bed mushroom (*Psalliata compestris*). *Ann. Agric. Suisse.* 59(10): 949–952.

Tewari, R.P. and Pandey, M. (2002). Sizeable income generating venture. In: *The Hindu Survey of Indian Agriculture*, pp. 165–167.

Toyama, N. and Ogawa, K. (1974). Comparative Studies on cellulytic and oxidizing enzyme activities of edible and inedible wood rotters. *Mush. Sci.*, 1: 760.

Upadhyay, R.C. and Vijay, B. (1991). Cultivation of *Pleurotus* species during winter in India. *Mush. Sci.*, 18: 533–536.

Wood, D.A. (1989). Mushroom Biotechnology. *Inter. Indust. Biotechol.*

©R.C. Ram & Asha Sinha (*Editors*)
***Microbial Diversity: Modern Trends*, 2007**
Daya Publishing House, New Delhi

Chapter 4

Biodegradation of Rice Residues by Soil Mycoflora in Rice Wheat Cropping System

Vibha and Asha Sinha

Department of Mycology and Plant Pathology, Institute of Agricultural Sciences, Banaras Hindu University, Varanasi - 221 005 (U.P.)

Introduction

The greatest challenge facing mankind in the 21st century specially in overpopulated Asian countries is to produce the bare necessities of life including food, feed, fiber and fuel from ever decreasing availability of cultivable land per caput, which in 2000 AD was 0.059, 0.077 and 0.161 ha in Bangladesh, China and India, respectively (Gill, 1994). These values will further decrease with time due to ever increasing population in these countries. India will need to produce an additional 90 million tonnes of food grain to feed anticipated population of about 1.4 billion by 2020. The only way to achieve this is to increase the productivity of each unit of land, which is a difficult task. This would necessitate optimization of our efforts in all dimensions, namely land utilization, soil and moisture conservation, crop and cropping system management with greater emphasis on residue management and efficient use of surface and underground water. Thus, what is needed is an integrated approach for the use of all the agricultural inputs.

Rice-wheat cropping system is extensively practiced in the state of Punjab, Haryana, Uttar Pradesh and Bihar covering an area of nearly 10 million hectare. Rice-wheat cropping pattern is very important in the eastern U.P. About. 90 per cent of wheat is planted after rice. Rice-wheat is the most common crop sequence of Varanasi region in UP. (0.68 m ha). The result of long term production potential obtained from experimentation with recommended package of practices on rice-wheat at Varanasi clearly indicate a declining trend in the yield of both rice and wheat over the last decade. Soil fertility declines combined with a reduction in quality and quantity of soil organic matter appears to be the major problem of most of the soils. The application of high dosage of chemical inputs has drastically changed the. nutrient status of the soil in the Indo-Gangetic Plains (IGP) of rice-wheat cropping system. During recent years the productivity trends indicate that any additional doses of N, P and K dose not give any impressive yield and the system is no longer sustainable. The continuous nutrient sucking capacity of high yielding hybrid rice-wheat varieties are resulting in depletion of organic matter and soil fertility.

Crop debris added to the soil after harvesting is usually a principle amendment applied to the soil system and plays an important role in maintaining soil fertility affecting both physical structure of the soil and its nutrient status. The most serious soil constraints are concomitant occurrence of moisture and nutrient stress. Both these characteristics are associated with soil organic matter (SOM) (Prasad and Power, 1997). Continuous practice, of intensive cropping systems such as rice-wheat, reduces SOM content (Hegde, 1998 and Nambiar, 1994). This coupled with reduced availability and application of farmyard manure, due to increased mechanization of agriculture, has further reduced SOM content in soil. Crop residues of wheat and rice, specially, the latter can be the major organic sources that can be added to soil and their application is reported to increase the SOM (Prasad *et al.*, 1999; Prasad and Mishra, 2001 and Silgram and Chambers, 2002). Addition of different amount and types of organic materials improves soil physical properties and yield of wheat grain and stubble (Barzegar *et al.*, 2002). Therefore, effective management of crop residues is imperative in sustaining the environment and crop productivity.

The crop debris is mainly made up of cellulose and lignin. Cellulose, the principal polysaccharide of plants, is by far the largest reservoir of biologically utilizable carbon on the surface of the earth. Lignin a major component of plants, is probably the second most abundant of all the continuously cycled organic materials on the earth. The biological decomposition of these two components is therefore, an important problem (Mukhopadhyay and Nandi, 1979). The major constituents of rice residue also are–Cellulose–15 per cent, Hemicellulose–35 per cent and Lignin–15 per cent (Harper and Lynch, 1985). Plant debris containing higher amounts of lignin such as straw, etc., are resistant to microbial attack and hence their mineralization proceeds slowly in soil. The lignin content of plant residues may serve as an index of the vulnerability of organic residues for microbial attack. The oxygen status of the soil (aerobic or anaerobic) is the most important factor in determining the qualitative and quantitative aspects of microbially mediated bio-degradation of plant remains.

Rice forms the principal food of nearly half of the world's population. Though, a large number of crops are growth in India, it is the second largest rice growing country in the world. Rice is probably one of the oldest food crops. It has been in cultivation for very ancient times in China, India, Java and probably North Africa. The root system in rice varies much between the different varieties. Short duration varieties have less extensive and less dense root development than long duration varieties. As a general rule, taller varieties have a greater penetration and lateral spread in the root system than the shorter varieties. Organic manures and phosphatic fertilizers promote root development.

After making use of economic parts, the remaining portion is mostly wasted except for a few crops. The adoption of mechanized farming in many advanced regions in the country has resulted in leaving a sizeable amount of crop stalk. These are either burnt or incorporated into the soil while preparing the field for next crop. Rice stubble can playa very important role in meeting the nutrients of rice and succeeding crops. However, their incorporation in soil is associated with immobilization of soil nitrogen, which may adversely affect the succeeding crop of wheat in rice-wheat rotation and application of some additional nitrogenous fertilizer is recommended to overcome the immobilization (Aulakh *et al.*, 2000). The rate of immobilization of nitrogen depends on the nature of soil microflora, soil temperature, fertilizer nitrogen status and the C/N ratio of organic matter added. Organic matter from diverse plant-tissues varies widely in their C/N ratio. Optimum level of C/N ratio in the range of 20–25 (1.4–1.7 per cent N) seems to be ideal for maximum decomposition. Rapid decomposition of incorporated rice residue is most important.

Among the principal crops in India residue yield as well as the nutrient potential (proportion of N, P and K) of rice is the highest. Rice straw, though not good in feeding quality, is also used for feeding cattle. However, in the regions where mechanical harvesting of wheat and rice is done, a sizeable quantity of rice residue is left in the field. Decomposition of this residue is pivotal for maintaining soil health and improving the crop productivity of the next crop of wheat. The problem of cellulosic material has been studied in two different ways. One is through the use of mineral nutrient and other by use of decomposing microorganisms. It is established that amount of nitrogen, contained in organic materials, controls the rate of decomposition (Nikhara, 1981; Debnath and Sinha, 1993). Application of nitrate or ammonical nitrogen accelerates rapid decomposition of rice straw (Brar *et al.*, 2000), if applied in suitable concentration under aerobic or slightly alkaline conditions, while its application in a soluble condition is temporarily immobilized. Despite addition of assimilate nitrogen, certain materials such as, saw-dust, rice husk, old bracken or coconut shell can not be decomposed while some others such as banana stem or hop bines take a long time for decomposition.

Microorganism decomposes the organic materials was established in 1850. Since then, bacteria have been given great deal of prominence and till the beginning of this century they were considered the sole agent responsible for disappearance of organic matter in soil. Soil microorganisms play an important role in increasing soil fertility and recycling of nutrients within soil. Different microorganisms including filamentous fungi, yeast, mycorrhiza, bacteria, cyanobacteria and actinomycetes possess the urease enzymes.

Recently, many workers have studied the role of fungus in relation to soil fertility and reported that their activities are not so varied as those of bacteria. Fungi are also important contributors to the ammonification, nitrogen transformation, cellulose decomposition and humification processes. The problem of decomposition of cellulosic materials is thus bound up not only with the study of the chemical constituents of the organic materials, but also with the microorganism's dynamics in soil during decomposition. The following are the genera of fungi which are most commonly encountered in soils and which can be isolated by conventional methods–*Acrostalagmus, Aspergillus, Botrytis, Cephalosporium, Gliocladium, Monilia, Penicillium, Scopulariopsis, Spicaria, Trichoderma, Trichothecium* and *Verticillium.* A number of fungi associated with the decomposition of rice residues are *Alternaria, Cladosporium, Pullularia, Cylindrocarpon* and *Fusarium* (Fungi Imperfecti); *Absidia, Cunninghamella, Mortierella, Mucor, Rhizopus, Zygorynchus* and *Pythium* (Phycomycetes); *Chaetomium* (Ascomycetes) and *Rhizoctonia* (Mycelia Sterilia, which fail to produce reproductive structures). The potential of crop residues of major cereals, pulses, oilseeds and commercial crops for recycling of valuable plant nutrients for sustained crop production is enormous.

Decomposition of rice residue in a rice-wheat ecosystem has crucial impact on germination, growth and development of succeeding wheat crop and is a major source of organic carbon in this system. In addition, its residue contains bacterial and fungal compounds and residues of micro and macro fauna. A large amount of rice residues left after harvesting in soil is metabolized by bacteria, actinomycetes, fungi, protozoa, nematodes, micro and macro orthopods, enchytraeids and certain worms. The decomposition of residue is accompanied by development of phylloplane microflora, colonization by saprophytic microorganisms, combination and ingestion by invertebrates resulting in incorporation of organic matter into soil and increase in the surface area. The microbial colonization and utilization of comminuted litter results in chemical degradation and utilization of tissues and production of complex phenolic polymers (humic acids, and formation of stable organo-mineral complexes (Woods, 1976).

Residue decomposition and nutrient cycling are important for maintaining soil health. Fungi play an important role in residue decomposition. However, information on their role in rice residue decomposition process is scanty in literature. An attempt has been made to review the literature on role of fungi in decomposition of rice residue in rice-wheat cropping system and other crops.

Decomposers Community

The interaction within and between the fungi, bacteria and invertebrate animals bring about decomposition of residues. Macfadyen (1963) and Reichle *et al.* (1975) emphasized that 80 per cent of the energy stored in organic matter passes through decomposers community. The transfer of minerals involves directly or indirectly the activity of various saprophytic groups. The residue after reaching the soil surface is immediately attacked by an array of soil organisms and animals. Swift *et al.* (1979) have proposed a conceptual model envisaging the decomposition as a function of catabolism, comminution and leaching. The process is regularized by the environmental factors, resource quality and decomposer community. The inorganic nutrients are released through the process of mineralization. The decomposition product of initial process differs in quality and thus in decomposition rate from the original resources. Berkenkamp *et al.* (2002) reported that decomposition of added organic matter started at the moment of residue incorporation.

Decomposition Process

Nandi *et al.* (1996) studied the two steps of composting process of converting rice straw into high humic substance content and reported that in first step lignin decomposing fungi *Phanerochaete chrysosporium, Polyporus versicolor* and *Ganoderma lucidum* in that order, were involved while in second step of humification general decomposting fungi were prevalent. Kuhad and Johri (1992) studied the colonization pattern and mechanism of action of white rot basidiomycetes fugus *Cyathus* sp. on rice straw and reported that fungal hyphae penetrated into the vascular bundles and their lateral branches penetrated the walls of adjoining cells thereby promoting decomposition. The highest decomposition of rice straw was obtained when it was inoculated with mixture of *Aspergillus, Helicomyces, Chaetomium* and *Streptomyces* (Limtong *et al.*, 1993). Sarojini and Mathur (1994) found rapid decomposition of paddy straw with *Trichoderma reesei, T. viride, Pleurotus sajor caju* and *Coplinum cinereus.* Chatterjee and Nandi (1980) isolated microorganism from the field soil associated with wheat stubble after six weeks of crop harvest and among these numerically dominant fungi *viz., Verticillium albo-atrum, Aspergillus wentii, Penicillium purpurogenum, A. flavus* and actinomycetes, namely *Streptomyces longisporus, S. griseoporeus*) and their mixture were used to study their degrading ability of wheat stubble. They reported that degradation of lignin was maximum by the mixture of fungi + streptomyces but individually *A. flavus* showed highest rate of lignin decomposition.

In a study, cellulolytic fungi *Chaetomium cellulolyticum, Rhizoctonia betaticola, Chaetomium globosum* and *Myrothecium verrucaria* were used to find out SCP production from wheat straw, it was found that *Chaetomium cellulolyticum* and *Chaetomium globosum* gave the highest percentage of protein and total protein (19.30 per cent, 15.03 per cent and 0.19 g/g of biomass, 0.15 g/g of biomass, respectively) while the highest carboxymethyl cellulase (CMC) activity was found in the case of *Rhizoctonia betaticola* (24.0 units/ml) followed by *C. cellulolyticum* (15.0 units/ml). *Myrothecium ven-ucaria* was poor producer (Kahlon *et al.*, 1980). Dhillon *et al.* (1980) studied bioconversion of the delignified rice straw by eight fungi *viz., Myrothecium verrucaria, Curvularia lunata, Sclerotium rolfsii, Fusarium equiseti, Helminthosporium maydis, Helminthosporium tetramera, Chaetomium globosum* and *Chaetomium cellulolyticum* and reported that *Chaetomium cellulolyticum* showed highest percentage of crude protein (32.1 per cent) of SCP per g of substrata followed by another species of same genera *Chaetomium globosum* 240 mg (30 per cent) per g of the substrate. In another study *Chaetomium thermophile* was found to produce maximum SCP when grown on delignified wheat straw (Sekhon, 1975). Choi *et al.* (1990) reported that with inoculation of cellulolytic microorganisms fermentation temperature and decomposition of rice straw increased rapidly. Compared to other treatments. Murayama (1984) reported that during decomposition of straw proportions of mannose, galactose, fiuctose, mammose and ribose increased consistently with time whereas the proportion of cellulosic glucose became consistently smaller. Choi *et al.* (2003) reported that the chemical agent and biological agent combined treatment was the most effective on the decomposition of rice straw.

Mixtures of MBK-336 (*Aspergillus*), MRY-586 (*Heliomyces*), MCT-794 (*Chaetomium*), ABK-372 (*Streptomyces*) and APK-425 (*Streptomyces*) were reported to causes higher production of compost from rice residue in the field (Limtong *et al.*, 1994). *Aspergillus* species isolated from paddy straw decomposition was most efficient fungal species isolated on MRP (Massoorie Rock Phosphate) (Akhaury and Kapoor, 1997). In a study mass loss of lupin residue was nearly twice that of barley (42 per cent) or wheat (38 per cent) straw after the first 90 days of decomposition. The difference in residue mass loss was also reflected in significantly higher rate of substrate induced respiration (SIR) and greater population of bacteria and fungal hyphae on lupin than on wheat or barley straw (Coockson *et al.*, 1998). Rate of decomposition of root residue in soil after incorporation remains higher in first arid second week, which gradually slowed down and finally becomes steady. The bacterial and fungal population increases as the amount of applied foots residue increase in soil (Sameni and Pour, 2001). In a study in New Zealand, the weight loss of AI (autum incorporated) barley straw average 33 per cent as compare with 18 per cent for the surface straw of SI (Spring incorporated). Irrigation increased straw decomposition during the cropping period by 68 per cent in AI treatment compared with only 37 per cent in SI treatment. The difference was related to the size of the residue-borne microbial population at the start of cropping period (Beare *et al.*, 2002). Hobbie (2000) suggested the anthropogenic deposition of N might increase leaf litter decomposition more in ecosystems characterized by low lignin litter than in those characterized by high lignin litter. In a study on decomposition of sun and shade leaves of chestnut, oak and beach (deciduous trees) Sariyildiz and Anderson (2003) observed that the decomposition rates were initially rapid and then platitude but there was significant differences in mass losses at 6, 12, 18 and 24 months. They suggested that the patterns of decomposition might be affected by influence of the culture conditions on the composition of fungal communities. The priming effect *i.e.*, the increase in soil organic matter decomposition rate after fresh organic matter input to soil, is often supposed to result from a global increase in microbial activity due to higher availability of energy released from the decomposition of fresh organic matter (Fontaine *et al.*, 2003).

Role of Fungi in Decomposition

Among the major groups of residue decomposing organisms, fungi are the most active. Although they have immense role in rice residue decomposition, there are very few studies on their structural and functional role during decomposition. Fungi have successive phases of colonization, exploitation and exhaustion of organic substrates during decomposition. Initially, the residue is colonized by weak parasite, which invade living tissues and pave the way for saprophytes. Later on the ground saprophyies colonize residue in successive and overlapping layers (Sadasivan, 1939 and Harper and

Lynch, 1985). Each fungus that colonizes the residue after the microhabitate paves the way for the next (Nandi *et al.*, 1996). Litter quality variables, such as total nitrogen, total phosphorus, total carbon, tissue nutrient concentrations explain most of the succession patterns and differences in the micro fungal assemblages of peat land plant litters (Thormann *et al.*, 2003).

Fungal Succession

The succession of fungi on different crop residue is scarcely understood in the tropics (Sharma and Dwivedi, 1972, Mehrotra and Aneja, 1979). Kskattriya (1990) studied the colonization and succession of fungi on various leaf litter and observed a specific pattern that resembled the one proposed by Hudson (1968). She reported that primary colonizers were deuteromycetes and ascomycetes. Somewhat similar successional pattern was observe on *Chenopodium album* by Mehrotra and Aneja (1979). Hudson (1968) from his detailed study on succession and colonization of fungi classified the fungi as:

1. Living parasites (Ascomycetes and fungi imperfecti *viz., Leptosphaeria sacchari, Gnomonia errabunda, Readeriella mirafilis.*
2. Senescent (Common primary saprophytes–Ascomycetes and fungi *imperfecti–Cladosporium herbarum, Epicoccum nigrum, Aureobasidium pullulance, Botrytis cinerea, Nigrospora* sp., *Curvularia* sp., and restricted primary and saprophytes–Ascomycetes and fungi imperfecti = *Leptosphaeria microscopiea, Fusicoccum bacillare,* Piggotia substellata) and
3. Secondary saprophytes (Ascomycetes, Basidiomycetes, Soil inhabitant *Mucorales, Penicillia*).

Beare *et al.* (1993) observed fungi including phyllosphere colonizer (*Alternaria alternata, Epicoccum nigrum*) and other common decomposing fungi (species of *Cylindrocarpon, Fusarium, Mucor* and *Phoma*) on sorghum residue. *Trichoderma, Fusarium* and *Penicillium* were identified as early decomposers whereas genus *Aspergillus* and class Mucorales (Zygomycotina) as late decomposer of pineapple leaf litter by Tiwari *et al.* (1994). Rosenbrock *et al.* (1995) found in initial phase of decomposition, the colonization of zymogenic mycoflora of genus *Mucor, Alternaria* and *Epicoccum* on the litter. This flora was progressively replaced by autochthonus species principally of the genus *Fusarium.* In Canada, in a study at different stages of decomposition approximately ten thousand micro fungal communities were isolated representing 292 species of filamentous micro fungi including 41 Ascomycetes, 21 Zygomycetes and 222 mitoscopic fungi. The most commonly isolated species were *Trichoderma viride, Rhinocladiella atrovirens, Penicillium pinophilum* and *Mortierella ramanniana* (Lumley *et al.*, 2001). Dominance of Dematiaceous, Hyphomycetes and Coelomycete were observed in early decomposing stage whereas Basidiomycetes reached a peak after six month but were absent from nitrogen rich pineapple needle. In later stage of decomposition soil fungi *Penicillium, Trichoderma, Absidia, Mucor* species were the most common (Santro *et al.*, 2002).

Factors Affecting Fungal Colonization on Residue

Primary fungal colonizer and successive invaders are governed by interacting factors of soil, climate and residues such as moisture, temperature, pH, depth, aeration and organic nutrients (Williams and Gray, 1974, Nikhra, 1981). Moisture is one of the factors that regulates the growth and activities of fungi on decomposing residue. During winter months, low temperature and moisture create an environment not conducive to fungal growth (Rai and Srivastav, 1982). The improved moisture during rainy season along with optimum temperature favours fungal growth. In rice-wheat ecosystems crops are grown in different seasons and the population and activity of fungi varies with a change in temperature and moisture. Increased temperature due to climatic change may increase the turn over of soil organic matter and hence reduce soil organic carbon (Webb *et al.*, 2003). The response of soil pH changes to the addition of organic matter, depend upon the type of plant materials and starting pH (Xu and Coventry, 2003).

Another factor that influences the fungal colonization is the C : N ratio of the organic residue. Majority of the fungi show optimum growth within a narrow C : N ratio (Nikhra, 1981). A C : N ratio of 60 was found most effective for fungal colonization and succession (Debnath and Sinha, 1993). Eiland *et al.* (2001) reported that the microbial community structure (measured as phospholipid fatty acid, PLFA profile) was affected by initial C : N ratio, with lower fungal/bacterial ratios in low compared to high C:N treatments after 12 months of composting of shredded straw of *Miscanthus*. Low initial C : N ratios caused a fast degradation of fibers during first three months of composting (Hemicellulose: 50–80 per cent, cellulose: 40–60 per cent), while high initial C : N ratios resulted in 10–20 per cent degradation of both hemicellulose and cellulose. Simoes *et al.* (2002) observed that the decomposition rates and nutrient release pattern were influenced by litter quality especially in initial N concentration and initial C : N. Dynamics of microbial biomass is closely related to N mineralisation (Vityakon *et al.*, 2000).

Devervre and Horwath (2000) studied carbon use efficiency by soil micro-organisms during rice decomposition in a paddy soil under aerobic and anaerobic (flooded) conditions at different temperatures (5, 15 and 25°C) in relation to nutrient availability in rice cropping and concluded that flooding had tendency to reduce 'C' mineralization and enhanced methane (CH_4) production while a decrease in temperature lowered methane (CH_4) production markedly. Microbial activities were similar under flooded and non-flooded conditions even though anaerobes decomposed less straw than aerobes. Significant amount of decomposition occurred under flooded condition but less straw was mineralized compared to aerobic condition. Tapper *et al.* (1981) also observed rapid decomposition of rice straw under aerobic conditions and marked increase in C in humic substances than under anaerobic condition. *Aspergillus* and *Penicillium* were the most dominant fungal species. Addition of mussorie rock phosphate was found to cause rapid loss in weight and decrease in C : N ratio owing to marked increase in fungal activities (Akhaury and Kapoor. 1997).

Mucor hiemalis and *Truncatella truncata* were significantly more abundant in the control (untreated litter) but their population decrease due to UV-B radiation indicating thereby their sensitivity to UV-B radiation while *Penicillium brevicompactum* was equally abundant in UV-B and control treatment (Gehrke *et al.*, 1995). *Alternaria alternata*, *Aspergillus* spp., *Cladosporium herbarum*, *Penicillium* spp. *Trichoderma* spp., *Cephalosporium* spp. and a filamentous yeast were the most dominant species on rubber litter during rainy season, their population decreased towards the end of season (Mathew *et al.*, 1998). Sources of carbon affects the predominance of various saprophytic fungi *Rhodotorula graminis* and *Fusarium oxosporum* stimulate higher proportion of sucrose while *Mortierella* isolates and FPC 341 (an unidentified isolate) lower. *Aspergillus flavus* and *Fusarium oxysporum* had higher metabolic activity on glucose containing disaccharide cellobiose than on glucose strongly suggesting preferential uptake of cellobiose compared to glucose and suggesting the potential ability to use the cellulose (Hobbie *et al.* 2003).

Residue Decomposition in Rice-Wheat Ecosystem

In rice-wheat cropping system wheat is grown just after harvest of rice and disposal of large amount of rice residue is a serious problem. One of the ways of rice residue disposal is *in situ* microbial decomposition. The detailed study of microbial decomposition of residue is important one. However, information on this is not available in literature, hence reports for the decomposition of other crop species is reviewed. Dkhar (1983) studied the colonization pattern of fungi on leaf stem and root residue of maize and that phycomyceta fungi were the pioneer colonizer but deuteromycetes such as *Trichoderma, Penicillium, Aspergillus* and *Alternaria* were predominant throughout decomposition processes. Similar was the finding of Kshattriya (1990). She found phycomycete *Mocur hiemalis, M. racemosus, Rhizopus stolonifer* and *Zygorhynchus* sp. pioneer colonizer while deuteromycetes such as *Penicillium chrysogenum, Trichoderma viride, Aspergillus flavus, A. niger* and *Fusarium oxysporum* were the secondary organism. The dominance of deuteromycetes and complete absence of phycomycetes and basidomycetes on herbaceous *Chenopodium album* litter was found by Mehrotra and Aneja (1979). Preponderance of *Fusarium, Mucor, Penicillium* and *Trichoderma* on decomposing wheat straw was recorded by Sadasivan (1939) while that of *Acremonium persicimum, Chaetomium globosum, Fusarium* sp. *Penicillium, Sordaria alcina, Trichoderma herzianum* and *T. langibrachatum* was reported by Harper and Lynch (1985). Abdel and El Said (2002) isolated 23 species of 11 genera of xylan decomposing filamentous fungi from decomposing rice straw, 16 species of 7 genera from wheat straw, and 20 species of 8 genera from sugarcane bagasse. Among these species *Aspergillus flavis, A. niger, Penicillium chrysogenum, P. corylophilum, P. funiculotum, P. oxalicum* and *Trichoderma harzianum* were the most prevalent species. Dominance of lignocellulolytic fungi *Cyathus stevecus* was reported by Halsall (1993) while that of *Cyathus olla, Copimus peicatiles* and *Panacolina foenigecu* by Baird *et al.* (1993) in the decomposition of residue in residue management practices. Green manure affects the yield and nitrogen content in rice-wheat agroecosystem (Aulakh *et al.*, 2000).

Incorporation of crop residues, like wheat or rice straw decreases the bulk density and increases the organic matter, cation exchange capacity, hydraulic conductivity, water holding capacity of the soil and also increases the grain and straw yield of rice (Das *et al.*, 2001). Verma (2001) recorded an increasing trend in wheat yield and organic carbon content in soil with the incorporation of crop residue. He also observed the positive improvement in soil fertility status with residue incorporation. Rice yields increased when wheat residue was incorporated with GM. Physiological efficiency was higher when rice straw was incorporated in wheat and when wheat straw plus GM were incorporated in rice. The long-term application of rice residue increased C accumulation in soil (Singh *et al.*, 2004a). Yield and soil nutrient changes in a long term rice-wheat rotation in India is studied by Bhandari *et al.* (2002). Soil C increases with the application of wheat straw and FYM in rice-wheat cropping system (Singh *et al.*, 2004b). Regular wheat-straw application results in the rise of grain yield by 0.31–0.5 t/ha for winter rye and by 0.23–0.52 t/ha for spring wheat. Straw application ensures soil sanitation from root pathogens, which also favours a rise in grain yield (Lazarev and Abrashin, 2000).

Stumpe *et al.* (2000) investigated the effects and residual effects of straw, farm yard manuring and mineral fertilization and observed that FYM and straw (with same amounts of dry matter) enhanced the organic carbon content and consequently the content of decomposable carbon. Jenkyn *et al.* (2000) studied the effects of incorporating different amounts of straw on growth, diseases and yield of consecutive crops of winter wheat. *Lantana* residues improve the nutrient status of soil and rice-wheat cropping system yield (Sharma *et al.*, 2003). Tajada and Gonzalez (2003) reported an increase in soil microbial activity, structural stability and soil porosity on the application of crushed cotton gin to the soil. Grandy *et al.* (2002) studied the organic amendment and crop rotation effects on the recovery of soil organic matter and aggregation in potato cropping systems.

Plant Nutrient Release

Fungi play an important role in the nutrient cycling in the soil. Many elements are mobilized from decomposing rice residue. Organic soil amendments initially increase the level of soil organic matter, microbial biomass, basal respiration and some enzyme activities related to C and N cycles. These values decrease but always remain higher than those of unamended soil (Ros *et al.*, 2003). Mukherjee *et al.* (1990) reported that all paddy straw application increased organic and nitrogen content. Similar effect was observed by Singh (2003) on the application of rice straw. Incorporation of wheat straw had also increased the available nitrogen and potassium status of soil but proved less beneficial than rice straw. Approximately 38 per cent N and 8 per cent P of total fertilizer used are provided by rice straw alone (Fores *et al.*, 1988). Shah *et al.* (1995) observed immobilization of ammonical nitrogen with addition of rice residue. In rice-wheat rotation application of organic material in soil increases organic carbon in both the soil and clay fraction (Ruixing and Dehui, 1988). Soil carbon (C) concentration increases significantly at higher fertilizer rate and rice stubble retention (Whitbread *et al.*, 2003). Adsorption of microbial product at clay stimulated reduces mineralization of carbon amendment and retention increases (Grant *et al.*, 1993). The soils with greater clay concentration have longer time period of initial N immobilization and mineralized less N from the added materials (Egelkraut *et al.*, 2000). With increasing temperature, crop residue and FYM diffusion coefficient of Zn increase due to presence of chelating agents (Prasad and Sinha, 1995). Cheshire *et al.* (1999) from their study using wheat straw in mesh bag buried for year reported that immobilization of nitrogen was not detected in first season owing to higher moisture content but it occurred thereafter. Soil organic material appears to play significant role in crop nutrient uptake in intensively cropped low-land rice soil (Olk *et al.*, 2000). The higher biomass and microbial activity is related to increase in pH and in the quantity of nutrient after incorporation of wood ash in to soil. The higher activity is linked to increase mineralization of organic matter, which follows by decreasing soil total C and N contents (Zimmennann and Prey, 2002). During decomposition of residue, carbon and potassium content decreases, nitrogen and phosphorus content increases and C : N ratio decreases with time. A high C : N ratio in decomposed wheat straw results in immobilization of nitrate nitrogen. Long-term straw application causes a rise in the contents of nutrients and humus in soils (Lazarev and Abrashin, 2000). Recycling of crop residues by incorporation or burning increase soil available K and organic carbon while total N content increases by rice straw incorporation (Surekha *et al.*, 2003). In the early stage, practically no mineralization takes place from the residue but thereafter mineralization proceeds gradually (Mishra *et al.*, 2001). Increase in relative phosphorus solubility due to straw incorporation owing to microbial activity was observed by Meena *et al.* (2002). Long-term effects of wheat residue management on some fertility indicators were studied by Du Preeze *et al.* (2001) and found only pH, K, P and Zn were significantly influenced by residue management practices after 11–12 years. The N mineralization is positively correlated with the total N and lignin content and negatively correlated with C : N ratio, cellulose, polyphenol and polyphenol : total N (Aderdiran *et al.*, 2003). They also observed that organic wastes had a positive influence on soil microbial biomass. The release of P, K, Ca, Mg, Zn, Fe and Cu was also studied. Incorporating green manure crops in to soil may increase bioavailability of phosphorus for succeeding crop (Cavigelli and Thien, 2003). Menan *et al.* (2004) observed an increase in C:N ratio when stubble were added to soil. Gyaneshwar *et al.* (2002) studied the role of microorganisms in improving the P nutrition of plants. Schulz *et al.* (2003) reported that more than 90 per cent of P released from the decomposing rice mill waste, however, around 60 per cent of the organic C and 45 per cent of N still remained.

The microbial biomass has a significant role to play in conservation of soil N and P in the stressed ecosystem (Arunachalam and Pandey, 2004). De Neve *et al.* (2004) studied the manipulating N mineralisation from high N crop residue using on and off farm organic materials. Nitrogen mobilisation increases with the incorporation of crop residues like groundnut haulm and maize stover in soil. The soil texture had little effect on N mineralisation in soil amended with crop residues (Mubarak *et al.*,

2001). Shammar *et al.* (2003) studied that during decomposition of leaf, bark and branch material of *Eucalyptus globus*, K was leached rapidly, Mg and P were released at similar rates to dry weight and Ca and N were released more slowly than dry weight.

In a mini-review on decomposition dynamics in mixed species leaf litter (Gartner and Cardon, 2004), what is clear is that natural mixing of litters has the potential to strongly affect dynamics of litter mass loss, litter nutrient content, and decomposer abundance and activity, and these interactions may well be very important for nutrient recycling and primary productivity in ecosystems.

References

Akhaury, Prema and Kapoor, K.K. (1997). Solubilization of insolubte phosphorus by fungi isolated from compost and soil. *Environment and Ecology*, 15(3): 524–527.

Anderdiran, J.A., De Bacts, N., Mnkeni, P.N.S., Kiekens, L., Muyima, N.Y.O. and Thys, A. (2003). Organic waste materials for soil fertility improvement in the Border region of the Eastern Cape, South Africa. *Biological Agriculture and Horticulture*, 20(4): 283–300.

Arunanchalam, A. and Pandey, H.N. (2004). Microbial C, N and P along a weeding regime in a valley cultivation system of North East India. *Tropical Ecology*, 44(2): 147–154.

Aulakh, M.S., Khera, T.S., Doran, J.W., Singh, K. and Singh, B. (2000). Yield and nitrogen dynamics in a rice wheat agro-ecosystem using green manure and inorganic fertilizer. *Soil Science American Journal*, 64: 1867–1876.

Barkenkamp, A., Priesack, E. and Munch, J.C. (2002). Modelling the mineralization of plant residue on the soil surface. *Agronomie*, 22(7–8): 711–722.

Barzegar, A.R., Yousefi, A. and Paryshenas, A. (2002). The effect of addition of different amount and types of organic material on soil physical properties and yield of wheat. *Plant and Soil*, 247(2): 295–301.

Beare, M.H., Pohlad, B.R., Wright, D.H. and Coleman, D.C. (1993). Residue placement and fungicide effects on fungal communities in conventional and no-tillage soils. *Soil Science Society of America Journal*, 57(2): 392–399.

Beare, M.H., Wilson, P.E., Fraser, P.M. and Butler, R.C. (2002). Management effects on barley straw decomposition, nitrogen release and crop production. *Soil Science Society of America Journal*, 66(3): 848–856.

Brar, S.S., Kumar, Sanjeev and Narang, R.S. (2000). Effect of moisture regime and nitrogen on decomposition of combine harvested rice (*Oryza sativa*) residue and performance of succeeding wheat (*Triticum aestivum*) in rice-wheat system in Punjab. *Indian Journal of Agronomy*, 45(3): 458–462.

Cavigelli, M.A. and Thien, S.J. (2003). Phosphorus bioavailability following incorporation of green manure crops. *Soil Science Society of American Journal*, 67(4): 1186–1194.

Choi, Y., Zongwen, L., Dehan, W. and Yurui, L. (2003). Decomposition effect of chemical-biological decomposing combined technique on rice straw and its fourier transform infrared spectroscopy. *Yingyong shengtai Xuebao*, 14(8): 1269–1272.

Choi, Y.H., Lee, S.B., Park, K.H. and Hong, J.S. (1990). Studies on the decomposition of rice straw by microorganism. I. Isolation and selection of cellulolytic microorganism and decomposition of rice straw. *Soil and Fertilizer*, 32(2): 42–50.

Chatterjee, S.K. and Nandi, B. (1980). Biodegradation of wheat stubbles by some soil microorganism. *Proceedings of RRAI Symposium*, PAU, Ludhiana, India, pp. 275–280.

Cheshire, M.V., Bedrock, C.N. and Williams, B.L. (1999). Effect of climate and soil type on the immobilization of nitrogen by decomposing straw and southern Europe. *Biology and Fertility of Soil*, 28(3): 306–312.

Clements, F.E. and Shear, C.L. (1954). *The Genera of Fungi.* Hafner Publishing Co., New York, USA.

Cookson, W.R., Beare, M.H. and Wilson, P.E. (1998). Effect of prior crop residue management decomposition. *Applied Soil Ecology*, 7(2): 179–188.

Das, K., Medhi, D.N. and Guha, B. (2001), Recycling effect of crop residue with chemical fertilizers of physio-chemical properties of soil and rice (*Oryza sativa*) yield. *Indian Journal of Agronomy*, 48(1): 8–12.

De Neve, S., Saez, S.G., Daguilar, S.S. and Hofman, G. (2004). Manipulating N mineralization from high N crop residue using on and off farm organic material. *Soil Biology and Biochemistry*, 36(1): 127–134.

Debnath, M. and Sinha, N.B. (1993). Effect of different C : N ratios and microbial culture inoculations on decomposition of rice straw. *Environment and Ecology*, 11(1): 1–6.

Devevre, O.C. and Horwath, W.R. (2000). Decomposition of rice straw and microbial carbon use efficiency under different soil temperatures and moistures. *Soil Biology and Biochemistry*, 32(11–12): 1773–1785.

Dhillon, G.S., Kalra, Singh, A, Kahlon, S.S. and Kalra, M.S. (1980). Bioconversion of the delignified rice straw by cellulolytic fungi. *Proceedings of RRAI Symposium*, PAU, Ludhiana, India, pp. 77–80.

Dkhar, M.S. (1980). Studies on ecology of edaphic microbial population and their activities in maize fields. *Ph.D. Thesis*, North-Eastern Hill University, Shillong, India.

Du Preez, C.C., Steyn, J.T. and Kotze, E. (2001). Long-term effects of wheat residue management on some fertility indicators of a semi-arid Plinthsol. *Soil and Tillage Research*, 63(1–2): 25–33.

Egelkraut, T.M., Kissel, D.E. and Cabera, M.L. (2000). Effect of soil texture on nitrogen mineralized from cotton residue and compost. *Journal of Environmental Quality*, 29(5): 1518–1522.

Eiland, F., Klamer, M., Lind, A.M, Leth, M. and Baath, E. (2001). Influence of initial C/N ratio on chemical and microbial composition during long term composting of straw. *Microbial Ecology*, 41(3): 272–280.

Fontaine, S., Mariotti, A. and Abbadie, L. (2003). The priming effect of organic matter: A question of microbial competition. *Soil Biology and Biochemistry*, 35(6): 837–843.

Forge, E., Menendze, M, and Comin, F.A. (1988). Rice straw decomposition in rice field soil. *Plant and Soil*, 109(1): 145–146.

Gartner, T.B. and Cardon, Z.G. (2004). Decomposition dynamics in mixed species leaf litter. *Oikos*, 104(2): 230–246.

Gehrke, C., Johanson, U., Callaghan, T.V., Chadwick, D. and Robbinson, C.H. (1995). The impact of enhanced ultraviolet B radiation on litter quality and decomposition process in vaccinium leaves from the Subarctic. *Oikos*, 72(2): 213–222.

Gill, K.S. (1994). Sustainability issues related to rice-wheat production system in Asia. In: *Sustainability of Rice–wheat Production System in Asia,* (Eds.) R.S. Panda, T. Woodhead and R.B. Singh. ROAP, FAO, Bangkok, pp. 36–60.

Grandy, S.A., Porter, G.A. and Erich, M.S. (2002). Organic amendment and rotation crop effects on the recovery of soil organic matter and aggregation in potato cropping system. *Soil Science Society of America Journal,* 66: 1311–1319.

Grant, R.F., Juma, N.G. and McGill, W.B. (1993). Simulation of carbon and nitrogen transformations in soi1 mineralization. *Soil Biology and Biochemistry,* 25(10): 1317–1329.

Gyaneshwar, P., Kumar, G.N., Parekh L.J. and Poote P.S. (2002). Ro1e of soil microorganisms in improving P nutrition of plant. *Plant and Soil,* 245(1): 83–93.

Harper, S.T.H. and Lynch, J.M. (1985). Colonization and decomposition of wheat straw leaves, internodes and nodes. *Journal of the Science of Food and Agriculture,* 32: 1057–1062.

Hegde, D.M. (1998). Effect of integrated nutrient supply on crop productivity and soil fertility in rice-wheat cropping system in semi arid and humid ecosystem. *Indian Journal of Agronomy,* 43: 7–12.

Hobbie, E.A., Watrud, L.S., Maggard, S., Shiroyama, T. and Rygiewicz, P.T. (2003). Carbohydrate use and assimilation by litter and soil fungi assessed by carbon isotopes and BIOLOG (R) assays. *Soil Biology and Biochemistry,* 35(2): 303–311.

Hobbie, S.E. (2000). Interactions between litter lignin and soil nitrogen availability during leaf litter decomposition in a Hawaiian montane forest. *Ecosystem,* 5: 484–494.

Hudson, H.J. (1962). The ecology of fungi on plant remains above the soil. *New Phytology,* 67: 837–874.

Hudson, H.J. (1971). *Fungal Saprophytism.* Studies in Biology, No. 32, London, Edward Arnold.

Jenkyn, I.F., Gristian, D.G., Bacan, E.T.G., Gutteridge, R.J. and Todd, A.D. (2001). Effects of incorporating different amount of straw on growth, disease and yield of consecutive crops of winter wheat growth on contrasting soil types. *Journal of Agricultural Sciences,* 136(1): 1–14.

Kuhad, R.C. and Johri, B.N. (1992). Fungal decomposition of paddy straw: Light and scanning electron-microscopic study. *Indian Journal of Microbiology,* 32 (2): 255–258.

Lazarev, A.P. and Abrashin, Y.I. (2000). The influence of applied wheat straw on the properties, biological activity and fertility of chernozems. *Pochvovedenic,* (10): 1266–1271.

Limtong, P., Vagani, S., Leaungvutivirong, C., Piriyaprin, S., Sunanthapongsuk, V. and Thunyadee, P. (1994). *Proc. of SABRO Seventh International Congress and WSAA Symposium,* Taipei, Taiwan, 35: 905–911.

Lumley, T.C., Gignac, L.D. and Currah, R.S. (2001). Microfungus communities of white spruce and trebling aspen logs at different stages of decay in disturbed and undisturbed sites in the boreal mixed wood region of Alberta. *Canadian Journal of Botany,* 79(1): 76–92.

Macfadyen, A. (1963). A contribution of the microfauna to totals soil metabolism. In: *Soil Organism,* (Eds.) J. Doeksen and J. Van. Der Drift. North Holland Pub. Co., Amsterdam, pp. 3–16.

Mathew, N.M., Kuruvilla, Jacob, C., Licy, J., Joseph, T., Meenattoor, J.R. and Thomas, K.K. (1998). Decomposition of leaf litter by *Hevea brasiliensis* Muell. Arg. Under subtropical condition of Meghalaya. *Proceedings of the 12th Symposium on Plantation Crops,* PLACROSYM XII, Kottayam, India, pp. 214–219.

Meena, S.L., Surendra Singh and Shivay, Y.S. (2002). Effect of time of nitrogen and phosphorus application on decomposition of wheat residue and nutrient uptake by rice. *Annals of Agricultural Research*, 23(2): 325–327.

Mehrotra, R.S. and Aneja, K.R. (1979). Microbial decomposition of *Chenopodium album* litter. I. Succession of decomposers. *Journal of Indian Botanical Society*, 58: 189–195.

Menan, D.P., Chan, K.Y. and Knigh, P.G. (2004) Long term impact of rotation, tillage and stubble management on the loss of soil organic carbon and nitrogen from achromic Luvisol. *Soil and Tillage Research*, 76(1): 59–68.

Mishra, B., Sharma, P.K. and Bronson, K.F. (2001). Decomposition of rice straw and mineralization of carbon, nitrogen, phosphorus and potassium in wheat field soil in western Uttar Pradesh. *Journal of Society of Soil Science*, 49(3): 419–424.

Mubarak, A.R., Rosenani, A.B., Siti, Z.D. and Anuar, A.R. (2001). Nitrogen mineralization in tropical soil amended with crop residues. *Tropical Agriculture*, 78(3): 165–173.

Mukherjee, D., Ghose, S.L. and Das, A.C. (1990). A study on the chemical and microbial changes during the decomposition of straw in soil. *Indian Agriculturist*, 34(1). 1–10.

Mukhopadhyay, Debdas and Nandi, B. (1979). Biodegradation of rice stumps by soil mycoflora. *Plant and Soil*, 53: 215–218.

Nambiar (1994). *Soil Fertility and Crop Productivity Under Long Term Fertilizer Use in India.* Indian Council of Agricultural Research, New Delhi, pp. 144.

Nandi, N., Hazra, J.N. and Sinha, N.B. (1996). Microbial synthesis of humus from rice straw following two step composting process. *Journal of Indian Society of Soil Science*, 44(3): 413–416.

Nikhra, K.M. (1981). Studies on fungi from Jabalpur soils woth special reference to litter decomposition. *Ph.D. Thesis*, Jabalpur University, India.

Olk, D.C. and Senesi, N. (2000). Properties of chemically extracted soil organic matter in intensively-cropped lowland rice soils. In: *Proc. of the Workshop on Carbon and Nitrogen Dynamics in Flooded Soils.* Los Banos, Philippines, 19–22 April, 1999, 2000, pp. 65–87.

Prasad, B. and Sinha, S.K. (1995). Effect of recycling of crop residues and organic manure on capacity factor and diffusion rate of zinc in calcareous soils. *Journal of Nuclear Agriculture and Biology*, 24(3): 185–188.

Prasad, R. and Mishra, B.N. (2001). Effect of addition of organic residues, farmyard manure and fertilizer nitrogen on soil fertility in rice-wheat cropping system. *Arch. Acken. Pjl. Boden*, 46. 455–463.

Prasad, R. and Power, J.F. (1997). *Soil Fertility Management for Sustainable Agriculture.* CRC, Lewis, Boca Raton, U.S.A., pp. 356.

Prasad, R., Gangiah, B. and Aife, K.C. (1999). Effect of crop residue management in rice-wheat cropping system on growth and yield of crops and cropping system. *Experimental Agriculture*, 35: 427–435.

Rai, B. and Srivastava, A.K. (1982). Decomposition of leaf litter in relation to microbial populations and their activity in tropical dry mixed deciduous forest. *Pedobiologia*, 24: 151–159.

Reichle, D.E., Mc. Brayer, J.F. and Ausmus, B.S. (1975). Ecology energertics of decomposer invertebrates in deciduous forests and total respiration budget. In: *Progress in Soil Zoology*, (Ed.) J. Vanvek. Academic Publishing House of Slovak Academy of Science, Prague, pp. 283–292.

Ros, M., Memandez, M.T. and Garcia, C. (2003). Soil microbial activity after restoration of a semi-arid soil by organic amendment. *Soil Biology and Biochemistry*, 35(3): 463–469.

Rosenbrock, P., Buscot, F., and Munch, J.C. (1995). Fungal succession and changes in the fungal degradation potential during the initial stage of titter decomposition in a black alder forest *Alnus glutinosa* (L.). *European Journal of Soil Biology*, 31(1): 1–11.

Ruixing, D. and Dehui, L. (1988). Decomposition and transformation of organic materials in bleached paddy soil. *Journal of Narjing Agricultural University (Nanjing Nongye Daxue LYucbao)*, 11(2): 67–72.

Sadasivan, T.S. (1939). Succession of fungi on decomposition of wheat straw in different soils with special reference to *Fusarium culmorum*. *Annals of Applied Biology*, 26: 497–508.

Saha, N., Das, A.C. and Mukherjee, D. (1995). Effect of decomposition on organic matter on the activities of microorganisms and availability of nitrogen, phosphorus and sulphur in soil. *Journal of the Indian Society of Soil Science*, 43(2): 210–215.

Sameni, M. and Pour, A.T. (2001). Effect of root residue of liquorice on different characteristics of soil from southern regions of Iran. I. *Aficrobial Characteristics Communications in Soil Science and Plant Analysis*, 328(19/20): 3259–3275.

Santro, A.V. de, Rutigliano, F.A., Berg, B., Fioretto, A. Puppi, G. and Alfuni, A. (2002). Fungal mycelium and decomposition of needle litter in three contrasting coniferous forests. *Acta Oecologia*, 23(4): 247–259.

Sariyitdiz, T. and Anderson, J.M. (2003). Decomposition of Sun and Shade leaves from three deciduous tree species, as affected by their chemical composition. *Biology and Fertility of Soil*, 37(3): 137–146.

Sarojini, K. and Mathur, R.S. (1994). Biodegradation of paddy straw with cellulolytic fungi and its application on wheat crop. *Bioresource Technology*, 47(92): 185–187.

Schulz, S., Tian, G., Oyewole, B. and Bako, S. (2003). Rice mill waste as organic manure on a degraded Alfisols. *Agriculture Ecosystem and Environment*, 100(2–3): 221–230.

Sekhon, A.K. (1975). Studies on utilization of wheat straw for production of cellulose and protein by thermophillic fungi. *M.Sc. Thesis*, PAU, Ludhiana, India.

Shammar, K., O'Connell, A.M, Grove, T.S., McMurtric, R., Damon, P. and Rance, S.J. (2003). Contribution of decomposing harvest residue to nutrient cycling in a second rotation *Eucalyptus globus* plantation in South–Western Australia. *Biology and Fertility of Soil*, 38(4): 228–235.

Sharma, P.D. and Dwivedi, R.S. (1972). Succession of fungi on decaying *Setaria glauca*. *Tropical Ecology*, 13: 183–201.

Sharma, P.K., Ladha, J.K., Verma, T.S., Bhagat, R.M. and Padree, A.T. (2003). Rice-wheat productivity and nutrient status in a lantana (*Lantana* Spp.) amended soil. *Biology and Fertility of Soil*, 37(2): 108–119.

Shaukat, S.S., Siddiqui, I.A. and Mehdi, F.S. (2003). *Avicennia marina* (mangrove) soil amended changes the fungal community in the rhizosphere and root tissue of mungbean and contributes to control of root Knor nematodes. *Phytopathologia Mediterranea*, 42(2): 135–140.

Silgram, M. and Chambers, B.J. (2002). Effects of long term straw management and fertilizer nitrogen addition on soil nitrogen supply and crop yield at two sites in Eastern England. *Journal of Agricultural Sciences*, 139(2): 115–127.

Simoes, M.P., Madeira, M. and Gazarini, L. (2002). Decomposition dynamics and nutrient release of *Cistus salvifolius* L. and *Cistus ladanifer* L. leaf litter. *Revista de Ciencias Agrarias*, 25(3–4): 508–520.

Singh, D.D. (2003). Management of crop residue in summer rice and its effect on soil properties and crop yield. *Crop research (Hissar)*, 25(1): 191–193.

Singh, V., Singh, B., Ladha, J.K., Khind, C.S., Khera, T.S. and Bueno, C.S. (2004a). Effect of residue decomposition on productivity and soil fertility in rice-wheat rotation. *Soil Science Society of American Journal*, 68: 854–864.

Singh, V., Singh, B., Ladha, J.K., Khind, C.S., Khera, T.S. and Bueno, C.S. (2004b). Long term effect of organic inputs of yield and soil fertility in rice-wheat rotation. *Soil Science Society of American Journal*, 64: 845–853.

Stumpe, H., Wittenmayer, I. and Merbach, W. (2000). Effect and residual effects of straw, FY manuring and mineral fertilization at field F of the long term trial in Halle (Saale) Germany. *Journal of Plant Nutrition and Soil Science*, 163(6): 649–656.

Surekha, K., Kumari, A.P.P., Reddy, M.N., Satyanarayana, K. and Cruzsta, P.C. (2003). Crop residue management to sustain soil fertility and irrigated rice yields. *Nutrient Cycling in Agroecosystem*, 67(2): 145–154.

Swift, M.J., Heal, O.W. and Anderson, J.M. (1979). *Decomposition in Terrestrial Ecosystems*. Blackwell Scientific Publications, Oxford.

Tapper, E.Z., Sidorenko, O.D. and Mohamed, Azazi (1981). Rice straw as a source of humus and mycoflora involved in its decomposition. *Biology Bulletin*, 8(1): 82–90.

Tejada, M. and Gonzalez, J.L. (2003). Effect of application of a compost originating from crushed cotton gin residue on wheat yield under dryland conditions. *European Journal of Agronomy*, 19(2): 357–368.

Thormann, M. N., Currah, R.S. and Bayley, S.E. (2003). Succession of microfungal assemblages in the decomposing peat land plants. *Plant and Soil*, 250(3): 323–333.

Tiwari, S.C., Tiwari, B.K. and Mishra, R.R. (1994). Succession of microfungi associated with the decomposing litters of pineapple (*Annanas comosus*). *Pedobiologia*, 38(2): 185–192.

Verma, K.P. (2001). Effect of crop residue incorporation and nitrogen of succeeding wheat (*Triticum astivum*). *Indian Journal of Agronomy*, 46(4): 665–669.

Vityakon, P., Meepech, S., Candish, G. and Toomsan, B. (2000). Soil organic matter and nitrogen transformation mediated by Plant residue of different qualities in sandy acid upland and paddy soils. *Netherland Journal of Agricultural Science*, 48(1): 75–90.

Webb, J.P., Bellomy, P., Loveland, P.J. and Goodlass, G. (2003). Crop residue returns and equilibrium soil organic carbon in England and Wales. *Soil Science Society of American Journal*, 63(3): 928–936.

Whitbread, A., Bliar, G., Konboon, V., Lefroy, R. and Naklong, K. (2003). Managing crop residue, fertilizers and Leaf differs to improve soil, nutrient balance, and the gain yield of rice and wheat cropping system in Thailand and Australia. *Agriculture Ecosystem and Environment*, 100(2–3): 251–263.

Williams, W.T., and Gray, T.R.G. (1974). Decomposition of the litter on the soil surface. In: *Biology of Plant Litter Decomposition*, (Eds.) C.H. Dickinson and G.J.P. Pugh. Academic Press, London, New York, 2: 611–632.

Wood, T.G. (1976). The role of termites (Isoptera) in decomposition process. In: *The Role of Terrestrial and Aquatic Organisms in Decomposition Process,* (Eds.) J.M. Anderson and A. Macfaclyen. Blackwell Scientific Publication, Oxford, pp. 145–168.

Xu, R.K. and Coventry, D.R. (2003). Soil pH changes associated with lupin and wheat plant materials incorporated in a red-brown soil. *Plant and Soil,* 250(1): 113–119.

Zimmermann, S. and Prey, B. (2002). Soil respiration and microbial properties in acid forest soil: Effect of wood ash. *Soil Biology and Biochemistry,* 34(1): 1727–1737.

©R.C. Ram & Asha Sinha (*Editors*)
***Microbial Diversity: Modern Trends*, 2007** **63–78**
Daya Publishing House, New Delhi

Chapter 5

Diversity in Stripe Rust Flora and Matching Genes in Wheat: The Indian Scenario

V.C. Sinha*\#, Rajiv K. Sharma* and Sanjay Kumar**

***Division of Genetics, Indian Agricultural Research Institute, New Delhi - 110 012**
****IARI Regional Station, Tutikandi, Shimla, H.P.**

ABSTRACT

All the three wheat rusts *viz.*, black, brown and yellow rust infect wheat crop in this country. The brown and yellow rust survives in the northern Himalayas particularly in central Nepal, while black and brown rust also survive in Nilgiri and Palney hills in South India. In our country, alternate hosts do not play any role in the perpetuation of rusts. These rusts survive on self sown plants, ratoon tillers and grasses in the Himalayas in the north and Nilgiri and Palney hills in south at appropriate heights. The new races/pathotype arise as a result of somatic hybridisation or mutation or through introduction from neighbouring countries. Whenever the new resistance sources are used in breeding programme to incorporate resistance against particular pathotype, the new pathotypes evolve to break the resistance. Suggested strategies for the management of rust include use of unutilized genes, adult plant resistance or slow rusting.

Keywords: *Diversity, Stripe Rust Pathotypes, Matching Genes, APR.*

Introduction

Wheat production in Asia especially in South east Asia increased dramatically during the past couple of decades. One of the key factors involved in the wheat revolution was the introduction of high

Corresponding Author: E-mail: vcssinha@rediffmail.com

yielding semi dwarf varieties that are resistant to lodging, responsive to fertilizer and widely adapted. Disease resistance has been one of the important features of new varieties that contributed to their spread. Over 27 m hectare is currently occupied by wheat and barely crop in India. In the past decades annual collection of about 400 samples of black and brown rust and 30-40 of yellow rust is made for identification of races which however, not expected to give a true picture of their occurrence and distribution of race flora. In Indo Gangetic plains, the main wheat producing tract of the country, all the three wheat rusts are prevalent. Yellow rust is more destructive in the hills and plains of Punjab and Western UP whereas it is almost absent in Central and Peninsular India (Singh *et al.*, 1978). Leaf rust is a problem of the country as a whole and stem rust is restricted to Central and Peninsular India, although it occasionally does appear in Northern India, but too late to cause any damage.

Rusts, particularly yellow and brown are the most important foliar diseases of wheat since prehistoric time. The importance of genetic resistance for the control of rust diseases was first demonstrated by Biffen (1905). Since then efforts have been made to identify new sources of resistance to rust disease and their utilization in rust resistance breeding. The most important requirement of a long term strategy for developing resistant cultivar is to search for novel diverse genes for resistance. Stripe rust caused by (Yellow rust) *Puccinia striiformis* West has been confined mainly to cooler parts *viz.*, plains of north Western India including foot hills of Himalayas, north eastern plains and northern parts of Central India and Nilgiri hills in south India (Figure 5.1).

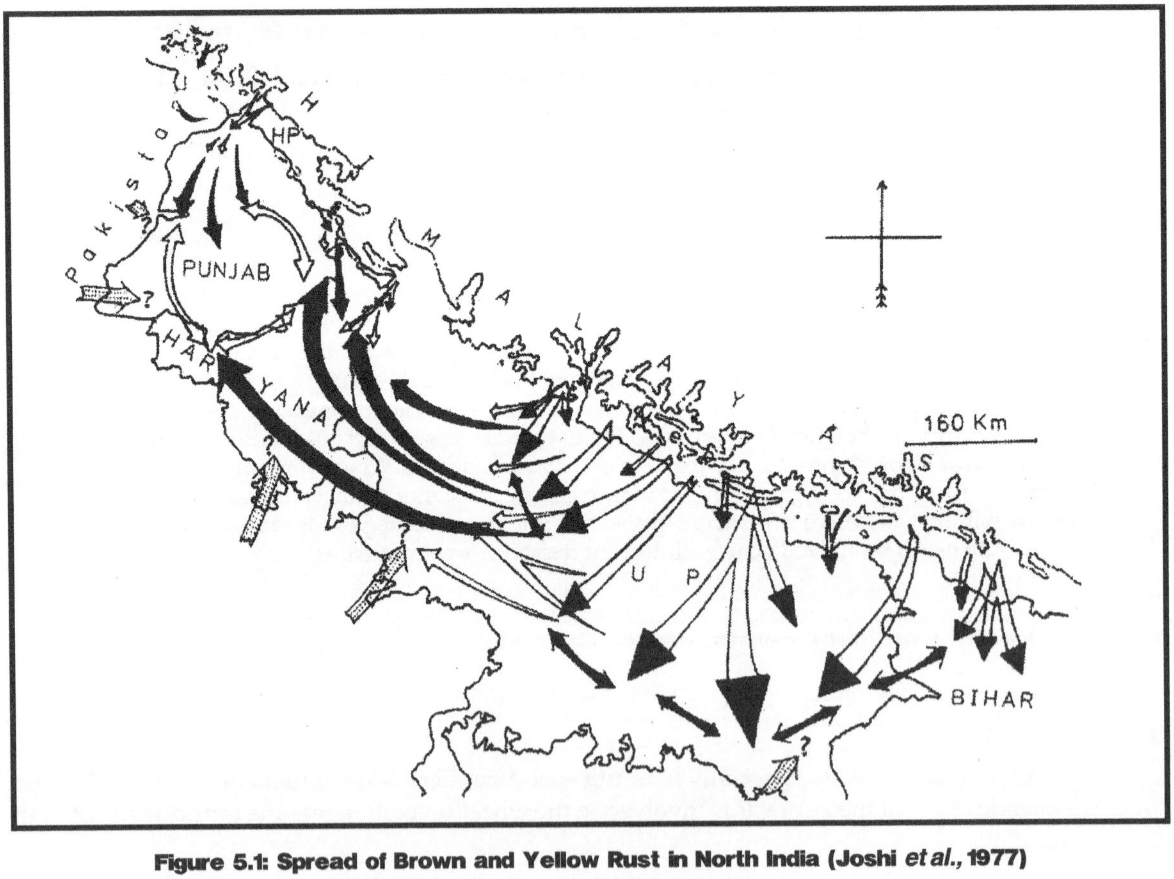

Figure 5.1: Spread of Brown and Yellow Rust in North India (Joshi *et al.*, 1977)

Stripe rust has the potential to cause more damage than that caused by leaf rust. The knowledge base for stripe rust is relatively poorer, however, it would seem that a similar mechanism is involved. As our knowledge grows it may be possible to totally manage rust diseases through genetic means in future. However the task of incorporating durable resistance in a cultivar that has all desirable characteristics acceptable to the farmers is still a formidable task. The lack of durability and diversity in resistance especially in remote and inaccessible regions of the world presents mother nature, an opportunity to continue her uninterrupted evolution of rust races.

Prevalence and Distribution of Wheat Rust Physiological Races in India

Information on prevalence and distribution of races of rust in nature is important for wheat breeders in their programme on breeding for rust resistance. Studies to obtain information on rust races occurring in nature are in progress since 1932 and the information has been compiled from time to time (Joshi *et al.*, 1976). The data on the occurrence of rust races has been recast area wise rather than state wise as done in the past. The states have been grouped arbitrarily into different areas as under:

Area–A: Northern Hills (H.P.), J&K, Assam, Nepal, Kumaon Hills.

Area–B: Punjab, Haryana, Delhi and Chandigarh.

Area–C: Uttar Pradesh, Bihar, Orissa and West Bengal.

Area–D: Madhya Pradesh, Rajasthan and Gujarat.

Area–E: Karnataka, Maharashtra, Andhra Pradesh and Tamil Nadu.

Area–F: Southern Hills (Nilgiri and Palney Hills).

New races 14A, 20A and 38A were identified during 1970–71 crop season (Sharma *et al.*, 1972) and were the main threats to the cultivation of Kalyansona in Punjab and Uttar Pradesh. Another race 'I' (Sonalika race) was detected during 1970–71 crop season (Sharma *et al.*, 1973) but was never found in nature in subsequent years. For stripe rust, however two distinct sources of infection are operative. The northern hills are effective in causing infection in northern India and southern hills are effective in the southern India only. During the last two years 56.8 per cent of the sample collected had the most virulent race 46S119 followed by 13.5 per cent of 38S102. These two pathotypes has distinct geographic distribution as 46S119 is dominating pathotypes of northern part while 38S102 is restricted to only southern parts of India. Other reported pathotypes were 46S103 (6.2 per cent), 67S64 (4.8 per cent), 46S102 (1.7 per cent) and 46S103 (1 per cent). Samples collected from the showed different pathotypes, identity of which is yet to be established. Against barley major pathotypes were ISO followed by OSO-1 Kumar *et al.*, 2005).

Survival and Multiplication of Yellow Rust Inoculum at the Source

In cereal rust the availability of inoculum is a limiting factor in the growth of an epidemic and therefore it becomes the main and sole objective of forecasting research. In forecasting cereal rust out breaks in the plains of India, the first step would be to study the survival and multiplication of inoculum at the source, since it is known that low rust severity at the source reduces the danger at the target area. In India most of the wheat is grown in Indo Gangetic plains. Mehta (1940) has shown that there is no local source of infection in the plains and every year inoculum blows down from certain foci in the hills to the plains. When the inoculum from the self sown plants passes to the regular crop it is soon multiplied and the spores are blown into the foot hills and near by plains by daily ketabatic winds. Ketabatic winds are flown from the hills every night in the clear weather. Rainfall is also an important factor in the establishment and subsequent multiplication of inoculum on the regular crop in the hills. Frost of –3°C does not harm the rust but moderate frost reduces the rust development and thus frost retards the rust multiplication at the source (Suryanaryana *et al.*, 1976).

Role of Western Himalayas in the Annual Recurrence of Yellow, Rust in Northern India

Studies were undertaken to identify pockets where wheat rusts, particularly yellow rust could survive all the year round in Shimla hills. The various places in the hills where yellow rust can survive all through the year around 8000 ft (2500 m asl). At higher altitude above (2500 m asl) yellow rust could not be found all the year round as reported by Mehta (1940) because of very low temperature. It is reasonable to presume that there are probably few pockets (foci) which are capable of harbouring yellow rust all the year round, but generally yellow rust can survive at an altitude range of 6000 to 8000 ft (2000–2500 m asl) (Joshi *et al.*, 1976).

Yellow Rust Epidemiology

Yellow rust occurs between Christmas and early January along the foothills where the rivers enter the plains. After establishing such Centres in Punjab, Haryana and parts of Western UP, it moves southwards and become omnipresent in this zone by the end of February. Spread of yellow rust in this zone is much faster in the initial stages when compared with brown rust. Though stray infection is seen on wheat in east and central UP and Bihar, it seldom reaches epidemic proportions. Very often the infection remains isolated (Joshi *et al.*, 1977).

Changed Virulence Pattern

Puccinia striiformis races 14, 20, 38 prevailed in the country before the cultivation of dwarf wheats. But after the cultivation of Mexican wheats three variants 14A, 20A, 38A predominated during 1975–80 in NWPZ. Race I which was picked up from Wellington is presently the only predominant virulence in the Nilgiri hills. Since 1982 race K prevailed with high frequency in NW India till the emergence of new races N, P, *Yr9* and PBW 343 virulence (Nayer *et al.*, 2001). These new variables in rust pathogen rendered the genes ineffective that are currently in use. Thus, in all 8 new races and biotype of *Puccinis striiformis* system were met with. These are 14A, 20A and 38A all virulent on Kalyansona. Race I which matches Sonalika and Strubes dickkopf and *Yr1*, race L, N and P has additional virulence for Hybrid 46 (*Yr3*, *Yr4b*), Chinese 166 (*Yr1*), *Yr3a* and *Yr48* (Table 5.1). Recently *Yr9* virulence has emerged in NWPZ where PBW 343 is the major variety having gene *Yr9* to which it is virulent but due to possible intervention of some unknown gene, PBW343 maintained field resistance to stripe rust in India (Nayer *et al.*, 1997b). However this did not last long, another race that has been detected that infects PBW343 and is tentatively designated as 78S84 (Nayer *et al.*, 2002). The possible emergence of this new race is a result of monoculture of the genotype having vertical resistance (VR) gene over a large area leading to boom and bust cycle. Exploring wheat cultivars with high yield potential and VR genes over large geographical region forced the pathogen to evolve which led to the development of more virulent races. The stripe rust virulence for *Yr9* was first detected in east Africa and the *Lr26* virulence in south western Russia and Balcon countries. In 1990 the *Yr9* virulence was isolated from Syria and subsequently in Iran. This country experienced major stripe rust epidemic in 1993. In 1994 the *Yr9* virulence which can attack *Very* 5 type cultivars like Pak-81, Pirsaback-85, Seri-82 was observed to be wide spread in north west frontier province of Pakistan. Since the movement pattern resemble that of 8156 of stripe rust migration which appears to have arrived in Indian sub continent from middle east in stages associated with weather pattern called western disturbances. This is the second case of stripe rust migration from middle east to sub-continent of south Asia caused by mother nature through wind and weather (Nagarajan and Saari, 1995).

Table 5.1: Population Shift in Stripe Rust Flora (*Puccinia striiformis* West) Over the Years and Break Down of Important Wheat Cultivars

Sl.No.	Period	Prevalent Races	New Races Build Up	Cultivars Break Down/Genes
1.	1960–1965	13, 19, 20, 31A	38A	NP770, NP792, NP824, NP710, NP200, NP201, NP202, K65, C591
2.	1966–1975	14, 19, 20, 38A	14A, 20A, 38A	Yr2: Kalyansona, Lerma Rojo, Sunara64
3.	1976–1985	20, 38, 14A	I, K	Yr2(KS): WL711, Kalayan sona, HD2009
4.	1986–1990	K, 14A, A	L, N	HD2329, HD2285, DL 153-2
5.	1990–1996	N, K	P, Yr9, vir.	Yr9: WH542, CPAN3004, UP2338
6.	1997–2000	Yr9	PPBW343 vir.	PBW343, PBW373

Evolution of Pathotypes

In India the alternate hosts for black and brown rust (*Berberies* and *Thalictrum sps.*) don't play any role in the perpetuation of both the rusts. The alternate host of yellow rust is unknown till date in the world. The rust survives in asexual cycle (uredo to uredo) in the hills both in north and south. Arisal of new pathotypes is either through somatic hybridisation, mutation or introduction (intercontinental dissemination). So far most of the variation in stripe rust that has been picked up, appear to be either through introduction or mutation (Nayer *et al*, 2001).

In the late 1970, yellow rust race identification procedure was modified in India by adopting in total the new world set of differentials (Table 5.2) proposed by Johnson *et al.* (1972). Also a practical method was adopted for identifying genes in resistance sources through multiple pathotype testing (Sinha *et al.*, 2003). Gene can be postulated from the correlation of the responses of selected resistance sources with those of control. However, this classic application of the Gene-for-Gene relationship (Flor, 1942) has involved problems, one of which is that the '*Yr*' tester lines are not isogenic lines and the assumption that they carried known genes is not proved against Indian pathotypes. Therefore, the postulations done so far may be considered provisional till complete genetical information on the genes in the tester lines is obtained.

Table 5.2: Constitution of Differential Set 0, A and B for the Detection of Pathotype of Stripe Rust

Wh147	Heines VII (Yr1)	Hybrid 46(Yr4)
Barely Local	Lee (Yr7)	HeinesVII (Yr2+)
WH416	Heines Kolben (Yr6)	Compar (Yr8)
HD2329	Vilmorin23(Yr3)	T. spelta album (Yr5)
HD2667	Moro (Yn10)	Tc*/Lr26(Yr9)
PBW343	Strubes Dickkopf	Sonalika (Yr2+)
HS240	Suwon 92Xomar	Kalyansona (Yr2 KS)
Anza	Riebesel 147/51 (Yr9)	
A 9-30-1		

Mechanism of Evolution of New Races

Average Virulence Factor (AVF)

The average virulence factor *i.e.* average number/per isolate is determined for stem, leaf and stripe rust according to Luig and Watson (1970) and it was 4.0, 2.1 and 2.2 respectively in 1935. But after the incorporation of vertical resistance in the cultivars, AVF in case of stem rust decreased to 3.0 in 1955, increased to 4.4 in 1965 and further declined to 3.1 in 1980. For leaf rust AVF increased to 3.0 in 1955 and remained consistent to 4.6 during 1965 to 1980 while in stripe rust AVF varied from 1.9 to 2.2 in 1935 to 1975 and declined to 1.5 in 1980 (Bahadur 1980). After 1980, 13 new pathotypes were added in stripe rust flora in India (Table 5.5).

Adaptation of Virulence

Virulences not only have specific host range but also differ in adaptation to ecological conditions. Joshi and Nagarajan (1977) divided Indo-Gangetic plains in three epidemiological zones for leaf rust management. Later on Bahadur and Nagarajan (1985) divided India into four ecological zones for stem rust management, six ecological zones for leaf rust management. Nagarajan *et al.* (1984) divided India into three ecological zones for stripe rust management.

Nature of Virulence Change

The population of plant pathogen remain in dynamic state. The shift may be due to either change in frequency of existing component in the population or due to infusion of new characters into single strain (Watson 1970). Upto 1965 variants of stripe rust were adapted to tall cultivars. These were replaced by races 14, 20 and 38 of stripe rust on dwarf wheat. According to Vander plank (1982) mutations arise in the population of the pathogen on susceptible host and may be more frequent in nature than revealed experimentally.

Mutation

It has been well documented that mutants are important source of variability and rates of mutation from a virulent to virulent varies greatly from locus to locus. Parlevelit and Zadocks (1977) for leaf rust gave an estimate of 1000,000 mutants per day in a single hectare of wheat with a mutation frequency of 10^{-6} for a given locus. In the field, variants of stripe rust races 14, 20, 38 were monitored after the incorporation of vertical resistance besides a few new races. Persoon *et al.* (1976) suggested that change in the pathogen population are due to selection pressure in the host population and potentially virulent mutants would be monitored in the large asexually reproducing population of the pathogen.

Virulence and Fitness Survival

Strains present before the cultivation of resistant wheat were wild or primitive (Watson, 1970). Races 13, 19, 31, A and G were the earlier races reported by Mehta (1940) in a very low frequency due to their survival ability and could be considered as primitive races. As a result of incorporation of VR in the cultivars the pathogen flora changed and complex races originated (Vanderplank 1982). In this context in India during more than past two decades pathogen evolved strains I, K, 46S119, and 78S84 which are more virulent. Vanderplank (1982) theorized that as the pathogen gains virulence for plants carrying strong genes for resistance, it looses fitness. However there are evidence that revealed that races with more genes for virulence predominates over those with less genes.

The Wheat Rust Resistant Genes

The discovery of rust resistant genes and the near isogenic lines in wheat–*Puccinia* system has opened a new dimension of disease resistance breeding work. McIntosh 2003 has listed 51 *Lr*, 41 *Sr*, and 32 *Yr* gene to confer resistance to rust flora of *Puccinia* system. Stripe rust resistant genes that have been utilized in the Indian wheat breeding programme are given in Table 5.3.

Table 5.3: Gene Postulated in the Improved Wheat Lines

Gene	Source	Release Wheat Varieties
Yr2	Heines VII	HD2009, HD2189, HD2285, HD2278, HD2329, HD2380, H1977, HD077, HP1209, HP1633, HS86, HUW234, HW741, HW971, IWP72, K8030, N15439, PBW175, PBW222, Raj2184, Raj3077, Sonalika, Swati, UP262, VL421, VL616, WH283
Yr2 (KS)	Kalyan Sona	BW11, GW173, HD2402, HD2182, HDR77, HD123, HP1102, K7410, K8027, Lok-1, PBW65, WL711
Yr3	Vilmorin	HS295
Yr9	Secale cereale	CPAN1922, CPAN3004, DL803-3, HD2610, HS207, HS240, HUW206, K8804, MACS2496, PBW299, PBW343, UP2338, WH542
Yr18	Lr34 Source	C306, DWR39, HD2189, HD2329, HD2610, H11077, HP1209, HPW42, HS295, N15439, PBW175, UP2338, WH147

However the exploitation of gene in commercial varieties exert selection pressure on the pathogen population as has occurred with *Lr26* in the recent past. The wide scale growing of Veery selection (1B/1R) translocation in various countries has resulted in the sudden evolution of different virulence matching *Lr26*. Similarly *Yr9* and *Sr31* gene respectively of yellow and stem rust are tightly linked to *Lr26* have also fallen pray to the new virulence evolved in Mediterranean and travelled to India after 1995 (Nagarajan and Saari 1995).

Identification of Yellow Rust Resistant Genes

32 potentially useful genes are known to condition resistance against stripe rust pathogen (McIntosh *et al.*, 1995, 1998). All the genes except *Yr11, Yr12, Yr13, Yr14, Yr16,* and *Yr*18 were classified for resistance in seedling and adult stages. Of the *Yr* genes identified *Yr8* and *Yr9* have been shown to be race specific; other genes (*Yr1– Yr7; Yr10, Yr19* to *Yr27*) express at seedling stage also. Expression of *Yr15* and *Yr17* is yet to be studied. Among the genes expressing at adult plant stage gene *Yr16* and *Yr18* do not show race specificity.

Pathogenic variability for stripe rust population has been reported worldwide. Many differential genetic stocks carry yet to be identified stripe rust resistance gene in addition to already known genes. For example the differential wheat Hybrid 46 that was known to carry *Yr3b* and *Yr4b* was found to carry 4b and a new gene that was not at *Yr3* locus. The promising *Yr* genes possessing resistance against Indian stripe rust flora are listed in Table 5.4.

Inheritance studies in the differential varieties such as *T. spelta* var. *album* (Luthra *et al.*, 1989), Moro (Prabhu *et al.*, 1991) and Heins VII (Prabhu and Luthra 1992) with Indian Pathotypes indicated that these wheat carried more than one named genes for stripe rust resistance.

In India 28 stripe rust races and their biotypes (Table 5.5) have been reported (Nayar *et al.*, 2001) out of which 14 have been picked up from Nilgiri. Most of the genes conditioning resistance to stripe rust have originated from *Hexaploid* wheat itself and very few genes have been transferred from wild species. The rust reaction recorded on the stocks carrying specific genes and the details about such genes conferring resistance to stripe rust are presented below.

Table 5.4: Proposed *Yr* Genes to be Pyramided in Indian Wheat Cultivars for Resistance to Stripe Rust

Sl.No.	Gene	Source	Characteristics
1.	Yr1	Chinese166	Susceptible to Barley Races
2.	Yr4	Hybrid 46	Low level of Vir in India
3	Yr5	Tri.Spel.Alb.	Rare vir. in snowy countries
4.	Yr6	Heins Kolben	High resistance at low Temp.
5.	Yr7	lumillo durum	High Frequency of Vir. Worldwide
6.	Yr8	Tri. Comosum	Resistant not durable
7.	Yr10	Moro 178383	No vir known in India
8.	Yr11	Jos cambier	APR
9.	Yr12	Mega	APR
10.	Yr13	Maris Huntsman	APR
11.	Yr14	Hobbit	APR
12.	Yr15	Tri. turgidum var. diccoides	Virulence unknown
13.	Yr16	Cappelle desprez	APR proved durable
14.	Yr17	T. ventricosum	More susceptible at low Temp.
15.	Yr18	Terenzio	APR

Based on the evaluation of wheat stocks carrying stripe rust resistance genes the *Yr* gene can be categorized into three groups.

Highly Effective Genes

Yr3a, Yr3b, Yr4b and their combination, *Yr5, Yr9**, Yr11, Yr12, Yr13, Yr14, Yr15* and *Yr16.*

Moderately Effective Genes

Yr1, Yr6, Yr10, Yr18 and *Yr28.*

Ineffective Genes

Yr2, Yr7, Yr8, Yr20, Yr21, Yr23, Yr21 and *YrA.*

Yr1

This gene is present in wheat variety Chinese-166. Consistently resistant to stripe rust pathotypes, but soon overcome by Vir., K in 1982 (Nagarajan *et al.*, 1984). However some of the stocks line like Maris ranger and Maris templer both carrying *Yr1* along with *Yr3a* and *Yr4a* showed resistance. This gene has not been commercially exploited for wheat improvement.

Yr2, Yr3 and *Yr4*

These genes have their origin in common wheat and show low infection type to Vir., K, otherwise all the three genes exhibited high degree of resistance to stripe rust races occurring in Wellington. Wheat cultivars Sonalika and Kalyansona carrying *Yr2 Ska* and *Yr2Ks* are susceptible to stripe rust while cultivar C306 carrying *Yr2* showed resistance indicating that it carries additional gene for stripe rust resistance.

Table 5.5: Pathotypes of *P. striformis* Reported in India

Designation		Detection Details	
Old	New	Year	Place
13	67S8	1937	Palne Hills
14	66S0	1965	Kangra
14A	66S64	1970	Punjab
19	70S0–2	1936	Rawal Pindi
20	70S0	1937	Nilgiri Hills
20A	70S64	1970	Punjab
24	0S0–1	1965	–
31	67S64	1936	Shimla
38	66S0–1	1965	Himachal Pradesh
38A	66S64–1	1970	Punjab
57	0S0	1965	–
A	70S4	1937	Gurdaspur
G	4S0	1965	–
G–1	4S0–3	1990	Nilgiri Hills
I	38S102	1973	Nilgiri Hills
K	47S102	1982	Punjab
L	70S69	1988	Dalang Maidan
M	1S0	1987	Shimla
N	46S102	1988	Dalang Maidan
P	46S 103	1990	Dalang Maidan
	46S 119	1996	Gurdaspur
Q	5S0	1990	Bhowali
P	47S103	1992	Nilgiri Hills
U	102S100	1992	Nilguri Hills
CI	–	1993	Leh
CII	–	1993	Leh
CIII	–	1993	Leh
	78S84	2001	Punjab

Source: Manual of Current Techniques in Wheat Rusts. Research Bulletin No. 2, 32 pp, Regional Station, DWR, Flowerdale, Shimla – 171 002.

Yr5

This gene is present in *Triticum spelta* var *album*. It imparts resistance to races of stripe rust present in Nilgiri hills including race K and I. The resistance of this gene also has not been deployed for commercial use in wheat improvement because a matching virulence appeared after its introduction. This gene has been reported to be allelic or closely linked to *Yr5* and *Yr1* (Johnson 1986).

Yr6

Originated from common wheat, present in Heins kolban, is moderately effective against stripe rust flora of Nilgiri hills. This gene was reported sensitive to light and temperature. The stripe rust population in India is appeared to be fixed for virulence (Kumar *et al.*, 1988). This gene is very common in winter and spring wheats. The resistance expressed by *Yr6* is influenced by the presence of *Yr2* gene (Johnson and Minchin, 1992).

Yr7

Present in Thatcher derived from *T. durum cv lumillo*. This gene has been ineffective against 13 Indian races of stripe rust both at seedling and adult plant stages (Upadhyaya and Kumar 1979). This gene also does not confer resistance to races land K (Kumar *et al.*, 1993). Apparently the gene *Yr1* is totally ineffective to race flora present in northern and southern hills (Tomar *et al.*, 1988).

Yr8* and *Sr34

The gene *Yr8* is present in variety compir. It is present independently along with *Sr34*. This gene *Yr8* exhibits resistance to stripe rust races prevailing in the southern hills, but due to the appearance of race K and I it became ineffective. Compair has shown resistance against ten races *viz.*, 13, 14, 14A, 20, 20A, 38, 38A, L, M and P of stripe rust. *Yr8* has not been deployed in commercial cultivars.

***Yr9/Sr31/Lr26*/PM8**

Many winter and spring wheats either in cultivation or used as genetic stock possess gene *Yr9* which is linked with *Lr26, Sr31* and also often with Pm8. This IB/IRS translocation has not only provided multiple disease resistance but also provide wider adaptability particularly in spring wheat. Gene *Yr9* was effective world wide with very low infection but virulence for this gene has appeared. The 1 BL/1RS translocation is the most successful wheat alien translocation and is in use for the improvement of common wheat world wide. In India all the wheat zones have cultivars carrying gene *Yr9*. When the area covered by varieties carrying *Yr9* increased in NWPZ a pathotype designated 46S119 virulent on *Yr9* evolved and was first picked up from Punjab in 1996 (Nayer *et al.*, 1996a). Many of the wheat lines carrying *Yr9* showed resistant to *Yr9* virulence which indicate that these varieties carry additional factor for resistance to *Yr9* virulence (468119). Winter wheats Kavkaz, Aurora, Clement and Skorospelka 35 (all with 1 BL/1RS) showed resistance to yellow rust at Wellington, indicating no virulence attacking *Yr9* is present in Nilgiris.

Yr10

This gene has its origin in Turkish bread wheat has high degree of resistance. No virulence for *Yr1* a has been detected in India. This has been used in limited stock in U.S.A. and Australia, however virulent pathotype emerged soon after the deployment of *Yr10* in USA. It appears to be one of the potential source of wheat improvement.

Yr11, Yr12, Yr13, Yr14, Yr16* and *Yr18

McIntosh (1988) and McIntosh (1993) classified these genes as APR. These genes except *Yr18* which is widely prevalent among winter and spring wheat show low infection type and slow rusting behaviour and closely linked to *Lr34* and leaf tip necrosis (Itn), a gene for tip nicrosis (Singh,1992, McIntosh, 1992). The gene frequency of *Yr11, Yr12, Yr13, Yr14* and *Yr16* in Indian wheat is negligible, *Yr18* accounts for around 20 per cent prevalence in Indian wheat. It has a tight linkage with *Lr34* and is therefore quite important. The durability of such type of wheat carrying *Lr34* and *Yr18* has been demonstrated in wheat variety Kundan (Sawhney, 1995).

Yr15

This gene has its origin in *T. turgidum* var. *dicoccoides.* This gene exhibits APR to stripe rust races prevailing in Lahaul spiti (H.P.) in northern hills. Virulence for this gene has not been reported and has not been used commercially. Vinod *et al.* (2004) have transferred *Yr15* to some high yielding wheat cultivars.

Yr17/Sr38/Lr37

The line RL6081 carrying *Yr17* shows low infection with moderate intensity of stripe rust at adult plant in Wellington. It is closely linked with *Sr38* and *Lr37* and located on short arm of chromosome 2A.

Wild Relatives as Donors of New *Yr* Genes

A large number of Yr genes have been derived from alien sources *viz., Secale cereale* (*Yr9*), *Aegilops comosa* (*Yr8*), *T. dicoccoides* (*Yr15*), *T. tauschii* (*Yr24, Yr28*), *T. turgidum* (*Yr26*), *T. spelta* (*Yr5*) and *T. vavilovii* (*Yr10*). Additionally, several other wild relatives *viz., Aegilops squarrosa* (*DD1*), *Aegilops markgrafi* (*CC1*), *Triticum Spelta* (*AABBDD3*), *Agilops ventricosa* (*DDMM2*), *Triticum dicoccoides* (*AABB2*) are other potential sources of novel Yr genes.

Inheritance of Yellow Rust Resistance

Studies on the genetics of stripe rust resistance have been carried out at IARI, Regional Station, Tutukandi, Shimla with emphasis on the Indian races of pathogen to understand the genetic action of the host genotypes. In a number of wheat cultivars of Northern Hills Zone of India, such as VL-421, HS 86, Sonalika, Shailja, HS 240 and HS 295, more than one gene was found to function against pathotypes 70S64 and 66S64-1. It is evident from the data presented in Table 5.6 that each *Yr* line possesses several genes. Riebsel 47/51 reported to possess single dominant rye gene (*Yr9*) against European races appears to possess 5 dominant and one recessive gene pairs when exposed to only three Indian pathotypes, 38S102, 47S102 and 70S64. The studies with present day pathotypes are in progress and preliminary data showed similar type of results.

Dominance of genetypes with 1 BL/1RS translocation has raised concern from rust management angle due to breakdown of *Yr9*-gene and only hope was 'Attila' lines (PBW343 and PBW373) with major coverage in Northern India. Consistent screening at IARI Regional Station Shimla resulted in identification of a number of winter lines with resistance against pathotype 46S119 virulent on *Yr9*. These genotypes were Capple Desprez, Moro, Mega, Hobbit and Vilmorin. DWR Regional Station Shimla has also developed few resistant sources against 46S119 out of which only FLW3 is different from 'Attila' on the basis of parentage and reaction pattern. IARI Regional Station Shimla has also developed two resistant sources WBM1587 and WBM1591 and their resistance against pathotype 46S119 is different from 'Attila' reaction pattern which also confirm with its distant parental linkages and very low coefficient of parentage value. Inheritance pattern of all these resistant sources are in progress and initial response has clearly shown its diversity. Studies on inheritance of resistance against pathotype 46S119 of stripe rust in popular wheat variety PBW343 conducted at IARI Shimla Centre indicated that the resistance is governed by two recessive genes and this is the limiting factor in transmitting this type of resistant genes in normal breeding programme of the country (Kumar *et at.,* 2005).

Table 5.6: Studies on Genetics of Resistance of Stripe Rust

Cultivar	Race	Genetics of Resistance
Chinese 166 (Yr1)	38S102	One dominant gene pair
	70S64	One dominant and one recessive gene pair
	66S64-1	One dominant gene pair. Also impart resistance to 70S64
Heines VII (Yr2)	67S64	One dominant and one recessive gene pair
	66S64	One dominant and one recessive gene pair
	66S64-1	One dominant and one recessive gene pair
	70S64	One dominant and one recessive gene pair
Vilmorin (Yr3)	38S102	Two recessive gene pair
Opal (Yr4b)	38S102	One dominant and one recessive gene pair
	47S102	One recessive gene pair
	70S64	One dominant and one recessive
T. spelta album (Yr5)	38S102	Two complimentary genes
	47S102	One recessive gene pair
	70S64	One dominant gene pair
	46S102	One dominant gene pair
	67S64	One dominant and one recessive gene pair
	66S64	One dominant and one recessive gene pair
Heines Peko (Yr6)	66S64	One dominant and one recessive gene pair
	67S64	One dominant and one recessive gene pair
	66S64-1	One dominant and one recessive gene pair
Compare (Yr8)	66864-1	One dominant and one recessive gene pair recessive gene
	66S64	Recessive gene pair derived from Chinese Spring
	70S64	One dominant and one recessive gene pair recessive gene pair allelic to the gene in Chinese Spring Two complimentary dominant gene pair
Riebesel 47/51(Yr9)	38S102	One dominant gene pair
	47S102	One dominant and one recessive gene pair
	70S64	Three airs of dominant gene
Moro (Yr10)	38S102	One dominant and one recessive gene pair
	47S102	One dominant gene pair
	70S64	One dominant and one recessive air
Chinese Spring	66S64-1	One recessive gene pair
	66S64	One recessive gene pair

Breakdown of these lines (PBW343 and PBW373) with new pathotype of stripe rust, 78S84 has forced to find detailed response of this pathotype. Advanced lines of the IARI Regional Station Shimla were screened against pathotype 78S84 and 65 per cent lines were found to be resistant (Table 5.7). These lines were also screened against two major pathotypes 46S102 and 46S103. The majority of resistant genotypes against 78S84 (67 per cent) were resistant to both 46S102 and 46S103, 9 per cent resistant genotypes were susceptible with both the pathotypes while 24 per cent were resistant with either 46S102 or 46S103. Among susceptible, majority (56 per cent) were susceptible with both other pathotypes also. 19 per cent of susceptible category were resistant with both 46S102 and 46S103 while 25 per cent were susceptible with one of the pathotype. These results indicated that breakdown of 'Attila' lines is a serious concern, this pathotype is not very virulent in nature. A systematic approach with screening of genotypes against 78S84 will help in finding better source that can be used for further wheat improvement with diverse resistance gene pool.

Table 5.7: Seedling Response of Advanced Lines of Wheat Against, Different Races of Stripe Rust

Pathotype	Resistant Lines
78S84, 46S102, 46S103	WBM1504, WBM1614, WBM1659, WBM1675, WBM1704, WBM1705, WBM1706M, WBM1716, WBM1723, WBM1736, WBM1749, WBM1758, WBM1760, WBM1761, WBM1762, WBM1763, WBM1765, WBM1767, WBM1775, WBM1780, WBM1783, WBM1786, WBM1794, WBM1795, WBM1799, WBM1806, WBM1814, WBM1816, WBM1818, WBM1833, WBM1834, WBM1835, WBM1839, WBM1840, WBM1846, WBM1848, WBM1852, WBM1853, WBM1860, WBM1861, WBM1862, WBM1864, WBM1868
78S84 and 46S102	WBM1770, WBM1766, WBM1821, WBM1845
78S84 and 46S103	WBM1493, WBM1553, WBM1680, WBM1756, WBM1784, WBM1809, WBM1822, WBM1842, WBM1849, WBM1851, WBM1859, WBM1865
78S84	WBM1683, WBM1725, WBM1737, WBM1808, WBM1831, WBM1937

Epilogue

Disease development is not a single phenomenon. It is the result of complex events involving several dynamic living and non-living entities *viz.*, the host, the pathogen and several components of environment like humidity and temperature etc. Appearance of disease requires host to be susceptible and virulent pathogen. Several strategies to manage wheat rusts have been proposed, however the use of genetic resistance is the most durable and economically viable. Search for novel resistance genes has been a never ending exercise and some of them like *Yr5, Yr10, Yr15* etc., remain to be utilized in new wheat varieties. However, the role of other factors like sowing time etc. cannot be overlooked in managing wheat rusts (Sharma *et al.*, 2005).

References

Bahadur, P. (1986). Physiologic specialization in wheat rust. In: *Problem and Progress in Wheat Pathology in South Asia*, (Eds.) L.M. Joshi, D.V. Singh and K.D. Srivastava. Malhotra Publishing House, New Delhi, pp. 69–91.

Bahadur, P. and Nagarajan, S. (1985). Gene deployment for the management of stem rust of wheat (*Puccinia graminis* (Pers.) f. sp. tritici. Erikss. and Henn. In India. *Indian J. Agric. Sci.*, 55: 219–222.

Biffen, R.H. (1905). Mendel's laws of inheritance and wheat breeding. *J. Agric. Sci.*, 1: 4–48.

DWR (2001). Report of the Coordinated experiments–Vol. V 2000–2001 Crop Protection (Pathology, Nematology), (Eds.) A.K. Sharma, D.P. Singh, J. Kumar, A.K. Singh, M.S. Saharan, K.S. Babu and S. Nagarajan. AICWIP and BIP, Directorate of Wheat Research, Karnal – 132 001 (Haryana), 166 pp.

Flor, H. H. (1958). Inheritance of pathogenicity of Melampsora lini. *Phytopathology*, 32: 653–662.

Joshi, L.M., Goel, L.B. and Sinha, V.C. (1976). Role of Western Himalayas in the annual recurrence of yellow rust in northern India. *Cereal Rust Bull.*, 4: 27–30.

Joshi, L.M., Srivastava, K.D., Nagarajan, S. and Singh, D.V. (1976). *Wheat Disease Newsletter*, IARI, New Delhi, Vol. 9

Joshi, L.M. and Nagarajan, S. (1977). Regional deployment of *Lr* genes for brown rust management. In: *Genetics and Wheat Improvement*, (Ed.) A.K. Gupta. ICAR, New Delhi, pp. 87–92.

Joshi, L.M., Nagarajan, S. and Srivastava, K.D. (1977). Epidemiology of brown and yellow rust of wheat in Northern India. I. Place and time of appearance and spread. *Phytopath. Z.*, 90: 110–122.

Jhonson, R., Stubbs, R.W., Fchs, E. and Chamberlin, N.H. (1972). Nomenclature of he physiologic races of *Puccjnia striiformis tritici* infecting wheat. *Trans. Brit. Mycol. Soc.*, 58: 475–480.

Jhonson, R. (1986). Genetics interaction between plant and parasite. In: *Natural Antimicrobial System in Plants and Animals*, (Eds.) G.W. Gould *et al.*, Bath University Press, Bath, U.K., pp. 131–145.

Jhonson, R. and Minchin, P.N. (1992). Genetics of resistance to yellow (Stripe) rust of wheat in some differential cultivars. Vortage fiir. Pflanzenzuchtung, 24: 227–229.

Kumar, J., Nayer, S.K,. Bahadur, P., Nagarajan, S., Bharadwaj, S.C., Parashar, M. and S.B. Singh (1988). Virulence survey of Yellow rust of wheat (*P.S. fsp. triticl*) and Barely (*P.S. fsp. hordel*) during 1985–87. *Cereal Rust and Powdery Mildew Bulletin*, 16: 30–35.

Kumar, J., Nayer, S.K., Parasher, M, Bharadwaj, S.C and Bhatnagar, R. (1993). Virulence survey of *Puccinia striiformis* in India during 1990–92. *Cereal Rust and Powdery Mildew Bulletin*, 21: 17–24

Kumar, S. and Dharam Pal. (2005). Response of new virulent pathotype, 78S84 in hexaploid wheats. *Indian J. Agric Sci.* (Unpublished).

Kumar, S. and Dharam Pal. (2005). Inheritance of Stripe rust resistance, against pathotype 46S119 in wheat of Indian subcontinent. *Indian J. Genet. and Pl. Breeding*, (Unpublished).

Luig, N.H. and Watson, I.A. (1970). The effect of complex genetic resistance in wheat on the variability of *Puccinia graminis* f sp. *Tritici Proc. Univ. Soc. New South Wales*, 95: 24–42.

Luthra, J.K., Verma, R.P.S. and Prabhu, K.V. (1989). Inheritance of resistance to Stripe rust (*Puccinia striiformis* West.) in some yellow rust resistant tester of wheat (*Triticum aestivum* L.). *Indian J. Agric Sci.*, 59(10): 617–621.

McIntosh, R.A. (1988). Role of specific genes in breeding durable resistance in wheat and *triticale.* In: *Breeding Strategies for Resistance to Rust of Wheat*, (Eds.) N.W. Simmonds and S. Rajaram. CIMMYT, Mexico, pp. 1–9.

McIntosh, R.A. (1992). Close genetic linkage of genes conferring adult plant resistance to leaf and stripe rust in wheat. *Plant Pathology*, 41: 523–527.

McIntosh, R.A., Hart, G.E. and Gale, M.D. (1993). Catalogue for gene symbols in wheat (University of Sydney) Australia, pp. 1333–1500.

McIntosh, R.A., Willings, C.R and R.F. Park (1995). *Wheat Rust: An Atlas of Resistance Genes*. CSIRO, East Melbourne, Victoria 3002, Australia, pp. 200.

McIntosh, R.A., Hart, G.E., Devoes, K.M., Gale, M.D. and Rogers, W.J. (1998). Catalogue of gene symbols for wheat. In: *Proc. 9th Int. Wheat Genet. Symp.*, Saskatoon, Saskatchewan, Canada, 5: 129–139.

McIntosh, R.A., Yamazaki, Y, Devos, K.M., Dubcovsky, W.J. and Appels, R. (2003). Catalogue of gene symbols for wheat. In: *Proc. 10th Int. Wheat Genet. Symp.*, (Eds.) N.E. Pogna, M. Romano, E.A. Pogna and G. Galterio. Vol. 4. On CO Instituto Sperimentale per la Cerealcoltura, Rome, Italy,.

Mehta, K.C. (1940). *Further Studies on Cereal in India*. Indian Council of Agricultural Research, New Delhi. Monograph No. 14, pp. 1–224.

Nagarajan, S., Bahadur, P. and Nayer, S.K. (1984). Contemplating the brown rust resistance (*Lr*) genes to mitigate the spread of *Puccinia recondita* f. sp. *tritici*. *Indian Phytopathol.*, 37: 490–497.

Nagarajan, S., Bahadur, P. and Nayer, S.K. (1984). Occurrence of new virulence 47S102 of *Puccinia striiformis* west in India during crop year 1982. *Cereal Rust Bulletin*, 12: 28–31.

Nagarajan, S. and Saari, E.E. (1995). Repeat performance of yellow rust spread from Turkey to Indian sub-continent. *Indian J. Mycol. Pl. Pathol.*, 25(1–2): 68 Abstr.

Nayer, S.K., Parasher, M. and Bharadwaj, S.C. (1996a). Occurrence of *Yr9* virulence of yellow rust and its management: In Indian context. In: *2nd Int. Crop Sci. Congress*, New Delhi, November 17–24.

Nayer, S.K., Nagarajan, S., Parashar, M., Bharadwaj, S.C., Jain, S.K. and Dutta, O. (2001). Revised Catalogue of genes that accord resistance to *Puccinia* species in wheat. Res. Bull. No. 348 pp., OWR Reg. Stn., Flowerdale, Shimla.

Nayer, S.K. Parashar, M. and Bharadwaj, S.C. (2002). Pathotype distributation of wheat rust in India 2001–02. In: *Report of Coordinated Experiments, Vol. 3: Crop Protection*, (Eds.) A.K. Sharma, O.P. Singh, J. Kumar, A.K. Singh, M.S. Saharan, K.S. Babu and D.S. Chauhan, pp. 73–78.

Nayer, S.K., Parasher, M., Bharadwaj, S.C. and S.K. Jain (1997b). Wheat rust spectrum in 1996–97. 36th All India Wheat Research Workers Meeting, HAU, Hisar, August 19–22.

Parlevliet, J.E. and Zadocks, J.C. (1977). The integrated concept of disease resistance: A new view including horizontal and vertical resistance to plants. *Euphytica*, 26: 5–21.

Persoon, C., Groth, J.V. and Mylyk, O.M. (1976). Genetic change in host parasite population. *Ann. Rev. Phythol.*, 14: 177–188.

Prabhu, K.V. and Luthra, J.K. (1992). Genetics of resistance to Indian races of stripe rust (*Puccinia striiformis*) in tester Heines VII of wheat (*Triticum aestivum*). *Indian J. Agric. Sci.*, 62: 599–602.

Prabhu, K.V., Luthra, J.K. and Verma, R.P.S. (1991). Inheritance of resistance to yellow rust (*Puccinia striiformis*) in tester line Moro wheat (*Triticum aestivum*). *Indian J. Agric. Sci.*, 61: 928–929.

Sawhney, R.N. (1995). Genetics of wheat rust interaction. In: *Plant Breeding Reviews*, (Ed.) J. Jenick. 13: 291–243.

Singh, S.D., Goel, L.B., Sharma, S.K. Nayer, S.K., Chatterji, S.C., Sinha, V.C. and Bahadur, P. (1978). Prevalence and distribution of physiologic races of wheat rusts in India during 1969–74. *Indian J. Agric. Sci.*, 48(1): 1–7.

Singh, R.P. (1992). Genetic association between Gene *Lr34* for leaf rust resistance and leaf tip necrosis in bread wheat. *Crop Sci.*, 32: 874–878.

Sharma, S.K. Joshi, L.M., Singh, S.D. and S. Nagarajan (1972). A new virulence of yellow rust on Kalyansona variety of wheat. In: *Proc. 5th European and Medit and Cereal Rust Conference*, Czechoslovakia.

Sharma, S.K., Joshi, L.M., Singh, S.D. and S. Nagarajan (1973). Note on new record of Sonalika infecting race of yellow rust in India and sources of resistance. *Indian J. Agric. Sci.*, 43: 964–965.

Sharma, Rajiv K., Kumar, J., Sinha, V.C. and S. Nagarajan. (2005). Anticipatory breeding for resistance to *Puccinia* spp. in *Triticum* system. *Annu. Rev. Plant Pathol.*, Vol. 3. (Eds.) R.K. Khetrapal. Indian Society of Mycol. Plant Pathol., Udaipur (In Press).

Sinha, V.C., Parashar M., Bharadwaj, S.C. and Rajiv, K. Sharma (2003). Management of yellow rust of wheat. *Indian J. Mycol. Pl. Pathol.*, 33(3): 501–502 (Abstr.).

Suryanaryana, D., Goel, L.B. and Sinha, V.C. (1971). Epidemiological studies on wheat rusts in Simla hills: Survival and multiplication of yellow rust inoculum at the source. In: *Proc. Indian Nat. Sci. Acad.*, 37(6): 460–463.

Tomar, S.M.S., M. Kochumadhavan and P.N.N. Nambisan. (1988a). Evaluation of *T. timopheevi* wheats for resistance to rust and Powdery mildew. *Indian J. Genet.*, 48: 69–73.

Upadhyaya, M.K. and Kumar, R. (1979). Identification of genes governing seedling and adult plant resistance to Indian races of stripe rust of wheat. In: 5th *Intern. Wheat Genet. Symp.*, Vol. 2: 1049–1056.

Vanderplank, J.E. (1982). *Host Pathogen Interaction in Plant Disease*. Academic Press, New York, London.

Vinod, Sivaswamy, M., Parasher, M., Menon, M.K., Sinha, V.C. and S.M.S. Tomar. (2004). Evaluation of stripe rust resistance genes and transfer of *Yr15* in Indian wheats. *Indian J. Agric, Sci.* (In press).

Watson, I.A. (1970). Change in virulence population shift in plant pathogen. *Ann. Rev. Phytopathol.*, 8: 209–230.

©R.C. Ram & Asha Sinha (*Editors*)
***Microbial Diversity: Modern Trends*, 2007** **79–93**
Daya Publishing House, New Delhi

Chapter 6

Fungi and Grain Storage in Tropics

Jai Prakash Rai**, Alok Kumar Singh**, Raghavendra Singh** and Asha Sinha***

***Department of Plant Pathology, **Department of Agronomy,
Faculty of Agriculture, Udai Pratap Autonomous College, Varanasi – 221 002, U.P.
*** Department of Mycology and Plant Pathology, Institute of Agricultural Sciences,
Banaras Hindu University, Varanasi – 221 005, U.P.**

Introduction

The regions of the Earth lying between the latitudes 23.5°N and 23.5°S are called Tropics. The climate of these regions is defined by uniformly high temperatures throughout t the year coupled with high relative humidities for varying periods depending on the duration of the summer rainy season. The growth rate of population in the countries of this region is often high with higher incidence of debilitating human diseases, larger low-income groups and low productivity of food crops. Such a condition puts up before the man of the modern times its greatest challenge-to produce more food to fight hunger of the everyday growing population. Significant developments have been observed in producing more especially with irrigated wheat and rice. At this stage, however, the yield potential of food crops perhaps seems to reach a plateau where the rate of increase in the yield has become lower. Moreover, in our efforts to produce more, we have increased the vulnerability of the crops to many pests and diseases and losses due to these agencies have become quite considerable. It is, therefore, high time that the scientists must rethink over controlling the losses and leak-outs that may suffice to feed millions of the population every year. In other words, a balance between the production technology and protection technology must be sought to meet this problem with greater efficiency. Of the protection strategy, the most important discipline still stands to protect the seeds from diseases and pests for a damaged and diseased seed seldom produces a healthy plant with the yield proper to its genetic potential. Protecting seeds requires less inputs resulting in greater outputs. Since only a few crops grow throughout the year, seeds of most of our food grains are needed to be stored up to the proper time of their sowing. Since the inception of agriculture seeds and food grains are stored by man to meet the food demands in the off-times and to sow them in the season for their multiplication. Storage of seeds is needed mainly for the following three purposes (Agarwal, 1994):

1. Seeds are stored for about six months from one harvest to the next planting season.
2. Seeds may also be stored for 18–36 months to insure against a following crop of poor yield and low quality, fluctuations in price and market demand against shortages, during outbreak of war and when the produce could not be sold in the present year.
3. Long term storage of germplasms. Seeds carry a wide range of microorganisms either externally or internally and these microorganisms become active in favourable conditions and cause considerable damage to the seed and several diseases on the crop raised from them in various ways. The traditional Plant Pathologist limits his concerns only up to the point of harvest. The pathology of the harvested product still gains less importance. Although it has long been recognized that some of the seedborne mycoflora might reduce the germinability of seeds when planted, or result in disease in the growing plant, yet none of the attempts has met with the success in controlling these losses completely within the economy of the low capital farmers in the tropics. To check the spread of such microorganisms seed health testing procedures were sought for, and in accordance, plant pathologists all over the world developed seed health testing procedures in 1920. Later on, Late Dr. Paul Neergaard took up initiatives on this problem and in 1966 after a meeting, the International Rules for Seed Health Testing were recommended. The two methods, *viz.*, agar plate method and blotter technique as recommended by International Seed Testing Association (ISTA) are used worldwide for seed health testing. Some important organizations related with the seed health testing are (*i*) Food and Agriculture Organization (FAO), (*ii*) International Seed Testing Association (ISTA), (*iii*) Inter African Phytosanitary Commission (IAPSC), (*iv*) Plant Protection Committee for the South East Asia and Pacific Region (SEAPRPC), (*v*) European and Mediterranean Plant Protection Organization (EPPO) and Near East Plant Protection Commission (NEPPC). Due to developments in agriculture and deforestation the ecological habitats maintaining biodiversity were highly threatened at many centres of diversity. The International Biological Programme (IBP), therefore at its inception in 1964, set up a subcommittee to study ways and means for collecting and conserving plant genetic resources which were threatened by such a human interference.

Fungi of Seeds

In seeds fungi find a very good habitat rich in nutrition. The interrelation of seeds and fungi is quite defined-either fungi use seeds as their means of dispersal or they utilize the nutrients locked up in the seeds for their nutrition making seeds their ideal host. Thus fungi may be both-externally as well as internally seedborne. Among the first to recognize that the fungal agents might be involved in the deterioration of stored grains and seeds were Drs. P.E. Ramstad and W.F. Geddes in the Department of Agricultural Biochemistry of the University of Minnesota in the late 1930s. Later, many workers like Nagel and Semenuik (1947) and Christensen and Kauffman (1965; 1969) studied the biodeterioration of grains due to storage fungi. Although the problems of maintaining viability of seeds in storage have always been a matter of concern to the people associated with seeds, yet more recently it has become also important and relevant to a much wider group of interests because of developing importance of long-term storage systems in a number of disciplines. Fungi from the standpoint of their ecology–the conditions that permit them to invade seeds–can be conveniently divided into two groups, *viz.* (1) Field fungi and (2) Storage fungi.

Field Fungi

Grains under the process of development and those fully matured serve as rich substrates for fungi. The term *field fungi* applies to the fungi that invade seeds developing on the plants in the field or after the seeds have matured and plants are either still standing or are cut and swathed, awaiting threshing (Christensen, 1972). These fungi require moisture content in equilibrium with relative humidity of at least 90–95 per cent to grow. In some regions, in some years a combination of wet weather during harvest, lack of drying facilities or lack of sufficient drying capacity and lack of transport may result in much grain of high moisture content being piled on the ground. If such high moisture content is coupled with a temperature that allows micro flora to grow, it will deteriorate rapidly. The invasion of the developing grain is necessarily pathogenic. After maturity, however, the fungi may invade the seeds either pathogenically or saprophytically (in the seeds made vulnerable by pathogenic attacks of same or different species).

Pepper (1960) lists approximately 180 species of filamentous fungi and about 20 species of yeasts that have been reported from barley kernels. Christensen and Kauffman (1969) state, '.....from a single gram of malting barley, about 25 kernels, we have isolated tens of thousands of colonies of yeasts and several million colonies of bacteria.' Malone and Muskett (1964) described 77 species of seedborne fungi isolated from seeds of soybean, 32 species from maize, 34 species from rice, 29 species from sorghum and 28 species from seeds of wheat. The same publication lists *Alternaria* from seeds of more than 100 species and *Fusarium* from seeds of 200 species of plants.

The fungi associated with crop seeds play an important role in seed deterioration during storage (Singh *et al.*, 2001). Concerning the effects of field fungi, Christensen and Kauffman (1969) state, 'field fungi may discolour seeds, cause death of ovules, shrivelling of the seeds or kernels, weakening or death of the embryos, and development of compounds toxic to man and to other animals. In grading of some grains, this discolouration is referred to as "weathering"–a misnomer, since the discolouration is a product of growth of micro flora.

Reduction in seed germinability resulting from invasion of seeds by field fungi has been reported by several workers. Christensen and Stakman (1935) found a high correlation between increasing percentage of seeds infected with *Helminthosporium* and *Fusarium* and decreasing germination percentage. Christensen (1936) confirmed the finding for barley seeds. Machacek and Greaney (1938) carried investigations over 'black point' or 'kernel smudge' of wheat and observed that black point caused by *Alternaria* did not result in decreased germinability but that caused by *Helminthosporium* did. Even if kernels whose pericarps heavily discoloured as a result of invasion by *Alternaria* do not usually have the embryo invaded by the fungus. *Fusarium* may invade and kill the developing or mature embryo without causing any noticeable discolouration of either the pericarps or embryo- the seeds appear sound, but actually are diseased or even dead.

Williams and McDonald (1983) in their marvellous review has listed the field fungi of various crops grown in the tropics. A summarized list of the above is shown in Tables 6.1 and 6.2..

Storage Fungi

The only name for the studies on storage fungi over some decades has been that of C.M. Christensen of the University of Minnesota (USA). Alone and together with H.H. Kaufmann, he has published an array of review articles encompassing a vast area in the discipline. (Christensen, 1980; Christensen

Table 6.1: Field Fungi that are Prominent on the Tropical Crops

Crop	Fungus	Name of Disease	Nature of the Damage Caused
Maize	Aspergillus flavus	Aspergillus ear rot	Grain discoloured; black, green, yellow or tan growth between grains; grain rot; aflatoxin contamination
	Botryodiplodia theobromae	Black kernel rot	Erumpent pycnidia make grain brown and black; dry rot
	Cladosporium spp.	Cladosporium rot	Grain discoloured greenish black, blotched or stripped; store rot
	Colletotrichum graminicola	Anthracnose	Grain with whitish strips flecked with reddish black spots
	Diplodia spp.	Diplodia ear rot	Grain discoloured; husks bleached; cobs sunken and rotted
	Fusarium spp.	Fusariun ear/kernel	Grain discoloured salmon-pink to reddish-brown; pink mycelium on surface of grain; white streaks in pericarps of late infected grains
	Helminthosporium mqydis (race T)	Ear rot	Black felty mould covering grains; ear rot; seedling blight
	Penicillium spp.	Penicillium rot	Powdery-green or blue mould on and between grains; embryos damaged; seedling blight
	Rhizoctonia zeae	Rhizoctonia ear rot	Salmon-pink mould growth on ears which become dull grey with salmon to dark-brown sclerotia on husks; mycelium and sclerotia occur in grain
	Trichoderma viride	Trichoderma ear rot	Green fuzzy mould growth on and between grains
	Ustilago maydis	Common smut	Infected grains enlarged and form galls which contains masses of powdery dark olive-brown to black spores
Rice	Alternaria padwickii	Stack burn disease	Grain discoloured and shrivelled; Seedlings blighted
	Cochliobolus miyabeanus	Brown spot (leaf and glume)	Grain discoloured; seedling blight and leaves disease; yield severely reduced
	Curvularia spp.	Black kernel	Grain discoloured; seedlings blighted
	Gibberella fujikuroi	Bakanae disease	Grain discoloured; seedlings blighted, stem elongation; yield reduced
	Phyllosticta glumarum	Glume blight	Grain discoloured
	Pyricularia oryzae (Magnaporthe grisea)	Blast (leaf and neck)	Grain discoloured; leaves and panicles diseased; yield severely reduced
	Septoria oryzae	Speckled blotch	Grain discoloured
	Tilletia barclayana	Kernel smut	Individual grains transformed into black spore masses
	Ustilaginoidia virens	False smut	Individual grains transformed into greenish spore balls
Wheat	Alternaria spp. and other genera	Black point/ kernel smudge	Kernels discoloured; black point if confined to embryo region; smudge if more extensive; grain embryo end shriveled; mycotoxins produced; germinability decreased; market value reduced
	Fusarium spp. and Helminthosporium spp.	Scab or head blight	Poorly filled grains with mycotoxins; reduced grain palatability
	Neovossia indica	Karnal bunt	Kernels completely or partially converted to masses of black spore balls; kernels fragile, malodourous and partially darkened, with breakage or partial erosion at the embryo end
Soybean	Cercospora kikuchi	Purple stain of grain	Grain discoloured pink to dark purple, cracked, emergence reduced

Contd...

Table 6.1–Contd...

Crop	Fungus	Name of Disease	Nature of the Damage Caused
	Cercospora sojina	Frogeye leaf spot	Grain discoloured grey to brown, cracked
	Macrophomina phaseolina	Charcoal rot	Sclerotia in seed coat; germination reduced
	Peronospora manshurica	Downy mildew	Grain size reduced, grain cracked and encrusted with oospores
	Rhizoctonia solani	Aerial blight	Grains infected within pods; macro- and microsclerotia formed on infected parts
	Septoria glycines	Brown spot	Seed coat and cotyledons invaded by mycelium
	Thielaviopsis basicola	Root rot	Mycelium may invade grains
Groundnut	Aspergillus flavus	Yellow mould, crown rot, aflaroot	Grain discolouration and rot; concealed damage; aflatoxin contamination
	Aspergillus niger	Crown rot	Grain discolouration and rot; concealed damage
	Macrophomina phaseolina	Charcoal rot	Grain discolouration and rot; concealed damage
	Rhizoctonia solani	Rhizoctonia pod rot	Grain with gray to brown discolouration; grain rot
	Rhizopus spp.	Seed and seedling rots	Grain discolouration and rot, concealed damage
	Sclerotium rolfsii	Southern blight	Grain shriveled and rotted often covered in grayish-white mycelium; blue damage
	Thielavoipsis basicola	Blackhull	Brown discolouration of grain; grain rot

Table 6.2: Field Fungi that Grow on or in Grains of Pulse Crops

Fungus	Common Name of Disease	Pulse Crop Affected
Alternaria spp.	Leaf spot; floral blight	Chickpea, pigeonpea,
Ascochyta spp.	Ascochyta blight, anthracnose	Bean, broad bean, chickpea, pigeonpea
Botrytis spp.	Gray mould	Bean, broad bean, chickpea, cowpea
Cercospora spp.	Leaf spots and blotches	Bean, cluster bean
Cladosporium spp.	Leaf and stem blights	Bean, broad bean
Colletotrichum spp.	Anthracnose	Bean, broad bean, chickpea, cluster bean, lablab, pigeonpea
Corynespora spp.	Target spots	Broad bean, cowpea
Diaporthe spp.	Stem and pod blight	Bean, cowpea
Elsinoe phaseoli	Scab	Bean
Fusarium spp.	Wilts, root, stem and pod rots	Bean, broad bean, chickpea, cowpea, pigeonpea
Macrophomina phaseolina	Charcoal rot	Bean, broad bean, chickpea, cowpea, pigeonpea
Peronospora spp.	Downy mildew	Broad bean
Phytophthora spp.	Blight	Bean, velvet bean, pigeonpea
Pleospora infectoria	Seed rot	Chickpea
Pythium spp.	Blight	Bean, chickpea
Rhizoctonia solani	Root, stem and pod rots, web blight	Bean, broad bean, pigeonpea, mungbean, urdbean
Sclerotinia sclerotiorum	Sclerotinia rot	Bean, Broad bean, chickpea, pigeonpea
Stemphylium spp.	Blight	Chickpea

and Kaufmann, 1965; 1969; 1974). The major loss due to storage fungi is the killing and discoloration of the seed embryo since it is at a higher moisture content than the rest of the grain. 'Blue eye in maize' and 'sickness' in wheat are the examples. A grain with a killed embryo cannot be called a seed since it does not have any chance of germination. A grain with damaged embryo is a loss to the consumer. Other deleterious effects of storage fungi on the grains are discolouration, alteration in the chemical composition of the grain, total disintegration of the grain, contamination of the grain with potent mycotoxins, mustiness and caking, and heating even to the point of combustion. Developing countries do recognize the postharvest losses of food grains that amount to tens of millions of tons per year. However, there is little precise information available on the magnitude of the storage losses of specific crops in specific regions, or on the relative importance of the several causal agents. The collection and assembly of such specific information is urgently needed to define the problem precisely and to develop control strategies (Williams and McDonald, 1983). Since the most important factor determining the magnitude of loss in storage is the moisture content of the commodity to be stored, the postharvest drying has a rather greater role to play in occurrence of the agents responsible for storage losses. This is particularly true for the fungal agents (storage moulds) which require high moisture content for their increased activity. Farmers in the tropics depend on the weather (sun and dry warm air) to dry their grains; they are, thus, on the mercy of the weather which is highly unpredictable in these regions. In the more humid tropics where there is no distinct dry season and relative humidities remain high throughout the year, rice and maize farmers have a very difficult time drying their harvested grain. The availability of cement drying floors is very limited and farmers often utilize the paved highways as a substitute. Drying the grains of high yielding varieties, which are photoperiod-sensitive is another problem encountered by farmers in the tropical countries. Though improved drying techniques and equipment for this have been developed but these are limited only to the rich and educated farmers; the average farmer is still in want of proper drying tools and technology. Besides, the climatic, economic and social circumstances of the farmers do not always match to these improved grain dryers. The traditional storage techniques adopted by most of the farmers in the tropics does not suit to the newer high yielding varieties with different grain characteristics and increased grain volume. Still, the storage losses are reduced when the grains are stored in decentralized on farm storage systems. The losses are often high when large quantities of grains are stored in centralized warehouses (such as those of Government agencies). In August 1982, 90,000 tons of wheat stored by Food Corporation of India was declared unfit for human consumption because of severe fungal spoilage.

Storage fungi also affect international seed/grain trade. In the recent years, it has become a necessity for some of the less-developed tropical countries to earn foreign exchange through export of the food grains. Such countries much often lack the advanced facilities for safe storage and disposal of the grains meant for export and there are additional opportunities for the storage fungi to develop. The grain may lie exposed to high humidity and rains while on the docks awaiting loading and condensation problems can be severe in ship holds when the vessels pass from the high temperatures of the tropics to the colder seas and climates of the temperate regions. Thus the grains even initially free from moulds become invaded considerably by storage fungi at the arrival of its destination with consequent risk of harmful effects on the consumers or economic loss for the importers and/or the exporters (Williams and McDonald, 1983).

The storage fungi comprise chiefly several 'group species' of the genus *Aspergillus* and about an equal number of less-well defined species of *Penicillium*. The group species of *Aspergillus* that are involved, *viz., Aspergillus restrictus, A. glaucus, A. candidus, A. versicolor, A. ochraceus* and *A. flavus* (listed in order of increasing moisture required for growth) or quite distinct from.one another and relatively easy to identify with a high degree of certainty. *Aspergillus restrictus* and *A. glaucus* are considered to be associated with incipient deterioration. The other common species of storage fungi–*Aspergillus candidus, A. ochraceus, A. versicolor, A. flavus* and *Penicillium* develop only later, after the growth of *A. restrictus* and *A. glaucus* has increased the moisture content of grain mass or of a portion of it, sufficiently to permit the growth of these higher-moisture-requiring or less xerophytic species (Christensen, 1972).

Harmful Effects of Storage Fungi

As far as harmful effects of storage fungi on the stored grains and seeds are concerned, extensive work has been done in this field. The major harmful effects can be listed as–discolouration of seeds and embryo, reduction in seed germinability, biochemical changes in the seed, increase in moisture content and heating, reduction in processing quality, enzymatic activities of seeds and production of mycotoxins. The major harmful effects of storage fungi on the seeds/grains are discussed below:

Discolouration and Changes in the External Appearance of the Seeds

The invasion of seeds by storage fungi causes blackening of the germ. This invasion takes place normally in the storage, but the contamination of seeds and, in certain cases, invasion of seeds (wheat, cotton and groundnut) by some species of *Aspergillus* (*A. flavus* and *A. candidus*) and *Penicillium* takes place before harvesting of the crop. The invasion causes discolouration, shrivelling of seeds, prolonged dormancy and quality losses. Such types of seeds are termed as 'sick' or 'damaged'. Relationship of discolouration and odour of seeds with the infection by mycoflora has been a point of investigation since long. Although fungi like *Alternaria* spp. are considered primarily field fungi (not surviving in storage more than six months at room temperature), they have a pronounced effect in storage as well. Chahal (1981) observed that discoloured seeds of Indian mustard (*B. juncea*) having 84–92 per cent infection of *Alternaria brassicae* in April become free from it when tested in September after storage at room temperature. Limited period of survival of *Alternaria brassicae* in mustard seeds in storage has also been marked by Singh (1994). The author has further found *Alternaria brassicae* to survive in the seed coats of infected mustard seeds. *Alternaria brassicae* was more frequently isolated from rapeseeds with gray to dark gray discolouration (Vishunavat *et al.*, 1985; Singh, 1994). Kolte (1985) stated that discolouration of rape and mustard seeds is primarily caused by *Alternaria alternata, Alternaria brassicae, Alternaria brassicicola* and *Penicillium citrinum.* Ali and Deka (1996) listed some fungal genera associated with discolouration of rice grain including *Curvularia, Drechslera, Nigrospora, Trichothecium, Fusarium, Aspergillus* and *Penicillium.* Jain *et al.* (1997) showed association of various fungi with discoloured and disfigured seeds of cumin (*Cuminum cyminum* L.). Among them, *Alternaria alternata* and *Curvularia lunata* were dominant in shriveled and black coloured seeds. Goyal and Jain (1997) found association of *Aspergillus flavus, A. niger* and *Fusarium* sp. in shriveled and brown seeds of *Vigna mungo.* Kumar and Agarwal (1997) recorded *Bipolaris maydis, Botryodiplodia theobromae, Curvularia lunata, Fusarium moniliforme, Sclerotium rolfsii, Trichoderma harzianum* and *Aspergillus flavus* causing black streaks; brown to black shredding, black pointed dots; dark black streaks; white to dark pinkish discolouration and white streaks; dull gray with light brown band; and greenish spots and yellow greenish discolouration, respectively on maize seeds. Moreover, they found that various types of discolouration on seeds might result in reduction in seed viability, germination, seedling vigour and loss in protein and starch contents. Srivastava and Jaiswal (2000) state that *Alternaria brassicae* affects seed quality of *Brassica oleracea* (L.) var. *botrytis* by causing seed spotting and seed discolouration. Godika *et al.* (2000) have reported symptomatic sunflower seeds with grayish brown and brown to black discolouration to yield *Alternaria alternata* after incubation on moistened blotters.

Effects on Germination, Viability and Seedling Tracts

As far as effects of storage fungi on the viability, seed, germination and seedling emergence and mortality are concerned, the investigations carried out prior to 1980s include those by Chandra *et al.* (1981); Gupta *et al.* (1993) and Bankole (1993; 1994).

There seems to be definite relationship between discolouration and germinability of the seeds. Discoloured seeds have been proved to be light in weight with reduced germinability (Singh, 1994). Germination and thousand seed weight of *Brassica campestris* seeds were observed to decrease in the seeds from pods having deep lesions caused by *Alternaria brassicae* (Kadian and Saharan, 1983). Reduced seed weight and seed size and its germinability were also observed by Vishuanavat *et al.* (1985) in discoloured rapeseeds. In an experiment conducted in Chitagong (Bangladesh), 41 per cent of the *Brassica campestris* seeds infected with *A. brassicae* failed to germinate and 73.4 per cent of seedlings died after germination (Mridha and Safa, 1985). Fungi do have a role in causing reduction in seed weight. *Aspergillus flavus* and *Alternaria carthemi* are reported to cause maximum reduction in the seed weight of safflower in the temperature range of 25°C to 35°C (Sandikar *et al.*, 1992). The germinability of stored seeds is also largely affected by storage structures and period of storage. Porous structures that allow moist air to enter the seed lot often promote the growth of fungi in the seed lot by raising the moisture in the inter-seed atmosphere. If allowed for sufficient time, this creates ideal environment for increased fungal activity that often results in reduction in germinability of the stored seeds. Mukherjee *et al.* (1992) observed changes in germination of groundnut seeds after storage for one year in gunny bags in laboratory. Germinability and dry mass of resultant seedlings of infected maize seeds was found decreasing with increasing storage period (Mycock and Berjak, 1992). Bankole (1993) reported decrease in seed germinability with prolonged storage of melon (*Citrullus vulgaris*) seeds. The percentage seed germination of maize was observed to decrease with storage time to the lowest after twelve months (Bankole, 1994). There have been attempts to isolate fungi that are responsible for reduction in seed germination and seedling traits. Mahajan and Joi (1992) observed twenty-eight seedborne fungi on pearl millet adversely affecting seed germination. Sharma and Singh (1992) attempted to find out the fungal metabolite inhibitory to the process of germination. They reported that root and shoot growth inhibition was maximum in *Brassica juncea* cv. Varuna and minimum in *Brassica rapa* by a partially purified toxin preparation of *Alternaria brassicae* at comparable dilutions. Moreno-Martinez *et al.* (1994) reported loss in germinability of all the cultivars of *Phaseolus vulgaris* tested, except for two (Negro Chiapas and Negra Veracruz) after storage at higher relative humidity conditions. In another experiment, the authors concluded that barley seeds had superior storage capability to maize and that storage fungi do not play as important a role as in the loss of seed viability in barley as they do in maize. Jain *et al.* (1997) found *Alternaria alternata* to cause pre- and post-emergence mortality and to reduce seed vigour index of cumin (*Cuminum cyyminum* L.) seeds. Goyal and Jain (1997) reported seven fungi, *viz., Aspergillus flavus, A. nidulans, A. niger, Fusarium pallidoroseum, Fusarium* sp. *Macrophomina phaseolina* and *Rhizopus oryzae* causing pre- and post-emergence mortality and to reduce seedling vigour of *Vigna mungo* seeds. Lima and Araújo (1999) isolated six fungal species, *viz., Rhizopus* sp., *Macrophomina phaseolina, Aspergillus niger, A. flavus, Fusarium* sp. and *Penicillium* sp. from groundnut seeds stating them responsible pre- and post-emergence damping-off of seedlings. Bhutta *et al.* (1999) reported increase in pre- and post-emergence mortality of sunflower seedlings and decrease in yield and seed mass of the oilseed with an increase in the seed infection level.

Reduction in the germinability of the seeds has been attributed to the death of the embryo, and reduced seedling emergence to the pre-emergence mortality of germilings resulting from the invasion by storage fungi. The whole seed lot may be rendered useless for sowing and consumption purposes. Decrease in germinability of seeds is higher and rapid following their storage at high moisture levels.

Biochemical Changes in the Seeds

The biochemical changes in stored seeds brought about by storage fungi include changes in the three major nutritional components of seeds, *viz.*, carbohydrate, proteins and lipids (Prasad *et al.*, 1988; Saxena and Karan, 1991; Rai, 2002).

Development of storage fungi increases moisture content and temperature of seed lots due to increased respiration of seeds. The moisture so produced in one corner can diffuse to the other areas and render it vulnerable to spoilage. By this time that the fungi have grown enough to have raised the moisture content high enough so that seed respiration becomes measurable, the seeds are dead and decayed. Temperature and relative humidity values of the inter-seed atmosphere play a very important role in creating the environments under which biochemical changes in the composition of the grain/ seed accelerate due to increased fungal activity. Such changes may lead to the reduction in the processing quality of grains for milling as in wheat, corn and many other cereals. Musty odour, often produced during spoilage of grains persists even in the preparation of food and beverages from such grains. The storage fungi deteriorate the quality of seeds used for oil production and reduce the oil content in the affected seeds (Singh *et al.*, 2001). Oil from infected samples is discoloured with acrid smell. Abramson and Chelkowski (1991) noted the production of fungal odour in stored moist cereals indicative of the biochemical changes brought about by fungi.

Seed rotting has been observed a phenomenon associated with fungal invasion of the seed. Kumar *et al.* (1986) isolated *Fusarium* spp., *Rhizoctonia* spp. *Alternaria brassicae* and *Alternaria brassicicola* from rotten seeds of mustard. Stimulation of activities of pectic enzyme complex, Cx, amylase, invertase and protease by storage fungi in finger millet has been reported (Prasad *et al.*, 1988). Storage fungi are known to stimulate hydrolysis of starch and protein producing extracellular amylase and protease besides cellulase and lipase (Prasad, 1979).

Several fungi growing on stored seeds are well known to produce metabolites toxic to other organisms including domestic animals and man (mycotoxins). Production of these powerful toxins that contaminate food and feed by many field and storage fungi is perhaps the most disturbing and emotive aspect of the grain invasion by fungi. The diseases caused by consumption of mycotoxins are known as 'mycotoxicoses'. Aflatoxins are the most important of the known mycotoxins. Apart from being highly toxic to a wide range of animal species, these toxins are also extremely potent human carcinogens. These are produced chiefly by *Aspergillus flavus*, a common storage fungus. The problem of aflatoxin production is confounded by the fact that *A. flavus* is cosmopolitan and worldwide in distribution and can attack a wide variety of substrates under storage conditions. The production of aflatoxin is increased by higher temperatures and this may be the reason behind more incidences of aflatoxicoses in tropics than in the temperate regions. Aflatoxin may cause serious disorders in the human beings or animals, when consumed, including liver damage and even cancer. A large number of mycotoxins have now been identified, their chemical structures determined and biological effects studied. Since there are many suspected mycotoxins which have not yet been fully identified, it is likely that several new mycotoxins will be reported in the near future.

Apart from the largest group of mycotoxicogenic fungal genus, *Aspergillus,* some other fungi producing mycotoxins include *Penicillium rubrum, P. putpureogenum,* (rubratoxins); *Aspergillus ochraceus, Penicillum viridicatum, P. cyclopium* (Ochratoxins); *Aspergillus clavatus, A. terrrus, A. patulum, Penicillium cyclopium* (Patulin); *Penicillium cyclopium, P. palitans, Aspergillus flavus* (Tremortins); *Fusarium oxysporum,*

F. moniliforme, *F. roseum* (Zearalenone); *Penicillium citreoviride*, *P. toxicarinum* (Citreoviridin); *P. citrinum*, (Citrinin) and *P. cyclopium* (Penicillic acid). *Aspergillus* spp. (Sterigmatocystin); *Penicillium* spp. (Sterigmatocystin, Cyclocholrotine); *Fusarium* spp. (Tricothecenes and ATA-toxin). Several toxigenic fungal species have been listed both as field fungi and storage fungi of cereal and legume grains, the most important of these are 'intermediate' forms being *Aspergillus flavus* and *Fusarium* spp. There is much greater potential of mycotoxin contamination of foodstuffs and grains in the tropics than in the temperate areas particularly for those produced by species of *Aspergillus* and *Penicillium*. It is ironical that most of the research studies on mycotoxins have been conducted in temperate countries with few papers reporting mycotoxin research in the tropical regions. Aflatoxin, the toxin responsible for the historical Turkey X disease is still the most important mycotoxin. The producer fungus *Aspergillus flavus* had long been known as a storage fungus and it was only this fungus that opened the field of mycotoxin research by drawing attention of the researchers of that time. In a brief description of factors favouring pre- and postharvest invasion of groundnut by toxigenic strains of *Aspergillus flavus*. Dickens (1977) listed (*a*) damage to the developing pod in the ground during cultivation; (*b*) damage by pests (insects, mites, millipedes, nematodes, rodents); (*c*) the effects of pod-rotting fungi and (*d*) severe fluctuations in soil moisture. In the cases of delayed harvesting, the plants become senescent and pods droop down towards the ground, much often touching it. Here, the seeds inside the pods are infected by the soil fungi that may include the species deteriorative to the grains in storage.

Environmental Factors and Activities of Fungi in Storage

The activity of majority of the organisms associated with the seeds in storage is affected by temperature, moisture and gaseous environment. Fungi, the primary cause of reduction in viability, are the most affected by these parameters. These factors, to the maximum extent, determine the genera of fungi that will prevail at a certain set of environmental parameters. In other words, the quality of fungal flora changes with the changing values of environmental parameters. Since fungi are aerobic organisms, they cannot flourish in anaerobic environment brought about by the depletion of oxygen by other organisms. This is particularly important in hermetic storage of damp grains in which the production of anaerobic environment is accelerated by respiration of the micro flora. In such a condition, though the activity of the micro flora including fungi is reduced, yet the viability of seed is lost rapidly.

It has been repeatedly emphasized that the moisture and temperature are the main factors that affect the micro flora of the stored seed. The relationship of environmental conditions with seed micro flora has very well been discussed by Roberts (1972) in his marvellous review entitled, 'Storage environment and the Control of Viability.' The moisture content of the seed itself does not play a direct role in determining the mycoflora of storage, although it has an equilibrium relationship with the relative humidity of the inter-seed atmosphere (water activity, a_w), therefore affecting the storage fungi indirectly. All the storage fungi are known to be completely inactive below a relative humidity of 62 per cent as evidences presented by Semenuik (1954). Milner and Geddes (1954) reported that there is very little activity below around 75 per cent relative humidity and from 75 per cent relative humidity upwards there has been found an expotential relationship between amount of fungi in seed and relative humidity (Bottomley *et al.*, 1950).

Deterioration in grain and mycotoxin production due to increased fungal activity is, therefore, largely dependent on the temperature and inter-seed relative humidity in the storage. Since seed viability decreases rapidly at higher temperatures, the only means of controlling microbial activity by temperature alone is by deep freezing. Moreover, localized temperature increases speak of increased biological activity including that of insects. Production of hot spots in stored grains by microorganisms means that the grain is already deteriorated. It is, therefore, desirable that entire grain bulk must be suitably managed to prevent deterioration beyond the hot spots.

Management of Deteriorative Fungi

Field Fungi

Majority of farmers in the tropical countries have smaller land holdings and limited monetary resources to apply management practices that prove economical to their business. Largely, they have to seek for the methods that require less machinery, preferably cheap and easy to learn and adopt. For this, they rely primarily upon the inherent quality of the crop cultivar and their skill in growing them. Any substantial improvement in practical control of storage diseases of grain with depend on (*a*) the determination of losses induced by fungal growth (*b*) application of new and old technology (*c*) extensive educational programs and (*d*) better rewards for grain quality (Tuite and Foster, 1979).

A rapid and precise assessment of grain condition so as to predict its storability is perhaps the most important tool required to control losses during storage. Injuries and physical damage to the seed coat, foreign material, insect infestation, amount of microbial inoculum, with keeping the storage conditions in view, one can have an approximate idea of the grain condition after storage for a certain time span. Cultural techniques like seed plating and assessing the quantity and quality of fungi on/in seeds may provide further information about the fate of the grain after storage. The one most ignored point in control of seed fungi is the susceptibility of the cultivar to the insect damage. The more is the seed damaged, the more will be venues available for entry of the fungal inoculum. Physical characteristics of the grain, such as husks in the case of maize and pods for legumes also affect the infection and subsequent invasion of the same by fungi. Development of the cultivars that could resist/avoid fungal flora by their physical and biochemical characteristics is, therefore, one of the most promising aspects in controlling field fungi. Of course, these characteristics, apart from preventing the grain spoilage by fungi, must not be such as to be unacceptable to consumers. The intellectual manpower engaged in the breeding of crop cultivars must, therefore, be aware that increased production through crop breeding will be of no consequence without adequate provisions for preventing the spoilage by fungal agents. In other words, our efforts should not be only to produce more, but also to save more from the considerable losses that take place by fungi. By doing this one can make earnings by savings that will be far more economic to the poor farmers with limited resources since the inputs required in saving the seed will be far more cheaper than those required in increasing the production.

The farmer can reduce the postharvest problems of field fungi by careful selection of the cultivar recommended for his region, proper cultivation techniques and crop care and timely harvesting the crop giving no more chances of infection by field fungi on matured grains. Rapid and proper drying of the grain will reduce the initial inoculum available for postharvest infection of the matured grain. These precautions become almost unpractical particularly to the small scale tropical farmer who does not have tools and technology for improved methods of crop cultivation such as quick harvesting through machines and improved dryers. For these processes, they depend solely on the hand labour which is not only unsophisticated but also time consuming. Also the problem worsens if the weather, which is highly unpredictable in such conditions, does not favor the farmer.

Storage Fungi

Maintenance of the low moisture content in the grain is the most important factor that helps in reduced infection and subsequent losses of grain by storage fungi. Low temperature also creates the environment unfavourable for growth and activities of storage fungi. Protection against insect infestation will be another important measure that will help in reducing the losses by such agents. These are much more difficult to achieve in the tropics than in the temperate regions for many reasons, including differences in climate, resources available, attitudes and incentives (Williams and McDonald, 1983).

Selection of the seed lot suitable for storage, proper drying of seeds to bring their moisture content down up to the safe level and storage in air tight storage structures such as polythene containers will be helpful to reduce the losses in storage satisfactorily.

Large scale spoilage can well be prevented by the practices of decentralized storage or on farm storage in the tropical areas. Though, this is done by the farmer for conserving the seeds not only for the next planting but also to feed the family and livestock throughout the year. This type of storage with traditional crop cultivars such as sorghum and millets in comparatively dry areas of tropics provides reasonable storage security for the traditional farmers, with low risk of large scale spoilage (Williams and McDonald, 1983). However, there is an urgent need for development of technologies for safe storage of food grains like rice and wheat in the more humid tropics. These technologies must not only be effective in reducing the spoilage but also cost effective for wide adoption by farmer.

Although, chemical fungicides have been used since long for treatment of the seeds, they are well known for their non-target effects on the mankind and its properties. Chemicals with residual effects and those not easily biodegradable are worth mention in this regard. Chemical hazards have forced the scientists of the era to rethink of their use particularly with the theme of 'Sustainable Agriculture'. Alternatively, people associated with plant protection sciences tried to search for a suitable alternative of the agrochemicals, especially pesticides, in the higher plants and their products. Due to high potency of antimicrobial properties, nonphytotoxic nature, being eco-friendly and easy availability with no side effects, the higher plants and herbs have been used to control several diseases of human being and animals. Their application against phytopathogenic microorganisms in general and in storage are given much attention nowadays to control these microorganisms causing severe damage to the crop and food commodities. The volatile and nonvolatile fractions of higher plants have been given much emphasis due to their high fungitoxic properties against parasitic, saprophytic and pathogenic fungi. The volatiles do not leave any residue; hence there is least chance of residual toxicity in treatment of food commodities.

We have entered the era of 'Modern Agriculture' which is based on crop cultivation with proper attention on the maintenance of ecological balance. Our cultivars must yield more, but never on the cost of the environment. It is the time when even a cosmetic article is enquired after its effects on the ecology prior to purchasing. Unjudicious use of harmful and residual pesticides is now considered a thing of past. People are becoming more and more aware about the pollution, which has now become virtually a threat to our own existence. Recently, Tsunami in Indonesia and other south eastern Asian countries and Katerina in USA have caused deaths and devastation of a large group of human population. Such climatological disturbances have become common and they reflect the pressure of human activities being put on the nature for his own short term gain. This shortcut method to get more and more has led us to the point from where our own life seems to be in grave danger. We have made the mess of the whole thing and this is not the mischief of some time. It has started since the so-called development of agriculture and particularly introduction of chemical agriculture. Chemicals have been once thought to revolutionize our life. Now, that revolutionization has become dear to us and our dear planet. At this point of time, we, therefore, do not have any other alternative except for keeping the maintenance of the ecological, geological and environmental balances on top priority and how all the other businesses are managed, are the most serious challenges before the scientists of this era.

References

Abramson, D. and Chelkowski, J. (1991). Development of molds, mycotoxins and odours in moist cereals during storage. In: *Developments in Food Science Vol. 26: Cereal Grain: Mycotoxin, Fungi and Quality in Drying and Storage*, p. 119–147.

Agarwal, P.K. (1994). *Principles of Seed Technology*. Publication and Information Division, Indian Council of Agricultural Research, New Delhi.

Ali, M.S. and Deka, B. (1996). Role of fungi causing grain discolouration of rice and management under storage. *Indian J. Mycol. Pl. Pathol.*, 26: 79–85.

Bankole, S.A. (1993). Moisture content, mould invasion and seed germinability of stored melon. *Mycopathologia*, 122: 123–126.

Bankole, S.K. (1994). Change in moisture content, fungal infection and kernel germinability of maize in storage. *Int. J. Crop Plant Dis.*, 12: 213–218.

Bhutta, A.R., Bhatti, M.H.R., Ahmed, I. and Shakoor, C.A. (1999). A study of correlation between seed and field infection intensity for establishing disease tolerance limits in sunflower. *Helia*, 22: 137–142.

Bottornley, R.A., Christensen, C.M. and Geddes, W.F. (1950). Grain Storage Studies. IX. The influence of various temperatures, humidity and oxygen concentration on mould growth and biochemical changes in stored yellow corn. *Cereal Chem.*, 27: 217–296.

Chahal, A.S. (1981). Seedborne infection of *Alternaria brassicae* in Indian mustard and its elimination during storage. *Curr. Sci.*, 50: 621–623.

Chandra, S., Marang, M. and Srivastava, R.K. (1981). Changes in the associated mycoflora and viability of seeds of some oilseeds in prolonged storage. *Geobios*, 8: 200–204.

Christensen, C.M. (1972). Microflora and seed deterioration. In: *Viability of Seeds*, (Ed.) Roberts, E.H. Chapman and Hall, London, pp. 59–93.

Christensen, C.M. (1980). Needed research on storage molds in grains, seeds and their products. *Plant Dis.*, 64: 1067–1070.

Christensen, C.M. and Kaufmann, H.H. (1965). Deterioration of stored grains by fungi. *Ann. Rev. Phytopathol.*, 3: 69–84.

Christensen, C.M. and Kaufmann, H.H. (1969). *Grain Storage: The Role of Fungi in Quality Loss*. University, Minn. Press, Minneapolis, 153 pp.

Christensen, C.M. and Kaufmann, H.H. (1974). Microflora. In: *Storage of Cereal Grains and their Products*. (Ed.) Christensen, C.M. pp.158–192. St. Paul: American Association of Cereal Chemists, Inc., 549 pp. 2nd ed.

Christensen, J.J. (1936). Association of microorganisms in relation to seedling injury arising from infected seed. *Phytopathology*, 26: 1091–1105.

Christensen, J.J. and Stakman, E.C. (1935). Relation of *Fusarium* and *Helminthosporium* in barley seeds to seedling blight and yield. *Phytopathology*, 25: 309–327.

Dickens, J.W. (1977). Aflatoxin occurrence and control during growth, harvest and storage of peanuts. In: *Mycotoxins in Human and Animal Health* (Eds.) Rodricks, J.V., Hesseltine, C.W. and Mehlman, M.A. pp. 99–105.

Godika, S. Agrawal, K. and Singh, T. (2000). Incidence of *Alternaria alternata* in sunflower seeds grown in Rajasthan. *J. Mycol. Plant Pathol.*, 30: 112–113.

Goyal, S.K. and Jain, M.P. (1997). Mycoflora of blackgrarn seed (*Vigna mungo* L. Hepper) under storage conditions, their pathological effects and control. In: *Proceedings of the Indian Phytopathological*

Society, Golden Jubilee International Conference on Integrated Plant Disease Management for Sustainable Agriculture, November 10–15, New Delhi, India (Abs.).

Gupta, I.J., Schmitthenner, A.F., and McDonald, M.B. (1993). Effect of storage fungi on seed vigour of soybean. *Seed Sci. and Technol.*, 21: 581–591.

Jain, M.P., Kumar, K.C. and Jain, S.C. (1997). *Alternaria* spp. associated with cumin (*Cuminum cyminum* L.) seeds, their pathogenicity and control. In: *Proceedings of the Indian Phytopathological Society, Golden Jubilee International Conference on Integrated Plant Disease Management for Sustainable Agriculture,* November 10–15, New Delhi, India (Abs.).

Kadian, A.K. and Saharan, G.S. (1983). Symptomatology, host range and Assessment of yield losses due to *Alternaria brassicae* infection in rapeseed and mustard. *Indian J. Mycol. Pl. Pathol.*, 13: 319–323.

Kolte, S.J. (1985). *Diseases of Annual Edible Oilseed Crops Vol. II. Rapeseed Mustard and Sesame Diseases.* CRC Press Inc., Boca Raton, Florida, p. 57.

Kumar, K., Singh, D.P. and Kumar, K. (1986). Control of *Alternaria brassicae* infection in mustard and rapeseeds. *Pesticides,* 20: 22–23.

Kumar, M. and Agarwal, V.K. (1997). Investigations on seed discolouration of maize. In: *Proceedings of the Indian Phytopathological Society. Golden Jubilee International Conference on Integrated Plant Disease Management for Sustainable Agriculture,* November 10–15, New Delhi, India (Abs.).

Lima, E.F. and Araujo, A.E. de (1999). Fungi causing damping-off, borne and transmitted by seeds of groundnuts. [*Fungos causadores de tombamento, transporlados e transmitidos atralles da semente do amendoim*]. *Revisla de Oleaginosas e Fibrosas,* 3: 71–76.

Machacek, J.E. and Greaney, F.J. (1938). The 'black point' or 'kernel smudge' disease of cereals. *Can. J. Res. C.*, 16: 84–113.

Mahajan, P.D. and Joi, M.B. (1992). Documentation of different seed borne fungi of pearl millet and their effect on seed germination. In: *Proceedings of the 44th Annual Meeting of Indian Phytopathological Society,* Bhavnagar, December 30, 1991–January 02, 1992. *Indian Phytopath.* (Suppl.) 43/44: CX. Meetings of the Zonal Chapters, Western Chapter: November 01–02, 1991 (Abs.).

Malone, J.P. and Muskett, A.E. (1964). Seed-borne fungi-description of 77 fungus species. In: *Proc. Int. Seed Test. Ass.*, 29: 179–384.

Milner, M. and Geddes, W.F. (1954). Respiration and heating. In: *Storage of Cereal grains and their Products* (Eds.) Anderson, J.A. and Alcock, A.W. Amer. Assoc. of Cereal Chemists, St. Paul, pp. 152–220

Moreno-Martinez, E., Vazquez-Badillo, M.E., Navaratte, R. and Ramirez, Gonzalez, J. (1994). Seed viability of different varieties of bean (*Phaseolus vulagaris* L.) stored under low and high relative humidity. *Seed Sci. and Technol.*, 22: 195–202.

Mridha, A.U. and Safa, N. (1985). Prevalence and control of seedborne *Alternaria brassicae* in mustard. *Bangladesh J. Plant Path.*, 1: 9–12.

Mukherjee, P.S., Nandi, S.K. and Nandi, B. (1992). Deteriorative changes in groundnut seeds in storage. *J. Mycopathol. Res.*, 30: 113–119.

Nagel, C.M. and Semeniuk, G. (1947). Some mould induced changes in shelled corn. *Pl. Physiol.*, 22: 20–23.

Pepper, E.H. (1960). The micro flora of barley kernels: Their isolation, characterization, etiology and effect on barley, malt and malt products. *Ph.D. Thesis.* Department of Botany and Plant Pathology, Michigan State Univ., East Lansing (unpublished).

Prasad, B.K. (1979). Enzymatic studies of seedborne fungi of coriander. *Indian Phytopath.*, 32: 92–94.

Prasad, B.K., Shanker, U., Narayan, N., Dayal, S. and Kishor, A. (1988). Physico-chemical changes in food reserve of coriander seed due to storage molds. *Indian Phytopath.*, 41: 386–388.

Rai, J.P. (2002). Studies on seed Mycoflora of Mustard (*Brassica juncea* L.). *Ph.D. Thesis.* Department of Mycology and Plant Pathology, Institute of Agricultural Sciences, Banaras Hindu University, Varanasi, India, 119 p.

Roberts, E.H. (1972). Storage environment and the control of viability. In: *Viability of Seeds*, (Ed.) Roberts, E.H. Chapman and Hall, London, p. 14–57.

Sandikar, S.N., Panchal, V.H. and Umatale, M.V. (1992). Studies on fungal biodeterioration in safflower seeds. In: *Proceedings of the 44th Annual Meeting of India Phytopathological Society*, Bhavnagar, December 30, 1991–January 02, 1992. *Indian Phytopath.* (Suppl.) 43/44: CXXI. Meetings of the Zonal Chapters, Western Chapter: November 01–02, 1991 (Abs.).

Saxena, N. and Karan, D. (1991). Effect of seedborne fungi on protein and carbohydrate contents of sesame and sunflower seeds. *Indian Phytopath.*, 44: 134–136.

Semeniuk, G. (1954). Microflora. In: *Storage of Cereal Grains and their Products*, (Eds.) Anderson, J.A. and Alcock, A.W. Amer. Assoc. of Cereal Chemists, St. Paul, pp. 152–220.

Sharma, T.R. and Singh, B.M. (1992). Parameters of resistance to *Alternaria brassicae* in some *Brassica* spp. In: *Proceedings of the 44th Annual Meeting of Indian Phytopathological Society*, Bhavnagar, December 30, 1991–January 02, 1992. *Indian Phytopath.* (Suppl.) 43/44: LIX. Meetings of the Zonal Chapters, Western Chapter: November 01–02, 1991 (Abs.).

Singh, A.K. (1994). Studies on seedborne nature of *Alternaria brassicae* (Berk.) Sacc. and *Alternaria brassicicola* (Schw.) Wiltshire and their effect on seed quality in rapeseed-mustard (*Brassica* spp.). *M.Sc. (Ag.) Thesis*, Department of Plant Pathology, College of Agriculture, G.B. Pant University of Agriculture and Technology, Pantnagar, (U.P.) India.

Singh, A.K., Vishunavat, K., Rai, J.P. and Sinha, A. (2001). Effect of *Alternaria brassicae* (Berk.) Sacc. and *Alternaria brassicicola* (Schw.) Wiltshire on oil content of mustard seed. *Res. on Crops*, 2: 372–374.

Singh, M.T., Singh, M.S. and Pandey, R.R. (2001). Effect of storage fungi on electrolyte leakage from rice seeds. *Indian Phytopath.*, 54: 123–125.

Srivastava, J.P. and Jaiswal, H.L. (2000). Mode of survival and effect of *Alternaria* spp. on cauliflower seed quality, effect of seed treatment and packaging material. In: *Proceedings of the National Symposium on Important Plant Diseases of North Eastern India and their Management*, December 6–7, 2000, Department of Plant Pathology, N.D. Univ. of Agric. and Technology, Kumarganj, Faizabad (Abs.).

Tuite, J. and Foster, G.H. (1979). Control of storage diseases of grain. *Ann. Rev. Phytopathol.*, 17: 343–366.

Vishuanavat, K., Agarwal, V.K. and Singh, R.S. (1985). Relationship of discolouration of mustard (*Brassica campestris* var. *sarson*) seed infection by *Alternaria brassicae* and its longevity on seed. *Seed Res.*, 13: 53–56.

Williams, R.J. and McDonald, D. (1983) Grain molds in the tropics: Problems and importance. *Ann. Rev. Phytopathol.*, 21: 153–178.

©R.C. Ram & Asha Sinha (*Editors*)
Microbial Diversity: Modern Trends, **2007** **94–102**
Daya Publishing House, New Delhi

Chapter 7

Indigenous Soil Microbial Biodiversity: Key to Sustainable Agriculture

D.P. Singh and B.K. Sarma

Department of Mycology and Plant Pathology, Institute of Agricultural Sciences, Banaras Hindu University, Varanasi – 221 005, U.P.

Introduction

The relationship between agriculture and natural resources is so intrinsically interwoven that they cannot be separated at any stage. However, the developmental activities planned in abrupt and unscientific manner in the recent past have led to the exploitive agriculture, which is still being carried out. Such an agricultural system has failed to generate any positivity towards its sustainability. Continuously increasing population demands for regular hike in food production but continuously decreasing per capita availability of arable land, irrigation water and degraded soil fertility restrict crop productivity. Unfortunately, there are no dramatic technological breakthroughs on the horizon to stand parallel with increasing pace of crop productivity.

Of all issues in the world, access to food is the greatest motive for social action. In the course of evolution, therefore, agriculture with the aim to produce more per se was developed to replace the plants and animals that exist in nature by those among them that humans prefer for their food and other utilities. However, in absolute terms, neither crop nor animal production yields more biomass than the natural population it replaces does. The motive behind agriculture in biospheric terms is basically aiming at taking away for humans the biological production from the majority of living things.

Therefore, the mobilization of scientific efforts for raising biological productivity without associated environmental harms can constitute one of the major and prime approaches for obtaining sustainability in crop productivity, which will be in harmony with nature too. It is therefore, appropriate to refer to such scientific and technological developments in the farmers' field as an *'ever-green revolution'*, because it is rooted in the principles of ecology, economics, human health and social and gender equity.

Soil Biodiversity: Key to Sustainable Agriculture

Soil is a dynamic living matrix essentially important for the maintenance of most life processes on the earth and atmosphere. Being an essential component of the terrestrial ecosystem, it is a critical resource to agricultural productivity and food security.

Table 7.1: Specific Role of Organism Communities in the Soil

Organisms	Specific Role	Effect on Soil
Worms (Earthworm), termites, Nematodes (Called Soil Engineers)	Mix soil horizons, particles and organic matter, increase porosity	Improve soil structure, directly determine vulnerability to soil erosion and availability of soil profile to plants
Fungi	Enhance decomposition processes, release of pollutants in the environment e.g. leaching of nitrates, Bioremediation	Affect nutrient cycling, plant growth and production
Nematodes	Protect plants from pest and disease outbreaks	Effective against soil-borne diseases/pests
Bacteria	Regulate the rates of carbon sequestration, gaseous emission and organic matter transformation	Determine carbon cycle
Phosphate solubilizing bacteria (PSB)	Mobilize fixed phosphorus to plants	Enhance phosphorus availability
Rhizosphetic bacteria (Pseudomonas, Bacillus etc.)	Live in association with the plant root rhizosphere and enhance plant growth, have antagonist effect against soil borne pathogens and activate and passive resistance against diseases in plants	Mobilize nutrients and chelated minerals e.g. iron in the soil, antagonize with soil borne disease causing agents and promote plant growth
Algae, Blue Green Algae, Azolla	Live in symbiotic association with plants roots and fix atmospheric nitrogen	Responsible for biological nitrogen fixation in the soil
Mycorrhizal fungi	Mycorrhizal associations provide plants with various nutrients and water availability	Act as soil conditioner

Soil is a native home of enormous living organisms assembled in the complex and varied communities. An acre of living topsoil contains approximately 900 pounds of earthworms, 2,400 pounds of different fungi, 1,500 pounds of bacteria, 133 pounds of protozoa, 900 pounds of arthropods, nematodes and algae and even small mammals in some cases. The type of healthy living soil needed to support humans in now and far into the future should balanced in nutrients, high in organic matter content and humus with a broad diversity of living organisms. Soil Biodiversity reflects the variability among living organisms ranging from the myriad of invisible microbes including bacteria, fungi, nematodes, protozoan and algae to the most familiar macro-fauna such as earthworms and termites. Rhizosphere of plants can also be considered as a living matrix in view of its symbiotic associations and interactions with diverse microorganisms and other soil components. In the form of communities, these organisms interact with one another, with various plants and animals and with various living biochemical processes in the soil to form a complex web of biological activities. Since the whole Soil

Biological Communities and their activities are living, it is susceptible to environmental factors as well as to the anthropogenic actions such as agricultural and forestry management practices. Soil Biological Communities are an important natural resource like sun energy, air and water and an integral part of agricultural and forestry ecosystems. They play a critical role in maintaining ecological functions, soil health and crop production. Each and every organism has a specific role in complex web of life in the soil.

The interacting functions of the soil microorganisms and the effects of the human activities during land management for agriculture affect soil quality and health. Soil quality is basically the capacity of soil to function for sustaining plant growth, enhance air and water capacity and support human health within natural and managed ecological boundaries. The principles of soil health include the ecological attributes of the soil. The attributes that are chiefly associated with the soil biota, its diversity its food web structure and activities and the range of activities and the functions it performs are greatly associated with the soil health and quality and are characteristic properties that may be vital for the continued capacity of the soil to support crops.

Successful management of agricultural resources in order to satisfy human needs while maintaining or event improving ecological quality and conserving natural resources is in the prime principles of sustainable agriculture. Improvement in the agricultural sustainability, that primarily require alongside effective water and crop management, and the optimum use and management of physico-chemical properties and fertility of the soil, largely rely upon the biodiversity and the biological processes being operated within the soil.

Need for the Management of Soil Biodiversity

> The sustained use of the earth's land, water resources and thereby plant, animal and human health is dependent upon maintaining the health of the living soil biota that provide critical processes and ecosystem services.
>
> *Agriculture Biological Diversity, Note by the Executive Secretary, United Nations Environmental Program, 2001*

The present days escalating population growth and increasing demand for food, shrinking land size, increasing industrialization and urbanization and need for sustainable and viable agricultural systems are critical to the issues to food security and poverty alleviation in societies in almost all developing countries in the world. In fact, this has given birth to what is known today after the emergence of Green Revolution in 1968 as Exploitive farming or Intensive Cultivation. From the experiences in the past, it is now being realized that exploitive agriculture that has been carried out unscientifically and only with an immediate profit motive has posed great threats to environment and human health. The exploitive farming community in India is now greatly realizing that intensive cultivation of land without conservation of natural resources and soil fertility have ultimately led to the springing up of the deserts. The indiscriminate use of pesticides, fungicides, herbicides and chemical fertilizers have caused adverse changes in biological balance in the soil and led to the rapidly exhaustion of the wonderful capital resource in the soil in the form of its biotic wealth left to us through the ages of natural farming.

The goal of efficient agricultural system should be to develop agro-ecology with minimum dependence on high external inputs (including synthetic fertilizers, agrochemicals and energy) in which interactions and synergy among biological components provide the mechanisms for the system to function in its own way making additions to soil fertility and crop production. Management practices such as zero-tillage, crop rotation, use of plant litters and organic manures possess the capacity to change the habitats of the soils and the food web and improve soil quality.

The judicious use of land, air and water resources which ultimately provide sustenance to plant, animal and human health is largely dependent upon maintaining the health of the living biota which regulate critical physico-chemical processes and ecological services. The current technological and developmental efforts that support increased agricultural production have largely ignored this living critical and most, vital agricultural components in the soil. The improved management of soil biological diversity could play a major role in maintaining soil quality and in achieving the goals of sustainable land and natural resource management for increasing crop productivity and assuring food security to the society.

Sustainable agriculture can be viewed as ecosystem management of complex interactions among soil, water, plant, animals, climate and ultimately people. The prime goal is to integrate these entire factors into a production system that can satisfy human needs while maintaining or improving environmental quality and conserving natural resources for coming generations. Improving agricultural sustainability requires, alongside the effective management of water and crops, the optimal exploitation and judicious management of soil fertility and soil biodiversity. This largely demands for the widespread adoption of management practices that enhance soil biological activity and build-up long-term crop productivity and soil health.

Keeping in view the facts that the abruptly adopted agricultural practices such as excessive ploughing, poor vegetation cover, low organic matter content and excessive addition of synthetic chemicals reduce the capacity of the soil to perform in natural manner, there is a recognized need to blend experiences and the ideas on the management of soil biodiversity in agro-ecosystems. Nevertheless, the fundamental role of soil biodiversity in generating efficient agricultural system that can lead to sustainable crop productivity and in improving human health is still a largely neglected issue in the majority of environmental initiatives.

Managing microbial biodiversity in the soil will–

1. Enhance overall nutrient status of the soil to produce healthy plants.
2. Eliminate undue reliance on costly and often risky external agricultural inputs.
3. Represent a significant biological control potential.
4. Eliminate unfavourable environment within the soil for crop production due to chemical inputs.
5. Produce healthy food in healthy environment for healthy people.

Soil organisms also play an important role in controlling plant pathogens and therefore, represent significant biological control potential. There exist opportunities to develop viable, effective and economically feasible pest/ disease controlling strategies through the introduction of antagonists or by managing indigenous soil microorganisms to increase their activities. However, the efforts to manipulate and exploit their potential for crop benefits must be compatible with the existing microbial symbionts, plant growth promoting rhizospheric bacteria, beneficial fungi, mycorrhiza and cyanobacteria that are being promoted for the motive to support plant growth as well as plant protection.

Managing soil biological diversity in such a way that could lead both to plant growth and protection is clearly an area with great opportunities for obtaining sustainable productivity in future.

Challenges of Managing Soil Biodiversity

The issues of economics, cost-benefits and livelihood usually drive the motive of farming communities behind crop production. In general the major concerns with the communities that perform farming lies with sowing crops as per seasons, land management issues including irrigation and water availability external nutrient inputs, access to the sources of fodder, fuel and feeds. They hardly pay attention to the other important issues such as control of soil erosion and land degradation, desertification and climate changes including air, water and soil pollution. Among an, the issues like maintenance of soil fertility, management of soil biological diversity and conservation of natural resources is of least concern.

Soil biota retains the capacity to increase or decrease agricultural productivity depending on its nature, composition, diversity and different biological activities. Farming practices also modify the variety of life in the soil and thereby associated biological processes. It can therefore be realized that cultivation practices being adopted by the farmers can be beneficial or detrimental to the soil biota, its functioning and its regenerative capabilities. Creating a more favourable environment within the soil for soil biological communities is basically a 'state of art' and a scientific approach and a challenge for not only farming communities all over the world but for scientists also.

It has now been greatly realized that the global solutions to the problems created due to the intensive or industrial farming practices rely in developing farming practices that conserve soil biodiversity. Exploiting soil organisms for future benefits, as is being suggested again in present agriculture is also not a proper solution. Available scientific literature and ongoing research work indicates the reduction/loss soil biodiversity and its functioning in different agricultural systems but, largely under controlled research conditions. Such works have largely been driven both as a matter of pure research or by commercial or private sector interests and completely lack the interests of poorer, small farm holder farmers' needs and ultimately, the national interest. This is why such researches have not significantly been applied or even when applied, made little success, for the management of soil biological diversity in farmer's fields. Applying any research model that has been carried out in small boundaries or under controlled conditions in laboratories, usually speculates about the consequences that would arise due to findings but, its application in open field.

Capturing the benefits of soil biological diversity and its functioning for sustainable, viable and productive agricultural system requires a better understanding of the linkages among the soil life, biological processes, ecosystem functions, and the impact of human interventions. It becomes a challenging task therefore, to manage effectively the complex interactions among soil, plant and animal life, ecological factors and human activities in the form of an integrated system. It is rather more challenging to transform such management strategies into viable practices in the field of farmers in order to obtain beneficial results.

There are reports that farmers allover the world are making their own efforts to manage their farms on the basis of the available indigenous knowledge and utilize some unplanned methods that unknowingly that favour sustainable land productivity to some extend. Unfortunately, these efforts are being made only in unsystematic, unplanned and discrete manner. It is therefore, a challenging task to foster these efforts through scientific as well as network backup and peoples' participation.

Ecological Principles Behind Soil Biodiversity Management

Managing soil biological diversity is basically managing the soil ecosystem in favour of crop production. Therefore, the soil biota and their ecological concerns must be managed effectively in order to yield maximum productivity without disturbing environmental balances. Various factors that involve site-specific decisions on crop selection and rotations, tillage, planting practices, external input application, crop residues, livestock production and grazing influence ecological interactions and ecosystem functioning. People concerned for land management need up-to-date information which enable them to develop management strategies for soil stabilization, nutrient cycling, crop diseases, pest infestations and detoxification of contaminants. These strategies certainly require understanding of the microbial habitats, nutrient sources, host interactions and physical and chemical properties of the soil on soil biota. Ecological concerns regulating growth or suppression of beneficial and/ or detrimental organisms in the soil are needed to be clearly understood in order to harness or control their activities in agro-ecosystem. By doing so, experts may be able to able to promote viable, productive and sustainable agro-ecosystem.

If farmers become able to understand the effects of various farming systems practiced by them on key agricultural soil organisms and their functions, and if they can be become able to know how to observe and assess what is happening in the soil, they can more successfully develop and adopt beneficial practices, or even modify their ongoing practices to yield more benefits. However, besides biophysical factors, socio-economic considerations and ignorance are the other key issues that affect farmers' decision on the aspects of management Certain common constraints that restrict farmers to adopt soil biodiversity management practices include the cost factors for labour, duration of the procedure, monetary inputs, availability of potential planting materials and viable microbial inoculants etc. and the most important one, the social acceptance.

Benefits of Soil Biodiversity Management

Soil organisms can act as the primary driving agents for regulating nutrient cycling, dynamics of organic matter, carbon sequestration and green house emissions and thereby, influencing structure, nutrient status and water regimes in the soil ecosystem. All these effects are essential to the functioning of an ecosystem in natural manner and in a cumulative manner add effectively to plant and animal health and ultimately lead to food security to the human society. The services offered by the soil organisms can lead to fulfill three objectives of the sustainability:

Economic Sustainability

Soil Biological Diversity reduces input costs by enhancing resource use efficiency. Less fertilizer input and less fertilizer loss due to leaching will the direct effects due to an efficient nutrient cycling in the soil. Similarly, improved soil structures also favour availability of water, air and nutrients to the plants. Presence of a diverse set of pest and disease control organisms in the soil reduces the cost of chemical pesticides and fungicides. It is estimated that the value of such 'ecosystem services' obtained by managing soil biological diversity by each year in the agro-ecosystem worldwide may exceed US$ 1,542 billion*.

Social Sustainability

Decisions made on the farms have direct effects to the local communities. Management of soil organisms is an essential part of sustainable farming practices that enhance quality of life of those

----

*Source: Pimentel, D., et al. (1997), Bioscience, 47(11), 747–757.

who work on the farm. Such farms may support other businesses and families in the communities. Management of soil organisms is an integrated approach that involves team working through participatory mode of working. The whole approach is therefore, holistic one improving the quality of life at family as well as community level.

Environmental Sustainability

Many soil organisms detoxify hazardous chemical that cause pollution to the environment. They can also degrade the toxic chemicals *e.g.* pesticides, fungicides etc. or absorb heavy metals that would otherwise become pollutants after reaching groundwater level. The conservation and management of biota can prevent soil and water pollution, land degradation and can also minimize or even eliminate the use of agrochemicals. Excessive reduction in soil biodiversity and the endangering of key species with unique functions may have catastrophic ecological effects ultimately leading to the reduction in agricultural productivity.

Food Security

Managing organisms in the soil has direct positive effect on crop yield and quality of the produce. This comes through the plant growth promoting and pest/disease controlling activities of soil organisms. Below ground biodiversity enhances resource use efficiency, sustainability and resilience of low-input agro-ecosystems, which is highly desirable in order to ensure food security and healthy food in every community, especially poorer, in almost an parts of the world. At the same time, certain soil and aquatic organisms are a major source of proteins and other nutrients to poor communities.

Management and conservation of soil biotic environment and its diversity contributes to the utmost need of farmers, especially in maintaining or even enhancing productivity and minimizing labour costs and other external inputs. At the same time, maintaining a healthy ecosystem in terms of

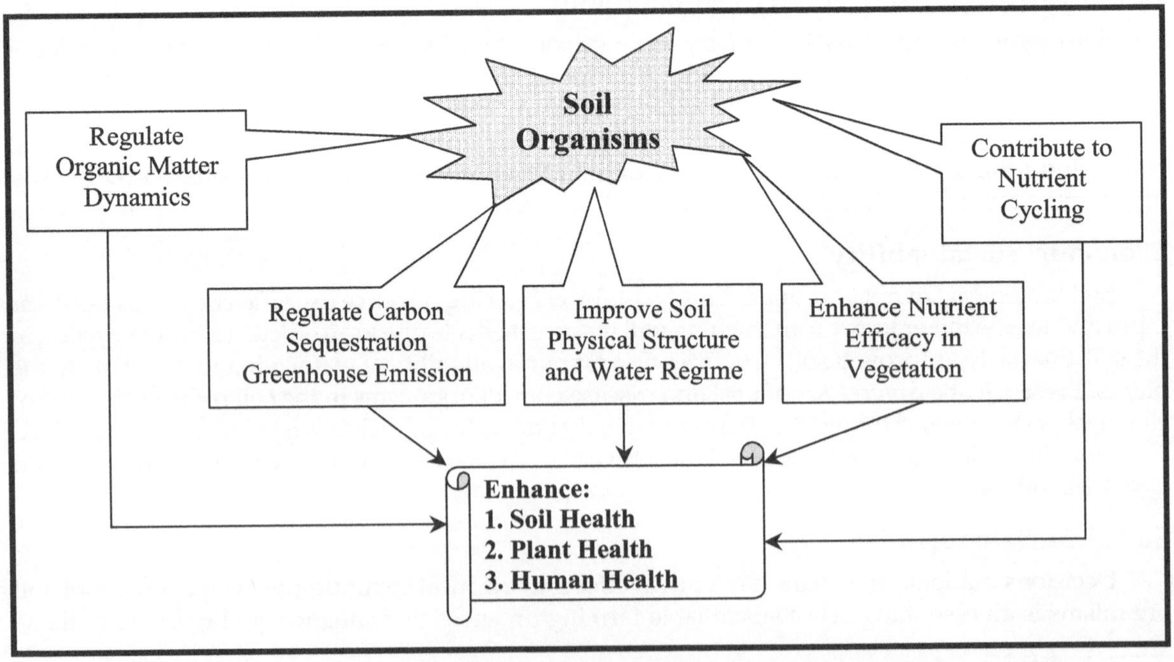

Figure 7.1: Schematic Presentation: Importance of Soil Organisms in Crop Productivity

water quality, prevention of land degradation and protection from soil erosion is in the national interest also. There is therefore, a need to recognize these multiple benefits of managing soil biota and to promote actions in this direction in order to enhance soil biological diversity and its vital and valuable activities.

An overview of potential benefits to nature and human kind due to the management of soil biological diversity and opportunities for exploitation of this relatively ignored research field in favour of sustainable crop productivity has been presented. Working with this area provides potential implications whereby farmers can actually manage soil biodiversity to enhance agricultural outputs while conserving natural resources. However, the adaptation such sustainable agricultural practices and technologies requires approaches with integrated efforts in natural resource and agro-ecosystem management in view of the complexity arising due to the soil biodiversity and the multiple bio-physical interactions. It is also important to mention that each approach and practice towards establishing a sustainable agricultural system has socio-economic as well as technical and environmental implications and only those options that would remain viable with time will gain social and cultural acceptability and be given due interest by the farming communities. The central focus of the investigator therefore, lies in developing such approaches and systems that are closure to both, the nature and to farming communities so that the farming communities balance socio-economics as well as ecological concerns to facilitate better adaptation and adoption of technologies.

Acknowledgement

DPS is grateful to the Department of Science and Technology, New Delhi, Government of India for Young Scientist Award.

References

Bai, Y., Xiaomin, Z. and Donald, L. S. (2003). Enhanced soybean plant growth resulting from coinoculation of *Bacillus* strains with *Bradyrhizobium japonicum*. *Crop Science*, 43: 1774–1781.

Burdman, S., Edouard, J. and Yaacov, O. (2001). Recent advances in the use of plant growth promoting rhizobacteria (PGPR) in agriculture. In: *Molecular Interactions in Agriculture and Forestry*, (Eds.) Subba Rao, N.S. and Dommergues, Y.R. 2: 229–250.

Cattelan, A.J., Hartel, P.G. and Fuhrmann, J.J. (1999). Screening for plant growth-promoting rhizobacteria to promote early soybean growth. *Soil Sci. Soc. Am. J.*, 63: 1670–1680.

Commission on Sustainable Development. (2000). United Nations: Economic and Social Council. Multi-stakeholder dialogue on sustainable agriculture (Note by Secretary-General), 8th Edition, 24 April–5 May, pp. 01–18.

DaSilva, E.J., Baydoun, E. and Bedran, A. (2002). Biotechnology and the developing world. *EJB Electronic J. Biotechnology*. ISSN: 0717-3458. Vol. 5(1) of 15 April, 2002, (PDF version).

FAO. (1996). *Success Stories in Food Security*. Vo1. 1. http:/ /www.fao.org/docrep/003/w2612e/ w2612e2a.

Neyra, C., Atkinson, A., Olubay, O., Sadasivan, L., Zaurov, D. and Zappi, E. (1997). Novel microbial technologies for the enhancement of plant growth and Biocontrol of fungal diseases in crops. CIHEAM-Options Mediterraneennes, 31: 447–456.

Pimentel, D. (1997). *Bioscience*, 47(11): 747–757

UNDP. (2001). *Agricultural Biological Diversity*, Note by the Executive Secretary, United Nations Environment Program.

World Food Program. (2001). WFP Hunger Map. http://www.wfu.org/countrybrief/hunger map/index.html#.

Zahir, Z.A., Arshad, M., and Williams, T.F. Jr. (2004). Plant growth promoting rhizobacteria: Applications and perspectives in agriculture. In: *Advances in Agronomy*, (Ed.) Donald, L.S. 81: 97–167.

©R.C. Ram & Asha Sinha (*Editors*)
***Microbial Diversity: Modern Trends*, 2007**
Daya Publishing House, New Delhi

103–127

Chapter 8

Role of Microbial Biocontrol Agents in Insect Pest Management

Nitin Joshi* and Manoj Kumar Pandey**

***Department of Entomology & Agricultural Zoology,**
****Department of Mycology & Plant Pathology**
Institute of Agricultural Sciences, Banaras Hindu University,
Varanasi – 221 005, Uttar Pradesh

ABSTRACT

The growing awareness of pesticidal hazards on man and environment, the immediate need for sustainable eco-friendly pest management has been acutely felt, providing a thrust for research and development of biological control tactics. Microbial control is an extension of biological control where only microbes (*i.e.*, virus, bacteria, fungi, nematode, protozoa) are employed to manage the pests. The microbial control of insect pests is directly related to human welfare and thus has attracted the attention of mycologist and entomologist in recent years. As a result, global interest has now been created in the development and use of microbial insecticides. This approach is sustainable, energy saving, non-polluting and ecologically sound. The major areas that offer opportunities for enhancing the impact of augmentation biocontrol agents are the utilization of their biodiversity, improvements in mass production and quality control besides promoting their integration with other components of Integrated Pest Management (IPM). Besides this, the use of biotechnology for insect pest management has produced a technological approach to IPM. For instance, if sustainable pest management is the ultimate target, we should perhaps be looking at way in which biotechnology could increase the persistence and numerical response of biocontrol agents in the fields, to make their effects more long lasting. Research should be centered on quantification of entomopathogens, selection of virulent strains, development of economically viable production technology, identification of toxins and their production on different media, standardization, timing and mode of field application.

Introduction

As man aggregated into villages and began to practice agriculture by aggregating his crops as well, he countered increasingly severe attacks by pests against himself and crops. Hence, for his survival, in addition to taking care of himself, he had to take care of plants health also. Even without knowing the causal nature of plant diseases or insects attack, man devised several methods to control them. His need for better methods were intensified during nineteenth and twentieth centuries which eventually led to the use of several organic and inorganic chemicals. These chemicals were indiscriminately used to control the pest, causing adverse effect on the non-target organisms, so that in the momentary span of less than 100 years, some forms of life became extinct, while many became threatened with extinction and the health and welfare of man himself came in dangerous/critical situation.

Insects cause enormous damage to our crops and forest trees. However, there are some friendly insects also like silkworm, honeybee etc. but the majority like borers, beetles, hoppers, moths, termites, aphids, scale insects etc. cause substantial amount of damage to crop plants in field condition as well as in storage godowns. To combat with these insect pest problems, the intensive and extensive use as well as misuse of insecticide took place causing wide spread damage to the environment. Not only that, the use of insecticides of chemical origin has also lead to serious consequence such as:

1. Pest resistance to insecticide.
2. Pesticide residual effect.
3. Pest resurgence.
4. Outbreak of secondary pest.
5. Disturbance in ecological harmony.

These ill effects have brought a complete change in strategy of pest control. The combined impact of all these. problems together with the rising cost of pesticide, provide the necessary feed back for control strategy, and led to development of the judicial and sound pest management programme, which is socially acceptable, economically feasible and ecologically sound to keep the pest population below Economic Injury Level (EIL). In quest of elimination/reducing these deficiencies or dangers of chemical crop protection, plant scientists developed an alternative–the biological control, a form of natural control.

Biological control can be defined as "The study and utilization of parasitoids, predators and pathogens for the regulation of pest population densities". In other words, biological control means reduction of the amount of inoculum of disease producing activity of a pathogen accomplished by or through one or more organisms other than man. If the organism used in biological control is of microbial type, it is called 'microbial control'. The mechanism generally involves: (*a*) Competition for same nutrients or commodity in limited supply but needed by the pathogen, (*b*) Antibiosis, which results from the liberation of an antibiotic or other chemical by one microorganism and harmful to insect, hyper parasitism, etc. Hyper parasitism indicates feeding of one organism on another, which is already a plant parasite.

Microbial Control of Insect Pests

Continuous search of substitutes for chemical pesticides, drudgery involved in registration of chemical pesticides and biosafety of microbials has made microbials attractive candidates for IPM:

This approach is sustainable, energy saving, non-polluting and ecologically sound. Therefore, global interest has now been created in the development and use of microbial insecticides.

The term 'microbial control' was coined in the year 1949 by Steinhaus. It can be defined as the exploitation of disease causing organism to reduce the population of insect pest below the damaging level. In other words, when microbial organism (pathogen) or their products are employed by man for the control of insect pest in a particular area is referred as microbial control. Microbes such as virus, bacteria, fungi, nematode, protozoa etc. are employed for the management of pest species. Microbial control of insect pests is directly related to human welfare and thus has attracted the attention of scientists involved in plant protection aspect in the last few years. As a result, the growth of microbial pesticides is at the rate of 10–25 per cent compared to 1–2 per cent of the synthetic pesticides throughout the world. However, most of which is dominated by *Bacillus thuringiensis* (*Bt*) products. Viewed in this context, there is a great potential for mass multiplication of other microbial pesticides too. Singh (1996) has reviewed the microbial agents employed for biological suppression. Useful recommendations have also emerged out of intensive research on microbial pesticides and their development as an effective biocontrol agent in field condition against insect pest on various crops (Table 8.1). These biopesticides are applied at the vulnerable/susceptible stages of crop growth to the insect pests, and further application is done after 10–12 days interval unlike chemical pesticides which are usually applied at 15 days interval.

Table 8.1: Field Recommendations of Microbial Biocontrol Agents for Insect Pest Management in Various Crops

Crop	Pest	Bioagent	Recommendation
Sugarcane	Chilo infuscatellus	GV	250 LE/ha and 0.05 per cent sandovit mixed in 200L of water and sprayed on 30th day of crop growth
	Leucopholis lepidophora	B. popillae	0.5 kg/ha for proper distribution of spores
Cotton	Helicoverpa armigera	HaNPV	450 LE/ha at ETL of 6–7 second instar larvae/20 plants, 2–4 sprays needed
	Spodoptera litura	SINPV	450 LE/ha at ETL of 20 second instar larvae/20 plants, 2–4 sprays needed
Tobacco	Spodoptera litura	SINPV	250 LE/ha mixed in 125 L of water, 2–3 sprays (Nursery treatment)
	Helicoverpa armigera	HaNPV	250 LE/ha on the inflorescence, 1–2 sprays
Pigeonpea	Helicoverpa armigera	HaNPV	500 LE/ha at pod initiation and full podding stage
Chickpea	Helicoverpa armigera	HaNPV	250 LE/ha at pod initiation and full podding stage
Mustard	Liapaphis eirsimi	V. lecanii	Spores at 10 × 10^{6}/ml and 0.05 per cent Teepol in water suspension are sprayed
Groundnut	Helicoverpa armigera	HaNPV	250 LE/ha, 2–4 sprays
	Holotrichia consanguinea	B. popillae	250–450 LE/ha, 2–4 sprays
Tomato	Helicoverpa armigera	HaNPV	250 LE/ha, 2–3 sprays
Coconut	Oryctes rhinoceros	M. anisopliae	Applied as water solution @ 10^{9}–10^{11} spores/ml after onset of monsoon/year
Citrus	Coccus viridis	V. lecanii	Spores at 10 × 10^{6}/ml and 0.05 per cent. Teepol in water suspension and 0.005 per cent quinalphos during the rainy season
Crucifers	Plutella xylostella	E. thuringiensis	0.3–0.5 kg/ha commercial powder/liquids of Bt var. kurstaki

Insect Viruses

Viruses are submicroscopic, intracellular, pathogenic entities. These are pathogenic to arthropods, belongs to at least eleven families. Viruses in the family Baculoviridae are the best known of all the insect viruses because the disease symptoms are easily recognized and they have the potential for development as microbial insecticides. Baculoviruses are double stranded DNA having rod shaped viruses. Important subgroups within the family are Nuclear Polyhedrosis Viruses (NPV) and the Granulosis viruses (GV). Work on insect viruses in India was started in 1968 with the report of NPV (Patel *et al.*, 1968). Since then, studies have progressed rapidly and several viruses reported to occur in insect pests belonging to orders of Lepidoptera, Diptera and Hymenoptera (Table 8.2). The Biocontrol Unit at Marathwada Agricultural University, Parbhani, has isolated NPV from *Helicoverpa armigera*, *Spodoptera litura*, *Spilosoma obliqua* and *Achaea janata* and standardized mass production techniques of these viruses (Anonymous, 1992). The studies conducted at various agricultural institutions regarding the feasibility of NPV as biocontrol agent has shown good responses. Sharma *et al.* (1997) evaluated NPV, *Trichogramma chilonis*, monocrotophos (Nuvacron® 36EC) and endosulfan (Thiodan® 35EC) for control of *H. armigera* on chickpea in Madhya Pradesh and. concluded that NPV gave the best control of the pest. Likewise, Neharkar *et al.* (1999) tested the comparative efficacy of insecticides against pod borer complex of pigeonpea in cv. BDN-2. They recorded the lowest pod damage (26.5 per cent) as well as highest yield (18.74 kg ha^{-1}) in plot treated with *H. armigera* Nuclear Polyhedrosis Virus (*Ha*NPV).

Table 8.2: The Major NPVs and GVs Recorded in India

Pest	Crops
	NPVs
Helicoverpa armigera	Pigeonpea, Chickpea, Cotton
Spodoptera litura	Tobacco and others
Spodoptera exigua	Tomato and others
Agrotis ipsilon	Potato and others
Plutella xylostella	Cabbage
Corcyra cephalonica	Rice
Mythimna seperata	Rice
	GVs
Spodoptera litura	Tobacco and Cabbage
Plutella xylostella	Cabbage
Agrotis ipsilon	Potato
Corcyra cephalonica	Rice
Chilo infuscatellus	Sugarcane

Mode of Action

It is a highly infective microbial biopesticide which can be used for management of *H. armigera*. The virus should be ingested to produce the disease. Due to alkaline gut juice, the virions are liberated from the polyhedral coat which attack the nuclei of cells of tissues *viz.*, fat body, tracheal matrix, haemocytes, sarcolemma, neurilemma and nerve cells of ganglion and brain. Within 3–5 days of spray, the dead larvae shows typical *Ha*NPV symptoms *i.e.* hanging of diseased larvae upside down. It is specific and selective viral biopesticide and hence kills only the target hosts and not other pests.

The infected larvae become pale and glossy and tissue get disintegrated and liquefied. The liquid which oozes out of the infected larvae (which hang upside down) contains millions of polyhedral occlusion bodies (FOB).

Method of Application

The *Ha*NPV is available only in liquid formulation which is diluted with water at the time of spraying. The spraying of *Ha*NPV is done in evening hours to avoid the harmful effect of Ultraviolet (UV) radiation on it. Tinopal @ 0.1 per cent is mixed in spray solution for protection against UV radiation. Besides, jaggery or sugar @ 0.5 per cent is also added in *Ha*NPV as phagostimulant.

Mass Production of NPV

The commonly employed methods for production of baculoviruses are: (*a*) insectary rearing of host insects on large scale, (*b*) collection of host larvae from infested crop and infecting them in laboratory and field collection of virosed larvae from the infested field after application of *Ha*NPV (Pawar and Thombre, 2001). The authors have developed a technique for rearing of host insects and mass multiplication of baculovirus in the laboratory. The important steps involved in mass multiplication are host insect rearing, infection of the larvae through artificial diet, harvesting, purification of diseased larvae, maceration, decantation, purification, standardization, formulation and storage of *Ha*NPV.

Advantages

1. NPV is class-specific *i.e.,* it is effective against the target pest only.
2. There is no waiting period between time of application and picking of fruits/pods etc.
3. The insects affected by NPV are also not considered dangerous to birds or other animals that may feed on the dead insects.
4. NPV is not known to cause injury to plants on which it has been applied and is also safer to the environment.
5. The treated insects have not yet shown any type of resistance against NPV.
6. NPV treatment can be integrated well with other control methods.

Disadvantages

1. NPV acts as a stomach poison, consequently good coverage is essential.
2. It is only effective against larval stages of Lepidoptera feeding on exposed plant surfaces. This limits its usefulness against insect pests that tunnel into the plants or plant parts.
3. Frequent applications are required compared with chemical insecticides as it has short residual effect.
4. NPV is also susceptible to Ultraviolet (UV) rays of sunlight.
5. Storage of NPV should be in cool and dry place away from direct sunlight *i.e.,* should be kept in refrigerators, whose availability is difficult one especially in villages.
6. The quality maintenance of NPV is a difficult task to achieve.

Bacteria

Several species of bacteria are associated with insects (Table 8.3). Transgenic plants incorporating crystal genes of *Bt* into their DNA so far have been created transgenic cotton, cowpea, tobacco, tomato, potato and cabbage have been developed. Several commercial products based on *Et* are also available which are quite promising, *e.g.* Dipel, Bactospeine, Delfin, Thuricide etc. Puntambekar *et al.* (1997) revealed that *E. thuringiensis* sub spp. *kurstaki* (NCIM 2514) at 10^8 spores/ml concentration caused more than 85 per cent mortality to the neonate larvae of the lepidopteran insects *Spodoptera litura* and *Phthorimaea operculella.* This strain at 10^{10} and 10^8 spores/ml concentration was also effective against *H. armigera* and *E. atomosa* in Maharashtra. They also recorded at least 1.5 times more grain yield than the untreated control. Prabhakara and Srinivasa (1998) evaluated the *Bt* formulations Biobit, Centari, Dipel and commonly used synthetic insecticides against *H. armigera* in pigeonpea. The results revealed that *Bt* formulations caused 58.72 per cent mortality of third instar larvae one day after application, while endosulfan and methomyl accounted for 82–90 per cent mortality. However, the residual activity of *Et* formulations decreased more rapidly compared to that of these synthetic chemicals. Nagrare and More (1998) revealed that *E. thuringiensis* subspp. *kurstaki* and *Ha*NPV alternated with *B. thuringiensis* subspp. *kurstaki* against *H. armigera* in Maharashtra were found to be effective with cost: benefit ratios of 1 : 8.80 and 1 : 4.95, respectively. Similar results were obtained by Kulat *et al.* (1999). They reported *B. thuringiensis* var. *kurstaki* @ 750 ml ha^{-1} to be most effective in minimizing larval population of *H. armigera* on chickpea in Maharashtra.

Table 8.3: Target Pests and Crops of *Bacillus thuringiensis*

Target Pest	Crop
Plutella xylostella, Crocidolomia binotalis, Pieris brassicae	Crucifers
Helicoverpa armigera	Tomato, Sunflower
Spodoptera litura, Helicoverpa armigera	Tobacco
Spodoptera litura	Beet root
Athalia lugens proxima	Radish
Papulo demoleus	Citrus
Achaea janata	Castor
Amsacta moorei, Helicoverpa armigera	Pulses
Acigona steniellus	Sugarcane
Earias vittella	Okra
Spoladea (= Hymenia) recurvalis	Amaranthus
Phthorimaea operculella	Potato

Mode of Action

They enter the body cavity through contaminated food and multiply. The other avenues of infection are injured integument of the insect due to mechanical damage, pricking of parasitoids and infections caused by protozoa or nematodes. Death occurs due to intoxication, sudden lack of oxygen, chemical changes in the gut or by toxins or crystals produced in the bacterial cells.

Method of Application of *Bt*

Bt can be applied as aerial spraying (forest crops), water treatment by aerial or ground equipment (vector control). Soil application by drip or overhead irrigation systems (field, fruit, and vegetable crops) and foliar application by spraying from knapsack sprayer (field, vegetable, and fruit crops) can be done. Wetting agents, surfactants, and stickers may be added to enhance effectiveness under some conditions.

Mass Production of *Bt*

Of the several entomopathogenic bacteria, the most common one *i.e., Bt* is commercially produced through fermentation technology. It can be easily produced on artificial media by adopting conventional fermentation, surface semisolid fermentation or submerged fermentation (Anonymous, 1993). Non-spore forming bacteria are eliminated by heating at 60°C for 15 minutes, growth of other spore formers is reduced by the addition of 50 g/ml polymyxin B to the nutrient agar medium. *Bacillus cereus* colonies are excluded by leaving colonies devoid of parasporal bodies.

Advantages of *Bt*

1. Different strains of *Bt* are class-specific *i.e.*, beneficial or non-target insects are not harmed. This high margin of safety recommends its use on crops where chemical pesticide use can cause adverse effects.
2. *Bt* is essentially non-toxic to humans, pets and wildlife.
3. There is no waiting period from time of application before re-entering the field and can be integrated well with other control methods.
4. The insects that ingest *Bt* and later die from it, are not considered dangerous to birds or other animals that may feed on those dead insects.
5. *Bt* is not known to cause injury to plants on which it has been applied and is not considered harmful to the environment.

Disadvantages of *Bt*

1. *Bt* acts as a stomach poison, consequently good coverage is essential as it must be eaten to be effective.
2. It is only effective against immature stages of insects of Lepidoptera, Diptera, and Coleoptera feeding on exposed plant surfaces. This limits its usefulness against insect pests that tunnel into the plants or plant parts.
3. It must be applied like a chemical insecticide as it lacks transmission and spreading mechanism/facilities.
4. Frequent applications are required compared with chemical insecticides due to short residual effect and most formulations persist on foliage less than a week following application.
5. The highly specific activity of *Bt* insecticides might limit their use on crops where problems with several pests occur, including non-susceptible insects like aphids, grasshoppers, etc.
6. *Bt* based products tend to have a shorter shelf life (one to two year) than other insecticides and storage should be in cool and dry place away from direct sunlight. The liquid formulations being more perishable than dry formulations.

Bt Products/Formulations

Available in different forms such as Water dispersible granule, dry flowable, aqueous suspension, granule, technical power, dust, wetable powder, emulsifiable suspension, aqueous flowable, bait and oil flowable. Manufacturers are experimenting with several techniques to increase the longevity of *Et* after application. One of these involves inserting *Et* toxic crystal genes into other species of bacteria which can better survive on leaf surfaces (*e.g.* M- Trak formulation of san diego strain). The agricultural products based on *Bt* subsp. *kurstaki* are Biobit @ Flowable Concentrate: *Bt* (15 per cent) and inerts (85 per cent); Biobit @ Wetable Powder: *Bt* (29 per cent) and inerts (71 per cent); Dipel @ 2X: *Bt* (6.4 per cent) and inerts (93.60/0); Dipel @ 4L: *Bt* (1.76 per cent) and inerts (98.24 per cent); Dipel @ ES: *Bt* (3.5 per cent) and inerts (96.5 per cent); Javelin @ Thuricide @ 328 and Thuricide @ 32LV: *Bt* (1.6 per cent) and inerts (98.4 per cent); Javelin @ WG: *Bt* (6.4 per cent), dispersing agent (15.0 per cent) and fermentation residue (78.6 per cent).

Fungal Pathogens

The term entomogenous fungi is derived from Greek word *Entomon* means insects and *Genes* means arising in *i.e.,* fungi arising in insects. An entomopathogenic fungus means fungi which cause disease in insects. So, these two have almost same meaning or synonymous to each other. Association of fungi with insects was noted long back in history. Insect disease and their symptoms have been recognized as far as back 2700 B.C. in China. Ancient Indian literature also refers to disease of honeybees and silkworm. In early 1835, Agostino Bassi popularly known as "Father of Insect Pathology" demonstrated the infectious nature of fungus *Beauveria bassiana* (the muscardine fungus) in silkworm and first suggested that an infectious microorganism might be used to control the insect pest. The fungal disease on insects are referred as mycosis. Fungi infect insects of most of the order such as Lepidoptera, Diptera, Coleoptera, Hymenoptera, Hemiptera, Isoptera, Orthoptera, etc. In some insect orders, nymphal or larval stages are often infected than adult stage. However, most of the entomopathogenic fungi attack adult stage of the insect. The characteristics of ideal entomopathogenic fungi are:

1. Good infection capability.
2. High post-infection virulence.
3. Adequate survival potential.
4. Abundant occurrence in nature.
5. Efficient dispersal mechanism.

Entomogenous fungi belong to the classes Phycomycetes, Ascomycetes, Basidiomycetes and Fungi imperfecti. The entomopathogenic fungi, *Beauveria* and *Metarhizium* are important because of their ability to infect a wide range of insects (Table 8.4). *B. bassiana* is best known as the causal agent of the disastrous muscardine in silkworms. It is the most widely distributed species of the genus in both temperate and tropical areas. Survival in soil is favored by darkness and reductions in both temperature and soil moisture; at 8°C and under dry conditions, 90 per cent of the conidia survive for over 635 days (Clark and Mandelin, 1965); on silica gel conidia survived up to 36 months of storage at –20°C (Bell and Hamalle, 1974). After spraying on the crops, conidia survival only 3 days on the same level of germination before application. Biological control experiments with it are numerous and have been directed particularly to the colorado beetle and other Coleoptera and Lepidoptera. Various species of mosquito larvae can be killed by dusting conidia on the water surface (Clark *et al.*, 1968); conversely, bees larvae were not susceptible. *B. bassiana* is employed on a large scale in the People's Republic of China to control pine caterpillars (*Dendrolimus punctatus*), green leaf hopper (*Nephotettix* spp.) and corn borer (*Ostrinia nubilalis*). *Verticillium lecanii* is a well documented entomopathogen of insect order Homoptera, most commonly aphids, scale insects and whiteflies in tropical and subtropical regions. Also, *V. lecanii* sometimes hyperparasitizes phytopathogenic fungi, mostly rusts and powdery mildews. The entomofungal pathogens with commercial potential are listed (Table 8.5).

Table 8.4: Important Entomopathogenic Fungi Recorded from India

Species	Insects Attacked
Beauveria bassiana	Helicoverpa armigera, Spodoptera litura, Chilo partellus,.Scirpophaga incertulas, Myllocerus spp.
Metarhizium anisopliae	Pyrilla perpusilla, Oryctes rhinoceros, Holotrichia consanguinea
Verticillium lecanii	Coccus viridis, Archips termias, Empoasca spp., Aphis gossypii, Tetranychus spp.
Verticillium aphidicola	Aphis spiraecola, Lipaphis erysimi, Brevicoryne brassicae, Myzus persicae
Nomuraea rileyi	Helicoverpa armigera, Spodoptera litura, Trichoplusia spp.
Hirsutella spp.	Nilaparvata lugens, Tetranychus spp. Erynia neoaphidis Lipaphis erysimi
Erynia neoaphids	Lipaphis erysimi
Paecilomces farinosus	Bemesia tabaci, Aleurocanthus woglumi

Table 8.5: Commercial Scale Production of Some Important Mycoinsecticides in the World

Fungus	Brand Name	Target Insect	Crop	Producer Country
Beauveria bassiana	Mycotrol	Whiteflies/Aphids/Thrips	Field Crops	Mycotech, USA
B. bassiana	Botanigard	Whiteflies/Aphids/ Thrips	Glass house tomatoes/Ornamentals	Mycotech, USA
B. bassiana	Naturalis	Sucking insects	Cotton, Glass house crops	Troy Bioscience, USA
B. bassiana	Conidia	Coffee berry borer	Coffee	Agr Evo, Germany
B. brongniartii	Betel	Scarab beetle larvae	Sugarcane	NPP (Calioppe), France
Metarhizium anisopliae	Bio-1020	Black Vine Weevil	Glass house Ornamental crops, Nursery	Bayer, Germany
M. anisopliae	Bio-Blast™	Termites	Houses	Eco Science, USA
M. flavpoviride	Green Muscle	Locusts, Grasshoppers	–	CABI, UK
M. flavpoviride	Bio Green	Red-headed cockchafer	Pasture/Turf	Australia
Verticillum lecanii	Vertalec	Aphids, Whiteflies and thrips	Glass house crops	Koppert, Holland
Paecelomyces fumosoroseus	PFR-97™	Whiteflies/Thrips	Glass house crops	Thermo Trilogy, USA
P. fumosoroseus	Preferal	Whiteflies		Biobest, Belgium

Mode of Action

Fungi do not have to be ingested to be infective; rather they are able to directly infect insects via the cuticle. This "contact activity" makes them unique among insect pathogens and they have a demonstrated capability to infect and suppress piercing-sucking insects. The "active ingredients " in fungal preparations are spores, also called conidia. These must contact the host to be effective, either directly at the time of spraying or later as the host moves over treated foliage. Good spray coverage is, therefore, essential. Once the spores are attached to the insect, they germinate and pierce the insect's body wall. In early stage of fungal infection, few necrotic spots develop at the invasion sites. Later on, insects become restless, sluggish, loose appetite and. coordination, malformed adult with reduced vigour, fecundity and longevity. Prior to death, internal tissue shows disintegration. The other diagnostic feature is the external fungal growth of mycelium and spores covers the host body with different colours and the disease is accordingly known as white, green, brown, red, yellow and even black muscardine.

M. anisopliae was found to be a more aggressive pathogen than *V. lecanii*, perhaps due to subtle differences in the mode of infection. *V. lecanii* often colonizes the host surface before invading the body, whereas *M. anisopliae* appears to penetrate and infect the insect soon after germination. Temperature influences infection and different pathogens work best at different temperatures. *M. anisopliae* strains were most virulent at 73°F (= 23°C). A drop in temperature of just 5–9°F (= 3–5°C) increased the time of death by one day, which could be critical in heavily infested greenhouses. Not all wood feeder thrips (WFT) developmental stages are equally susceptible to fungal infection. Eggs, for example, are laid in host plant tissue and are thus protected from contact with the pathogen. Larval and pupal stages are more resistant than adults, presumably because fungal inoculum is shed when the insects molt. Molting is an important factor in insect resistance to infection, especially in an insect like WFT in which the time between molts is short. Pupae are more susceptible to infection by *M. anisopliae* than are larvae. Adults are the most susceptible stage; however, infected adults continue to lay eggs for a few days. Re-treatment of an infested crop is, therefore, *V. lecanii* has been used to successfully control WFT on chrysanthemums and cucumbers.

Method of Application

High volume spray applications that produce a fine mist with good leaf wetting (but not to run-off) give better levels of control than low volume electrostatic sprays. Tank mixes of fungal biopesticides and other pesticides should only be prepared according to the manufacturer's guidelines because some materials are harmful to fungi. Some older insecticides are inhibitory, and if used should be applied separately. Fungicides are the most inhibitory, and need to be used with caution in any spray program in which beneficial fungi are used. However, fungi may be sprayed successfully either 48–72 hours before or after fungicides have been applied.

Mass Production of Fungi

Fungi can be mass multiplied on solid or liquid media. Coconut water and organic waste in copra industry have been used for production of *M. anisopliae* for control of red palm weevil. Fungal pathogens could be cultured on crushed moist sorghum grains or carrot broth. The pure culture of the fungus is maintained on PDA + 1 per cent yeast slant in tubes. Production may be with crushed grains or coconut water. However, there are some differences in the production procedure of some of the important fungi used in biological control.

Small Scale Production of M. anisopliae

M. anisopliae is possible to produce like local product in plastic fermenters. The isolate *M. anisopliae* must be passed through target host and then be cultivated on Sabouraud's liquid medium. As far as the concentration of conidia was very high and reached after 8 days 3.19×10^{10} conidia/g. Light and darkness have not any decisive effect on mass production of conidia. The temperature is most important for the yield of conidia. At first 6 days is the best temperature 24–25°C and then it is possible to decrease temperature to 22–20°C. Submersed culture produce through 3–4 days blastospores, which are suitable like seed culture with medium for plastic fermenters.

Harvesting process must be done by mixing it with Tween–80 or Triton–X–100 and then is used Siloxyd 125–150 g per litre of mud and dry during 24 hours. Formulation WP is suitable for conidia *M.*

anisopliae and shelf life during storage is good at 4–10°C. For standardization, 100 g of conidia powder (2.5×10^{10} conidia/g) must be mix with Siloxyd and diluent, so that the total weight will be 500 g. The recommended dose per ha is in this case is 500 g of formulated product (*e.g.* 2.5×10^{12} of conidia) which results in 55–80 per cent mortality after 2–3 applications.

Small Scale Production of B. bassiana

On agar media, conidiogenesis starts after six days, while in liquid's culture this takes only 3–4 days (Samsinakova, 1966). In stirred liquid cultures employed in mass production of this fungus, so-called "blastospores" develops, which are thin-walled, larger (3–5 × 2–3 mm) and less resistant than conidia (Mueller-Koegler and Samlioakovci, 1970); blastospores germinates in 6–10 hours at 18–24°C, while conidia require 15–20 hours. The germination of conidia requires a saturated atmosphere and the optimal temperature for growth is 25–30°C, minimum 10°C, and maximum 32°C apparently depending on the geographic origin of the isolate; no germination occurs either below 10°C or above 35°C. The thermal death point of conidia has been determined as 50°C for 10 minutes in water. The optimal pH for growth is 5.7–5.9 and 7– 8 for conidia formation. Good growth occurs on maltose and sucrose and, among others, on the nitrogen sources glutamic and aspartic acids, and ammonium oxalate, citrate or tartrate. The 2–3 times application of its formulation product @ 2×10^6 to 2.8×10^{12} spores per ha results in 49–85 per cent mortality of targeted insect pest.

Small Scale Production of V. lecanii

Laboratory culture is possible on all conventional mycological media, but the best growth is on Sabouraud's agar (Khalil *et al.*, 1983). The polyethylene cushions made of large thin walled, polyethylene tubing sealed into sections which were partially filled (1 cm high layer absolutely horizontal) with submerged culture of *V. Lecanii* after 2–3 days cultivation and inflated with sterile air. Product was harvested after 14–16 days by discarding the medium and retaining the mud. Yield from a 0.8 per cent peptone, 1 per cent sorbitol medium was 1.0×10^{13} conidia/m^2. The process of harvesting is similar to that of *M. anisopliae* and the product is packaged like water dispersible powders which is prepared for dilution with water into a final spray. The optimum dose of 2×10^7 to 2.8×10^9 spores per ha sprayed 2–3 times results in 73 to even 100 per cent mortality of the target insect pest.

Advantages of Using Entomopathogenic Fungi

1. No mammalian toxicity or no adverse effect on non-target pest.
2. Simple to produce and easy to apply.
3. High degree of specificity.
4. Absence of development of pest resistance.
5. Relatively cheaper when produce in mass.
6. Compatible with other pesticides.
7. No left over residue in the environment (eco-friendly).
8. Easily disseminable.

Disadvantages of Fungal Pesticide

1. Slow in action.
2. Storage problem.
3. Low adaptation by farming community.

4. Marketability and availability is not sufficient.
5. Effectiveness changes with environmental condition.

Nematodes

A certain group of nematodes that feed on insects by entering the insect body, releasing a lethal bacterium and reproducing in the insect cadaver are called insect pathogenic or entomopathogenic nematodes (EPNs). EPN comprises two words, *i.e.*, Entomo = insect and pathogenic = producing disease. Nematodes associated with insects are referred to as entomophilic, entomogenous or entomopathogenic are known to parasitize, cause disease and kill the insects. They belong to the families Steinemematidae and Heterorhabditidae, are the best candidates for biological control of insects among beneficial nematodes. These nematodes are symbiotically associated with the bacteria *Xenorhabdus* and *Photorhabdus* spp. which makes them highly virulent particularly against lepidopterous and coleopterous pests.

A large number of companies are in the commercial production of EPN. In USA, *Steinemema carpocapsae* EPN is being used against large number of pests belonging to Coleoptera, Lepidoptera and Diptera etc., those of white grubs, mole crickets, root weevils, bugs, borers on variety of crop plants and turfs. In UK, black vine weevil, strawberry root weevil are being controlled by s. *carpocapsae*, S. *riobrave* and *Heterorhabditis* spp. Four nematode species *viz.*, *S. carpocapsae*, *S. riohrave*, *S. feltiae* and *H. bacteriophora* are nematode industries success stories. Difference in the field efficacy for the two EPN *Steinemema* and *Heterorhabditis* were validated, *Steinemema* was found to be more effective against mole crickets whereas the latter was found very effective against white grubs. Besides, they hold promise against household pests, sucking pests and those of public and veterinary significance. Most of EPN producing companies are located in western world. The ease of use out in nematode products is constantly being improved through formulation research. The quality of the nematodes that survive the rigors of manufacturing process is analyzed by determining their shelf life and virulence. Nematode's shelf life is predicted from storage energy lipid reserves of infective juvenile nematodes; virulence being assessed using insect bioassays.

Mode of Action

The EPN naturally associated with specific bacteria are found as free-living in the soil, where they are in a non-feeding juvenile stage. Some species move through the soil in search of a fresh insect larva, while others wait for a larva to crawl by them. The second instar juveniles of nematodes enter the insect larva through the insect's natural openings *i.e.*, mouth, anus, spiracles and some species of EPN have teeth with which they can break through the insect's hard cuticle and enter directly in to the haemocoel of the insect. The non-feeding juvenile nematodes carrying the specific bacterium releases it in the haemolymph which kills the insect larva through septicemia. In the days following infection, the bacteria breakdown the insect into its molecular components so that they and the nematode can feed and multiply. When the nutrients in the insect are used up, the bacteria again take up residence in the nematode, the nematode stops growing at a non-feeding juvenile stage and the nematode-bacteria pair leaves the insect carcass (Sharma *et al.*, 2005).

Method of Application

Sprinkling infective juveniles directly onto the soil surface is the most common method. A spray volume of 750–1900 1/ha is usually sufficient for most nematode species to reach the target insects in soil. Nematodes can be applied with nearly all commercially available ground or aerial spray equipments. They can also be applied using drip and sprinkler irrigation systems. However, best results are obtained when applied after the fields/area are lightly irrigated. Foliar application to control insect pests on the foliage requires considerable efforts, because of rapid desiccation due to wind and solar UV radiation which creates inefficiency of EPNs to stay on the site. However, under favourable conditions *i.e.*, high humidity and calm atmosphere, EPN are very effective against foliar pests. EPN can also be applied as an IPM component to control insects of trap crops and can also been used successfully in traps designed to lure and kill insects (Sharma *et al.*, 2005).

Mass Production of Nematodes

The EPN are amenable for mass production. Mass production of EPN has evolved from large scale *in vitro* solid media (Glaser *et al.*, 1940), *in viva* production (Dutky *et al.*, 1964) and three-dimensional solid media (Bedding, 1981, 1984, 1986). For *in viva* production, infection of *Carcyra* is essential. This is followed by harvesting the nematodes with filter paper in a concave side up watch glass in large petridish. The infective juveniles are aspirated or decanted and then transferred to a storage container at 10–15°C. *In vitro* production technique has been developed by Bedding (1986) to collect large number of nematodes using chicken offal medium on a porous polyether polyurethane foam substrate. The successful mass multiplication of *Steinemema* and *Heterarhabditis* species has been standardized (David *et al.*, 1994). Wouts medium appeared to be best for all isolates followed by egg yolk and dog biscuit medium.

Advantages of EPN

1. EPN are highly virulent, possess chemoreceptors to search out their hosts and can be cultured easily under *in vivo* and *in vitro.*
2. EPNs are safe to vertebrates, plants and non-targets, have been exempted from pesticide registration in the USA.
3. They are easily applied using standard application equipment, are compatible with many chemical pesticides.
4. EPNs are amenable to genetic selection.
5. Insecticidal nematodes are virtually without competition from other biological agents for control of soil inhabiting and plant-boring insects and cost of application equals that of chemicals.
6. Unlike other bioagents, nematodes kill insects in quick span of time *i.e.*, within 24–48 hours.
7. EPNs spray does not lead to any residue problem or groundwater contamination.

Disadvantages of EPN

1. Till date not a single company is producing EPN in India.
2. Shelf life is not uniform and is predicted from storage energy reserves (fatty acids and lipids) of infective juveniles correlated with the type of diet used for production.
3. Storage of EPN is not easy especially of Steinemematid infective juveniles which are maintained at 1–2°C.
4. The identification of most potential strains against local pest problems has to be worked out.

Protozoa

Insects are parasitized by several protozoa, but only Sporozoa and Cnidospora cause death of parasitized host. The infection takes place through infective food and water within 72 hours of ingestion and also by transovarian transmission. Under moist conditions, microsporidia remain viable for 10–12 months. The important protozoan parasites are listed (Table 8.6).

Table 8.6: Protozoan Pathogens of Insect Pests in India

Species	Pests
Vairimorpha spp.	Helicoverpa armigera, Spodoptera litura, S. exigua
Farinocystis tribolii	Tribolium castaneum
Nosema spp.	Anadevidia peponis, Spilarctia obliqua, Adisura atkinsoni, Crocidolomia boinotalis
Mattesia dispora	Ephestia cautella
Bacillidium spp.	Thrips flavus
Microsporidium spp.	Spodoptera litura, S. exigua, Utethesia pulchella
Tetrahymena spp.	Chilo partellus

Mass Production of Protozoa

Large number of protozoan spores can be obtained from field collected insects. For greater certainty of species determination, production in the laboratory is a prerequisite, which is limited by amenability of its host to laboratory culture and our ability to transmit the pathogen to the host. The latter is a problem with many microsporidian spores that are produced in one host but infect an unknown intermediate host. A more convenient alternate host can often be used for production of microsporidia that do not have a high degree of physiological selectivity. Spores can be fed to the host on artificial diet for their natural food. Protozoa can be cultured *in vivo* and *in vitro*. The procedure involves extraction of spores from the host, purification, obtaining aseptic spores, transport and storage.

Augmentation Microbial Biocontrol

Augmentation biological control is a system that seeks to suppress the target pest through enhancing the numbers and/or activity of promising natural enemies in target agro-ecosystems. Very often native species/strains are utilized, as they tend to have potential for local adaptation. The common strategy is to multiply such promising biocontrol agents in large numbers and release them in the target crop system at the appropriate time in relation to pest infestation. Repeated releases or applications of the biocontrol agents or products may often be necessary within a crop season and this constitutes the 'inundative' approach in augmentation biocontrol.

Entomophages like parasitoids on eggs (*e.g. Trichogramma*) and larvae (*e.g. Bracon*) as well as predators (*e.g. Chrysoperla*) and entomopathogens like viruses, bacteria and fungi are potential candidates in augmentation biocontrol and can be integrated with other components of IPM. The potential for augmentation biocontrol appears to be greater in irrigated agriculture, especially in high value crops. The scope offered in horticultural crop systems would merit special focus (Thakur *et al.*, 1991; Sithanantham and Swamiappan, 1998). Augmentation biocontrol is found promising especially in vegetable crops (Vitdav *et al.*, 1985; Diwakar and Pawar, 1987; Ansari *et al.*, 1992). The target pest range of augmentation biocontrol agents should also be expanded wherever possible, so as to expand the demand and investment on mass production. Extending augmentation biocontrol to forest pests (Varma, 1999) and to mulberry crop system (Chandramohan, 1999) are among such possibilities.

The aspects of augmentation biocontrol, promising themes, recent progresses and utilization of the emerging opportunities as relevant to the developing world are emphasized below.

Improved Understanding of the Native Biodiversity of Microbial Biocontrol Agents

It is commonly observed that native species or strains of natural enemies that are deployed for augmentation biocontrol tend to perform better than exotic ones due to their greater ability to adapt to the target ecologies. Further, there are possibilities for identifying intra-species diversity in searching ability and/or efficiency of predation or parasitism among entomophages. For the same cost of mass production, it would be possible to achieve substantial gains in impact, if we can substitute with more adapted or efficient species/strains of these agents.

Focused surveys in contrasting ecologies are useful in discovering the existing native diversity among promising biocontrol agents. The routine approach of targeting biodiversity surveys on the basis of geopolitical demarcations should be replaced by integrating ecological criteria involving geographical information system (GIS). Efforts have been made to assemble strains of baculoviruses (NPV) on *H. armigera,* as well as bacterial pathogens (*Et*) and fungal pathogens (*M. anisopliae*) of importance to *P. xylostella* (ICIPE, 2001).

Adaptation to Physical and Chemical Stresses

Microbial agents such as viruses and bacterial products/formulations are often confronted with inactivation due to ultraviolet (UV) radiation in sunlight, while fungal pathogens may experience stress due to extremes of humidity and temperature. Besides this, all of them suffer from chemical stresses, largely due to pesticides used in the agro-ecosystem and may also encounter stress due to the chemistry of the plant surface on which they are deposited.

Improvements in isolation/identification of strains of *Bt* (Dhandapani, 1999a) and baculoviruses (Rabindra, 1999a) are also important basic themes that would encourage further research. Assessing the scope for genetic improvement of microbials such as *B. thuringiensis* (Dhandapani, 1999b) should be pursued. Fungal biopesticides must be held under appropriate conditions; at all times in transit and in storage on the farm. Fungi are sensitive to high temperatures and even warm conditions can be damaging if prolonged. Generally, if the products are kept cool 60°F (15°C) and dry, quality will be maintained for at least a year. Short-term storage (under two hours) at higher temperatures is unlikely to have a significant effect, but if fungi are held under warm conditions for extended periods, the spores are inactivated and the rate of germination in the survivors declines. In general, the higher the storage temperature, the more rapid the rate of inactivation.

Exploiting the Complementarity in Impact among Microbial Biocontrol Agents

The impact of the biocontrol agents could also be achieved by exploiting their complementarity during deployment. For instance, the debilitating effect of larval parasitism can enhance the rate of mortality caused by baculovirus or *Bt* among caterpillar pests. Synergistic effect of granulosis virus on the biological activity of NPV, as 'enhancer' has been reported recently with the gypsy moth, *Lymantria dispar* L. (Shapiro, 2000). Similar results have been reported with *H. armigera* by Whitlock *et al.* (1996). Granados *et al.* (2001) have demonstrated that 'enhancers' from granulosis virus of *Trichoplusia ni* can augment the toxicity of *Bt* to lepidopterous species. These findings could lead to exploiting the synergistic effect towards improved impact of baculoviruses.

The potential for use of different microbial agents such as *Vairimorpha* (Santharam, 1999a), *Nomuraea* (Vimaladevi, 1999), *Metarhizium* (Sundara Babu, 1999a), *Bacillus popilliae* (Easwaramoorthy, 1999a), *Bacillus thuringiensis* (Santharam, 1999b; Dhandapani, 1999a), granulosis viruses

(Easwaramoorthy, 1999b; Parameswaran, 1999) and other baculoviruses (Santharam, 1999b; Mohan, 1999) and entomopathogenic nematodes (Subramanian, 1999) has been explored and promising results have been obtained. Integrating their use by utilizing their selectivity and complementarily needs to be further pursued (Dent, 1997; Sundara Babu, 1999b). The efficacy range of microbial biocontrol agents can be manipulated by modifications in the formulations. Asano *et al.* (2000) have found that the culture medium of *B. thuringiensis* serovar *japonensis* Buibui enhances the insecticidal activities of the δ-endotoxins from *B. thuringiensis* serovars *kurstaki* and *aizawai* against lepidopterous pest insect pests.

Interaction of Host Plants with Performance of Microbial Biocontrol Agents

Augmentation biocontrol agents often express differential efficacy/impact depending on their interactions on different host plants. These interactions should be well understood, so that their deployment could be selective and visualization of their impact realistic. Even plant parts within the target host plants (pigeonpea and sorghum) have been shown to differ in their effect on parasitism of the same host eggs by *Trichogramma* spp. (Romeis, 1997). Likewise, plant (leat) surface chemistry of host plants has been known to affect the persistence of NPV (Farrah and Ridgway, 2000). Recent studies at ICIPE have shown that different cultivars of cowpea can affect the performance of the fungal pathogen (*M. anisopiiae*) in suppression of the legume thrips, *M. sjostedti* (ICIPE, 1997). Such understanding could help in selective deployment of the bioagent in favorable target crops.

Field Deployment Practices that could Enhance the Performance of Microbial Agents

The field release or application strategies may also be suitably chosen to enhance the performance of biocontrol agents, wherever appropriate. For baculoviruses and *Bt*, use of UV protectants and phagostimulants like jaggery or sugar as spray adjuvants offers scope for enhancing their performance (Rabindra and Jayaraj, 1987). Mixing boric acid 0.025 per cent or Neem Seed Kernel Extract (NSKE) 2 per cent could also enhance the microbial activity of *Bt*. Fungal biopesticides spray solutions should be used immediately because if they are held for several hours under warm conditions, spore viability will decline and the level of control will be reduced.

Integration of Microbial Biocontrol with Other Pest Control Practices

Efforts to incorporate biocontrol as a key component of IPM programmes must be sustained and supported. For example, Sundara Babu and Muthuswami (1995) reported that *B. thuringiensis* var. *kurstaki* (Delfin) combined with half the recommended doses of quinalphos or fenvalerate was most effective in reducing the larval population and damage on chickpea, pigeonpea and cotton. Similarly, Bijjur *et al.* (1994) carried out field experiments for the control of *H. armigera* on sunflower and pigeonpea with *Ha*NPV and showed that 2 sprays in sunflower and 3 in pigeonpea at fortnightly intervals with half the recommended dose of endosulfan + *Ha*NPV @ 250 LE/ha were as airective as endosulfan spray in reducing the larval population and plant damage. The virus infection increased the susceptibility of *H. armigera* to the insecticide, thereby giving better control. The use of baculoviruses and *Et* for management of *H. armigera* in chickpea in India (Lal *et al.*, 1998) in combination with host prant resistance and use of safer insecticides that conserve the entomophages in the crop system (Chandrasekaran, 1999), can also offer a model.

It would be useful to focus on the compatibility of augmentation biocontrol agents with the other crop protection practices. Compatibility between augmentation biocontrol agents and neem products needs to be further studied to promote their integration. Abdul Kareem (1999) has reviewed the scope for integrating neem use with microbial control.

Refinements in Mass Production and Delivery Systems

Back up research in the areas of mass production (Sivasubramanian, 1999b; Kennedy and Sathiah, 1999a; Rajasekaran *et al.*, 1999b) product formulation (Muthuswami, 1999; Kennedy and Santhiah, 1999a) quality control (Kennedy and Sathiah, 1999b; Swamiappan, 1999) and registration requirements (Sathiah, 1999) are also being pursued and need to be strengthened. Standardizing the bioassay procedures for testing the virulence of microbial agents to suit to local requirements (Santharam, 1999b; Rabindra, 1999a; Rabindra *et al.*, 1999) as well as for characterizing the strains (Udayasuriyan *et al.*, 1999; Rabindra, 1999a) to contribute to quality control should receive further emphasis. Refinements in mass production that can enhance the fitness or quality of the entomophages should also continue to receive attention. *In vitro* production of insect viruses (Rajasekaran *et al.*, 1999a) backed up by insect cell culture (Rajasekaran *et al.*, 1999b) should be explored.

Risk Assessment

This aspect is receiving much attention in recent years. Seigel (2001) has reviewed the mammalian safety of *B. thuringiensis* based insecticides. The US-EP A has registered 177 products containing viable *Bt* during 1961–1995 and there were only two recent reports of *Bt* infection in man in recent years. Nevertheless, these two individuals had experienced either extensive burns or blast injury, which predisposed them to *Bt* infection. Increase in human antibody levels following *Bt* exposure has also been reported but with no increase in incidence of asthma or other illnesses. It has been concluded that based on laboratory studies and field experience, *Bt* insecticides have an excellent safety record.

Interaction of microbials with non-target beneficial species, especially the entomophages is currently also being addressed in India (Balasubramanian, 1999; Sivasubramanian, 1999a). The potential for integrating augmentation biocontrol with pesticide use in the target system has received some, but limited attention.

Introduction of Insect Pathogens into Pest Habitat

We should exploit the possibility of inducing local epizootics of insect diseases by attracting the adults to devices that can contaminate them with the particular insect pathogen so that when they return to their breeding habitat, they transfer/transmit the pathogen to other individuals in the population. Jackson *et al.* (1992) has used light traps to attract adult *Helicoverpa* and to contaminate the moths with NPV, with the intention of creating epizootics in local populations. We need to experiment on systems that can enhance the efficiency and cost effectiveness of this approach. Light or food that can help also contaminate the adult females may be worth exploring (Sundara Babu, 1999a).

Biotechnological Approaches for Enhancing Microbial Agents Effectiveness

Advances in biotechnology can help to make IPM strategies more sustainable through increasing effectiveness of microbial agents. Engineering microbials, plants, recombinant DNA technology, genetic improvements of entomopathogenic fungi are some of the methods to enhance the effectiveness of microbial agents.

Biotechnology for plant protection can currently be characterized by three approaches to genetic engineering:

1. The engineering of natural enemies of pests to be more effective.
2. The engineering of plants or their microbials associates with genes that protect the from pests, and

3. The engineering of natural enemies or plants with genes for pesticide resistance, which increase option for pesticide use.

Thus, the engineering of microbial pesticides means for increasing the virulence and engineering the plant associated microorganisms. For example, *Agrobacterium radiobacter* was the first genetically manipulated organism to be released for pest management. Inoculation with this organism can protect a number of horticultural crops from infection by the crown gall bacterium, *Agrobacterium tumifaciens.* Later on, the microbials of carrying groups were identified and engineered to make effective pesticides.

Baculoviruses

Baculoviruses have very slow kill performance, which is characteristic of wild type viruses. Hence, a favourable portion in microbial control with baculoviruses would be to use recombinant baculoviruses which can kill insects more efficiently in shorter time. The baculovirus genome is amenable for genetic engineering and the knowledge of the molecular biology of these viruses has enabled cloning of several proteins of insecticidal value, *viz.*, toxins, juvenile hormone esterase, mellitin, trehalose, fungal insecticidal protease, scorpion and mite toxins into NPV. Some of the genetically engineered baculoviruses has been reported with increased speed to kill such as BmNPV has foreign gene from *Andractomus australis* found 40 per cent increase of speed of killing of *Bombax mori.*

The scorpion venom gene (AalT) has been cloned into the ecdysteroides-UDP-glucosyl transferase (EGT) gene-deleted NPV of *Autographa californica* (AcMNPV). The insect-specific neurotoxin gene Tox 34 from the itch mite, *Pyemotes tritici* has been cloned into the EGT–deleted HzNPV. This recombinant virus killed *H. zea* larvae faster than the wild type virus. The deletion of EGT gene results in early degeneration of malpighian tubules, resulting in increased speed of kill (Flipsen *et al.*, 1995).

The GV of *Trichoplusia ni* (TnGV) produces a protease termed as enhancin, which increases the susceptibility of insects to NPV by breaking down the peritrophic membrane. Such factor has also identified in GVs of *Helicoverpa armigera* (*Ha*GV) and *Pseudaletia unipuncta* (*Pu*GV). The genes encoding for the enhancins of *Tn*GV, *Pu*GV and *Ha*GV have been cloned and sequenced. Enhancin therefore appears to be of potential in genetic manipulations for enhancing the insecticidal activity of baculoviruses.

Bacteria

Efforts to improve *Bt* potency have focused primarily on increasing host range by combining the different genetic forms of the δ-endotoxin, which have different virulence within and between insect orders. The genetic manipulation of cry genes in *Bt* offers a promising means of improving the efficacy and cost effectiveness of *Bt* based bioinsecticide products. Certain combinations of cry proteins have been reported to exhibit synergistic toxicity towards lepidopteran and dipteran pests.

Fungi

Fungi have received little attention with respect to genetic improvement, partially due to the limited development of fungal based biopesticides, but also because the fungal genome is comparatively complex and the genetic basis of virulence and other key properties are poorly understood. Little work has. been done about manipulation of culture media and genetic improvement of entomopathogenic fungi. Genetic modifications in fungi is complicated by the fact that the leading candidates are deutromycetes with no sexual stage. However, chemical mutagenesis, parasexual cycle, protoplast fusion and direct genetic manipulation could be used. For example, the mutants of *M. anisophae* and

Paecilomices farinosus, have been generated which were significantly more virulent at low relative humidity than parentals.

Future Thrust Areas/Researchable Areas

There is a need to encourage mass production and use of NPV's and GV's by the government, public sector cooperatives, non-governmental organizations, private agencies and agricultural universities. This will also create job opportunities for unemployed youths, farmwomen and underemployed farmers in lean periods. The future prospects of *Bt* appear to be bright with the selection of new serotypes with spore negative and crystal positive traits, *in vitro* production of *Bt,* development of cost effective liquid fermentation technique and development of transgenic plants with superior *Bt* toxin. The past has provided appreciable scientific information on entomopathogenic fungi as potential microbial insecticides in the framework of IPM and biological control of agricultural pests. However, much of this work is still at a very seminal stage and a great deal of fundamental work with fungi still remains to be carried on. Till date not a single company is producing EPN in India. So, efforts should be made for their mass production suitable for different strains utilizing indigenous resources and scale up procedures. Research on adaptation of entomopathogens to climatic stresses such as temperature and humidity besides tolerance to sunlight or moisture stress needs to be explored further. Community mass production laboratories for the production and distribution of microbial agents for controlling the insect pests should be provided. Adequate research resources should be allocated to sustain research in biosystematics of the important biocontrol agents as a vital back up to our efforts to utilize biodiversity in natural enemies of importance in augmentation biocontrol. In near future, it seems likely that developments in biotechnology and genetic engineering may be used to manipulate desirable traits and to improve the field activity of the entomopathogens. The following issues are important for making a breakthrough in the use of microbial pesticides:

1. Quantification of microbial bio-diversity.
2. Characterization.
3. Development of virulent strains.
4. Development of economically viable mass production system.
5. Evolution of proper storage techniques.
6. Development of suitable formulation and biosafety.
7. Selection of suitable adjuvants.
8. Creating village level crop protection units for multiplication of microbial pathogens and their distribution to needy farmers on payment.

Efforts should be made to pool the resources and share the expertise by establishing a consortium for reference collections and identification assistance of the biocontrol agents.

Surveys should be backed by insectary services that can provide a cheap supply of uniform and adequate numbers of host/prey so as to help support live cultures of promising species/strains of entomophages discovered. Such repositories can serve as a gene bank and so provide access for drawing from these sources for different research needs.

Conclusion

In spite of great potential of microbial pesticides such as viruses, bacteria, fungal pathogens, entomophilic nematodes and protozoa in managing the insect pests, it is surprising to see that Plant

Pathologists and Entomologists in India have not emphasized this aspect of research which would otherwise have paid high dividends. However, the time has come to renewed interest due to the realization of the interesting role of such organisms in controlling insect pest without causing environmental damage and with economical advantage in comparison to prevailing use of chemicals pesticide. It is true that microbial control of insects will never supersede chemical treatment, but must be exploited as valuable alternative of chemical control. There are some problems with field efficacy and storage of microbial pesticides, the prospect of microbial pesticides seems to be promising. It is also important to strengthen research on the choice of species/strains of biocontrol agents based on their potential for controlling distinct geographical populations of the target pests. Assessment of the compatibility of entomophages to the pesticides commonly used on the target crops can offer improved scope for their integration. Efforts to improve efficiency in mass production and quality control as well as to generate bio-efficacy and biosafety data for facilitating product registration should be intensified. Given the right support and research and development back up, augmentation biocontrol could emerge as a vital component of IPM in many agricultural cropping systems of the developing world. There is need and scope to sponsor a more active partnership among the researchers, extensionists, developmental agencies, private enterprises and the farmers. In developing countries, there could be opportunities to involve communities in promoting the utilization of biocontrol agents. At last it can be said that "Adopt microbial control today for a better tomorrow".

References

Abdul Kareem, A. (1999). *Integration of botanicals in microbial control technology*. In: *Emerging Trends in Microbial Control of Crop Pests,* (Eds.) R.J. Rabindra, G. Santharam, N. Sathiah and J.S. Kennedy. Tamil Nadu Agricultural University, Coimbatore, India, pp. 173–180.

Anonymous. (1992). Final Technical Report of an ad-hoc Indo-US Project on *Microbial Control of Insect Pests of Pulses, Oilseeds and Vegetables.*

Anonymous. (1993). *Production Technology of Natural Enemies*. Project Directorate of Biological Control, Bangalore, India.

Ansari, M.A., Pawar, A.D. and Arunkumar, D. (1992. Possibility for biocontrol of tropical armyworm, *Spodoptera litura* on potato. *Plant Protection Bulletin,* 44: 27–31.

Asano, S., Ogiwara, K., Takahashi, M. and Indrajith, S.L. (2000). Culture medium of *Bacillus thuringiensis* serovar *japonensis* Buibui enhances the insecticidal activities of δ-endotoxins from *B. thuringiensis* serovar *kurstaki* and *aizawi* against lepidopterous pest insects. *Applied Entomology and Zoology,* 35(4): 575–582.

Balasubramanian, G. (1999). Interaction of microbials with entomophages, In: *Emerging Trends in Microbial Control of Crop Pests,* (Eds.) R.J. Rabindra, G. Santharam, N. Sathiah and J.S. Kennedy. Tamil Nadu Agricultural University, Coimbatore, India, pp. 185–190.

Bedding, R.A. (1981). Low cost *in vitro* mass production of *Neoplectana* and *Heterorhabditis* species (Nematoda) for field control of insect pests. *Nematologica,* 27: 109–114.

Bedding, R.A. (1984). Large scale production, storage and transport of the insect parasitic nematodes *Neoaplectana* sp. and *Heterorhabditis. Annals of Applied Biology,* 101: 117–120.

Bedding, R.A. (1986). Mass rearing, storage and transport of entomopathogenic nematodes. In: *Fundamental and Applied Aspects of Invertebrate Pathology,* (Ed.) R.A. Samson *et al.* Vepdhoven, Netherlands, Proceedings of the 4th International Colloquium on Invertebrate Pathology, pp. 308–311.

Bell, M.R. and Hamale, R.J. (1974). Viability and pathogenicity of entomogenous fungi after prolonged storage on silica gel at –20°C. *Canadian Journal of Microbiology,* 20: 639–642.

Bijjur, S., Kulkarni, K.A. and Lingappa, S. (1994). Utilization of NPV of *Heliothis armigera* (Hübner) management in sunflower and pigeonpea ecosystem. *Indian Journal of Entomology,* 56 (3): 241–245.

Chandramohan. (1999). Prospects of microbial control in mulberry, In: *Emerging Trends in Microbial Control of Crop Pests,* (Eds.) R.J. Rabindra, G. Santharam, N.I Sathiah and J.S. Kennedy. Tamil Nadu Agricultural University, Coimbatore, India, pp. 83–87.

Chandrasekaran, S. (1999). Insecticide testing for toxicity to natural enemies. In: *Ecology Based Pest Management.* (Eds.) R.J. Rabindra, S. Palaniswamy, P. Karuppuchamy, K. Ramanaju and R.P. Sridhar. Tamil Nadu Agricultural University, Coimbatore, India, pp. 241–243.

Clark, T.B., Kellen, W.R., Fukuda, R. and Lindegren, J.E. (1968). Field and Laboratory studies on the pathogenicity of the fungus *Beauveria basiana* to three genera of mosquitoes. *Journal of Invertebrate Pathology,* 11: 1–8.

Clark, T.B. and Mandellin, M.F. (1965). The longevity of conidia of three insect hyphomycetes. *Transaction of British Mycological Society,* 48: 193–209.

David, H., Easwaramoorthy, S. and Rajaguru, P. (1994). Entomophilic nematodes for biological control of crop pests in India: Retrospect and prospects. In: *Biological Control of Insect Pests,* pp. 53–66.

Dent, D. (1997). *Bacillus thuringiensis* as an insecticide. In: *Exploitation of Microorganisms,* (Eds.) D. Gareth Jones. Chapman and Hall, London.

Dhandapani, N. (1999a). *Bacillus thuringiensis* toxin binding sites and insect resistance management. In: *Emerging Trends in Microbial Control of Crop Pests,* (Eds.) R.J. Rabindra, G. Santharam, N. Sathiah, J.S. Kennedy. Tamil Nadu Agricultural University, Coimbatore, India, pp. 140–145.

Dhandapani, N. (1999b). Genetic improvement of *Bacillus thuringiensis.* In: *Emerging Trends in Microbial Control of Crop Pests,* (Eds.) R.J. Rabindra, G. Santharam, N. Sathiah, J.S. Kennedy. Tamil Nadu Agricultural University, Coimbatore, India, pp. 140–145.

Diwakar, B.J. and Pawar, A.D. (1987). Biological Control of tomato fruit borer, *Helicoverpa armigera* in Kamataka. *Indian Journal of Plant Protection,* 15: 57–62.

Dutky, S.R., Thompson, J.V. and Cantweft, G. E. (1964). A technique for the mass propagation of the DD-136 nematodes. *Journal of Insect Pathology,* 6: 417–422.

Easwaramoorthy, S. (1999a). Practical techniques with *Bacillus popilliae.* In: *Emerging Trends in Microbial Control of Crop Pests,* (Eds.) R.J. Rabindra, G. Santharam, N. Sathiah, J.S. Kennedy. Tamil Nadu Agricultural University, Coimbatore, India, pp. 279–285.

Easwaramoorthy, S. (1999b). Genetic manipulation in granulosis virus. In: *Emerging Trends in Microbial Control of Crop Pests,* (Eds.) R.J. Rabindra, G. Santharam, N. Sathiah, J.S. Kennedy. Tamil Nadu Agricultural University, Coimbatore, India, pp. 115–121.

Farrah, R.R. and Ridgway, R.L. (2000). Host plant effects on the activity of selected Nuclear Polyhedrosis Virus against the corn earworm and beet armyworm (Lepidoptera: Noctuidae). *Journal of Environmental Biology,* 29(1): 105–115.

Flipsen, J.T.M., Mans, R.M.W., Kleefsman, A.W.F., Knebel-Morsdorf, D. and Vlak, J.M. (1995). Deletion of the baculovirus ecolysteroid UDP-glucosyltransferase gene induces early degeneration of malpighian tubules in infected insect. *Journal of Virology,* 69(7): 4529–4532.

Glaser, R.W., McCoy, E.E. and Girth, H.B. (1940). The biology and economic importance of a nematode parasitic on insects. *Journal of Parasitology*, 26: 479–495.

Granados, R.R., Fu, Y., Corsaro, B. and Hughes, P.R. (2001). Enhancement of *Bacillus thuringiensis* toxicity to lepidopterous species with the enhancing from *Trichoplusia ni* granulosis virus. *Biological Control*, 20: 153–159.

ICIPE. (1997). *Annual Scientific Report* 1996–97. ICIPE Science Press, Nairobi, Kenya.

ICIPE. (2001). *Annual Scientific Report* 2000–01. ICIPE Science Press, Nairobi, Kenya.

Jackson, D.M., Brown, G.C., Nordin, G.L. and Johnson, D.W. (1992). Autodissemination of baculovirus for management of tobacco budworm (Lepidoptera: Noctuidae) on tobacco. *Journal of Economic Entomology*, 85: 710–719.

Kennedy, J.S. and Sathiah, N. (1999a). NPV formulations and quality control. In: *Emerging Trends in Microbial Control of Crop Pests,* (Eds.) R.J. Rabindra, G. Santharam, N. Sathiah, J.S. Kennedy. Tamil Nadu Agricultural University, Coimbatore, India, pp. 100–105.

Kennedy, J.S. and Sathiah, N. (1999b). Technology for mass production of *Helicoverpa armigera* nuclear polyhedrosis virus. In: *Emerging Trends in Microbial Control of Crop Pests,* (Eds.) R.J. Rabindra, G. Santharam, N. Sathiah, J.S. Kennedy. Tamil Nadu Agricultural University, Coimbatore, India, pp. 256–259.

Khalil, S.K., Bartos, J. and Taborsky, V. (1983). Effect of temperature, pH of the medium and sugar on the germination of spores, development of mycelium and sporulation of the entomopathogenic fungus, *Verticilium lecanii* (Zimm.) Vilgas. *Agric. Trop. et Subtr.*, Universitas Agriculture Praga, 16: 255–274.

Kulat, S.S., Nimbalkar, S.A, Radke, S.G. and Tambe, V.J. (1999). Evaluation of biopesticides and neem seed extract against *Helicoverpa armigera* on chickpea. *Indian Journal of Entomology*, 61(1): 19–21.

Lal, S. S., Ahmed, R. and Yadav, C.P. (1998). Integrated Pest Management strategies for *Helicoverpa armigera* infesting chickpea. *Plant Protection Bulletin*, 50(1–4): 17–19.

Mohan, K.S. (1999). Strategies for the management of *Oryctes* rhinoceros with baculovirus. In: *Emerging Trends in Microbial Control of Crop Pests,* (Eds.) R.J. Rabindra, G. Santharam, N. Sathiah, J.S. Kennedy. Tamil Nadu Agricultural University, Coimbatore, India, pp. 122–127.

Muller-Kogler, E., Samlioakovci A. (1970). Zur Massenkultur des Insektenpathogenen Pilzes *Beauveria bassiana*. *Experientia*, 26: 1400.

Muthuswami, M. (1999). Development of tank-mix formulations of microbial pathogens. In: *Emerging Trends in Microbial Control of Crop Pests,* (Eds.) R.J. Rabindra, G. Santharam, N. Sathiah, J.S. Kennedy. Tamil Nadu Agricultural University, Coimbatore, India, pp. 181–184.

Nagrare, V.S. and More, G.D. (1998). Economics in using bioagents against *Helicoverpa armigera* on pigeonpea. *Indian Journal of Entomology*, 60(2): 203–206.

Neharkar, P.S., Patange, N.R., Kendre, M.S. and Munde, A.T. (1999). Comparative efficacy of insecticides against pod borer complex of pigeonpea. *Journal of Soils and Crops*, 9(2): 178–181.

Parameswaran, S. (1999). Scope of utilizing granulosis viruses as microbial control agents. In: *Emerging Trends in Microbial Control of Crop Pests,* (Eds.) R.J. Rabindra, G. Santharam, N. Sathiah, J.S. Kennedy. Tamil Nadu Agricultural University, Coimbatore, India, pp. 106–114.

Patel, R.C., Singh, R. and Patel, P.B. (1968). Nuclear Polyhedrosis Virus of gram pod borer, *Helicoverpa armigera*. *Journal of Economic Entomology*, 61: 191–193.

Pawar, V.M. and Thombre, U.T. 2001. Mass production of *Helicoverpa* Nuclear Polyhedrosis Virus. In: *Proceedings of 3rd Training on Mass Production Technology of Biological Control Agents*, NCIPM, New Delhi, April 4–12, 2001.

Prabhakara, M. S. and Srinivasa, N. (1998). Residual activity or field persistence of *Bacillus thuringiensis* formulations against pod borer, *Helicoverpa armigera* in pigeonpea. *Indian Journal of Pulses Research*, 11(2): 95–100.

Puntambekar, U.S., Mukherjee, S.N. and Ranjekar, P.K. (1997). Laboratory screening of different-*Bacillus thuringiensis* strains against certain lepidopteran pests and subsequent field evaluation on the pod boring pest complex of pigeonpea (*Cajanus cajan*). *Antonie van Leeuwenhoek*, 71(4):319–323.

Rabindra, R.J. and Jayaraj, J. (1987). Evaluation of certain adjuvants for Nuclear Polyhedrosis Virus (NPV) of *Heliothis armigera* (Hbn.) on chickpea. *Indian Journal of Experimental Entomology*, 26: 60–62.

Rabindra, R.J. (1999a). Processing and standardization of nuclear polyhedrosis viruses for pest management. In: *Ecology Based Pest Management*, (Eds.) R.J. Rabindra, S. Palaniswamy, P. Karuppuchamy, K. Ramanaju, R.P. Sridhar. Tamil Nadu Agricultural University, Coimbatore, India, pp. 82–84.

Rabindra, R.J. (1999b). Identification of baculoviruses by restriction endonuclease analysis. In: *Emerging Trends in Microbial Control of Crop Pests*, (Eds.) R.J. Rabindra, G. Santharam, N. Sathiah, J.S. Kennedy. Tamil Nadu Agricultural University, Coimbatore, India, pp. 266–273.

Rabindra, R.J., Sathiah, N. and Kennedy J.S. (1999). Standardization techniques and bioassay for virulence of isolates of nuclear polyhedrosis virus. In: *Emerging Trends in Microbial Control of Crop Pests*, (Eds.) R.J. Rabindra, G. Santharam, N. Sathiah, J.S. Kennedy. Tamil Nadu Agricultural University, Coimbatore, India, pp. 260–265.

Rajasekaran, B., Rabindra, R.J. and Kennedy, J.S. (1999a). Insect cell culture techniques. In: *Emerging Trends in Microbial Control of Crop Pests*, (Eds.) R.J. Rabindra, G. Santharam, N. Sathiah, J.S. Kennedy. Tamil Nadu Agricultural University, Coimbatore, India, pp. 274–276.

Rajasekaran, B., Rabindra, R.J. and Kennedy, J.S. (1999b). *In vitro* production of insect viruses. In: *Emerging Trends in Microbial Control of Crop Pests*, (Eds.) R.J. Rabindra, G. Santharam, N. Sathiah, J.S. Kennedy. Tamil Nadu Agricultural University, Coimbatore, India, pp. 277–278.

Romeis J. (1997). Impact of plant characters and cropping systems on the searching behavior and parasitization efficiency of *Trichogramma* spp., egg parasitoids of *Helicoverpa armigera. Dissertation*, Universitat Hohenheim.

Samsinakova, A. (1966). Growth and sporulation of submerged cultures of the fungus *Beauveria bassiana* in various media. *Journal of Invertebrate Pathology*, 8: 395–400.

Santharam, G. (1999a). Potential of *Vairimorpha* spp. as microbial pesticide. In: *Emerging Trends in Microbial Control of Crop Pests*, (Eds.) R.J. Rabindra, G. Santharam, N. Sathiah, J.S. Kennedy. Tamil Nadu Agricultural University, Coimbatore, India, pp. 169–172.

Santharam, G. (1999b). Evaluation procedures or testing virulence of *Bacillus thuringiensis* strains. In: *Emerging Trends in Microbial Control of Crop Pests*, (Eds.) R.J. Rabindra, G. Santharam, N. Sathiah, J.S. Kennedy. Tamil Nadu Agricultural University, Coimbatore, India, pp. 294–296.

Sathiah, N. (1999). Registration of biopesticides: requirements. In: *Emerging Trends in Microbial Control of Crop Pests,* (Eds.) R.J. Rabindra, G. Santharam, N. Sathiah, J.S. Kennedy. Tamil Nadu Agricultural University, Coimbatore, India, pp. (197–203.

Seigel, J.P. (2001). Mammalian safety of *Bacillus thuringiensis* based insecticides. *Journal of Invertebrate Pathology,* 77: 13–21.

Shapiro, M. (2000). Effect of two granulosis viruses on the activity of gypsy moth (Lepidoptera: Lymantriidae) nuclear polyhedrosis virus. *Journal of Economic Entomology,* 93(6): 1633–1637.

Sharma, M.L., Rai, H.S. and Verma, M.L. (1997). Biopesticides for management of *Helicoverpa armigera* (Hubner) in chickpea. *International Chickpea and Pigeonpea Newsletter,* 4: 26–27.

Sharma, M.P., Hussaini, S.S. and Adholeya, A. 2005. Entomopathogenic nematodes–A potential future microbial biopesticide. *Agriculture Today,* May, 2005, pp. 45–47.

Singh, S.P. (1996). Biological control in India. In: *IPM System in Agriculture, Vol. II: Bio-control in Emerging Biotechnology,* (Eds.) R.K. Upadhyay, K.G. Mukerji and R.L. Rajak. Aditya Books Pvt. Ltd., New Delhi, India. pp. 216–328.

Sithanantham, S. and Swamiappan, M. (1998). Status and scope for research on utilization of *Trichogramma* sp. in horticultural ecosystem in India, pp. 177–183. *Mitt. Biol. Bundesanstalt. Land-Forstwirtsch,.* Berlin-Dahlem, H. 356 pp. 31–34.

Sivasubramanian, P. (1999a). Safety of microbials to entomophagous insects. In: *Emerging Trends in Microbial Control of Crop Pests,* (Eds.) R.J. Rabindra, G. Santharam, N. Sathiah, J.S. Kennedy. Tamil Nadu Agricultural University, Coimbatore, India, pp. 191–196.

Sivasubramanian, P. (1999b). Mass production of predators and parasitoids. In: *Ecology Based Pest Management,* (Eds.) R.J. Rabindra, S. Palaniswamy, P. Karuppuchamy, K. Ramaraju and R.P. Sridhar. Tamil Nadu Agricultural University, Coimbatore, India, pp. 104–110.

Subramaninan, S. (1999). Improved methods of mass production ofentomopathogenic nematodes. In: *Emerging Trends in Microbial Control of Crop Pests,* (Eds.) R.J. Rabindra, G. Santharam, N. Sathiah, J.S. Kennedy. Tamil Nadu Agricultural University, Coimbatore, India, pp. 299–302

Sundara Babu, P.C. (1999a). Development of *Metarhizium anisopliae* for the Management of *Oryctes rhinoceros.* In: *Emerging Trends in Microbial Control of Crop Pests,* (Eds.) R.J. Rabindra, G. Santharam, N. Sathiah, J.S. Kennedy. Tamil Nadu Agricultural University, Coimbatore, India, pp. 146–154.

Sundara Babu, P.C. (1999b). Advanced dimensions in the use of entomopathogenic fungi for pest management. In: *Ecology Based Pest Management,* (Eds.) R.J. Rabindra, S. Palaniswamy, P. Karuppuchamy, K. Ramaraju and R.P. Sridhar. Tamil Nadu Agricultural University, Coimbatore, India, pp. 85–92.

Sundara Babu, P.C. and Muthuswami, M. (1995). Field efficacy of *Bacillus thuringiensis* var. *kurstaki* against *Helicoverpa armigera* on chickpea, pigeonpea and cotton. *Presented in Symposium on Integrated Pest Management and Environment* at Madras, Tamil Nadu, India, February 2–4.

Swamiappan, M. (1999). Ethics in the production of microbial pesticides. In: *Emerging Trends in Microbial Control of Crop Pests,* (Eds.) R.J. Rabindra, G. Santharam, N. Sathiah, J.S. Kennedy. Tamil Nadu Agricultural University, Coimbatore, India, pp. 88–93.

Thakur, J.N., Rawat, V.S. and Pawar, A.D. (1991). Successful introduction of *Trichogramma chilonis,* an egg parasitoid of Anar butterfly, *Deudorix epijarbas* in Kullu valley. *Journal of Insect Science,* 4(2): 163–164.

Udayasuriyan, V., Sivakumar, N.C. and Chitra, V. (1999). Characterization of *Bacillus thuringiensis* (*Bt*) toxins. In: *Emerging Trends in Microbial Control of Crop Pests,* (Eds.) R.J. Rabindra, G. Santharam, N. Sathiah, J.S. Kennedy. Tamil Nadu Agricultural University, Coimbatore, India, pp. 288–290.

Varma, R.V. (1999). Microbial control of forest insects pests in India. In: *Emerging Trends in Microbial Control of Crop Pests,* (Eds.) R.J. Rabindra, G. Santharam, N. Sathiah, J.S. Kennedy. Tamil Nadu Agricultural University, Coimbatore, India, pp. 78–82

Vimaladevi, P.S. (1999). Potential use of *Nomuraea* as a microbial pesticide in India. In: *Emerging Trends in Microbial Control of Crop Pests,* (Eds.) R.J. Rabindra, G. Santharam, N. Sathiah, J.S. Kennedy. Tamil Nadu Agricultural University, Coimbatore, India, pp. 155–168.

Whitlock, V., Fleischak, C.A. and Whitlock, C.E. (1996). Isolation and assay of *Helicoverpa armigera* GV enhancement factor. *Proceedings of Annual Meeting of Social Inventebrate Pathologists,* Cordopo, Spain. Academic, Orlando, FIUSA.

Yadav, D.N., Patel, R.C. and Patel, D.S. (1985). Impact of inundative release of *Trichogramma chilonis* Ishii against *Heliothis armigera* in Gujarat, India. *Journal of Entomological Research,* 9(2): 153–159.

©R.C. Ram & Asha Sinha (*Editors*)
***Microbial Diversity: Modern Trends*, 2007** **128–151**
Daya Publishing House, New Delhi

Chapter 9

Induced Resistance: A Biotechnological Approach for Induction of Defense-related Molecules in Plant Disease Management

Anuj Kumar, Vinod Kumar and R.N. Kharwar*

Mycopathology Laboratory, Centre of Advanced Study in Botany, Banaras Hindu University, Varanasi – 221 005, India

ABSTRACT

Each plant microbe's interactions are initiated at the stage of the cell. It is vital to build up appropriate and well-organized method of disease management and its large-scale use. Recently, the induced resistance has increased popularity as investigative tools in plant cell biology in spite of certain limitations in part, because of parallel improvement of some refined result on biological task of chemical inducers, PGPR, siderophores, PR-proteins and phytoalexins are reviewed.

We gave special accent on reimbursement; negative aspects and future considerations as well as challenges and advance use of chemical and biological elicitors on induced resistance in conventional agriculture. This review describes regarding some of these approaches, deliberations and how they are pushing as a step closer to inspection the intricate complexity with which plants regulates their own function down to sub-cellular level.

Keywords: *Induced Systemic Resistance (ISR), Systemic Aquired Resistance (SAR), Plant Growth Promoting Rhizobacteria (PGPR), and Chemical inducers, PR-proteins, Siderophores.*

---

*Corresponding Author: E-mail: rnkharwar@yahoo.com

Introduction

Plants and microorganisms in cooperation allied to each other, plant comprises a tremendous bionetwork for microorganisms and its interactions. Microbes interact with plant tissues and cells with different stages of their liberty. The most remarkable point of microbes with plant is beneficial and pathogenic interactions. On the basis of constitutive and inducible responses, plants have evolved refined defense response against microbes. These responses may localize or spread throughout the whole plant system, through the signal transudations, detecting intra-cellular protein of the pathogen based on a gene-for-gene interaction recognition system or triggering a biochemical attack or programmed cell death (PCD). The specificity of plant-bacteria interactions provides numerous implications for the managing of plant diseases. New biotechnological eco-friendly approaches are currently being developed on exciting of plant defense response, and on the use of plant-beneficial bacteria for biological control of plant disease and growth promotion. Plant approves several mechanisms to defense themselves from pathogens including beneficial or detrimental interactions. Surveillance that plants can recognize pathogens was the one of the most striking discoveries in plant pathology. (Chester, 1933; De wit, 1997). Each plant has a range of pathogens to which it is prone, and another to which it is resistant. The hypersensitivity reaction (HR) occurs in non-host plants of a given pathogen group build up a type of blocking necrosis against these pathogens when their tissues are invaded. The induced resistance ensuing HR contrast with the attuned reaction of a pathogen on its host plant, which, consists of infection and invasion of plant tissues or organs, and often plant death. Plant defense mechanisms are characterized by a permutation of constitutive and inducible responses. General barriers or preexisting biochemical defenses includes constitutive responses and instead of this inducible responses can be localized or systemic and are more refined because they engage the recognition of the pathogen by the host plant, signal transudations, and the expression of several genes. In a localized response, plant tissues retort beside pathogens by a sort of programmed cell death (PCD) consisting effectively of electrolyte leakage from the cytoplasm and an oxidative burst. But in case of systemic defense, spread of signal from the site of interface as a response to chemicals, microorganisms, mechanical damage or stress. The signal is intercede by several molecules, which utility is as messengers in plants, *i.e*, Salicylic and Jasmonic acid, or even volatiles such as Nitric oxide and Ethylene (Baker, 1997). These messengers activates the Pathogenesis-Related (PR) proteins genes in response to pathogens aggression by interacting with specific binding proteins which produce several enzymes, *viz.* peroxidase, lipo-oxygenases, super-oxide dismutases, and phenylalanine ammonia lyase (PAL), which involved in the secondary metabolism of plants and specially in the formation of phenolic compounds, or phytoalexins, glucanases and chitinases, with anti-microbial activity. The inducible defense response in plants produces a restricted range of biochemical compounds with low specificity for the target pathogen. Though, the specificity of the response in plants deception in the fact that they perceive intracellular definite types of proteins secreted by the pathogen (elicitors), using an interaction recognition system, which triggers the defense response. It is already known that bacteria can create close links with plants or endophytes in a more delicate affiliation with the plant. The non-symbiotic, valuable plant bacteria embrace plant-growth-promoting rhizobacteria (PGPR) and biological control agents (BCA). PGPR are inhabitants of the rhizosphere, the volume of soil under the immediate influence of the plant root system. In the rhizosphere, discharge of macrobiotic compounds by the plant favored large, active microbial population. Along combination with plants, some genera *Pseudomonas, Serratia Azospirillum* and *Bacillus,* often limits the growth of certain soil-borne plant pathogens and enhance growth of the root system and as well as entire plant, (Kloepper, 1991). Siderophores that chelate iron, in some cases a correlation has been found with biological control of soil-borne plant pathogens or deleterious microorganisms, or even with induction of the systemic defense response. More rarely, BCA induce systemic acquired resistance (SAR) in the plant by eliciting, to a few extents, the HR response devoid of infection, or by causing some sort of stress to the plant pathogen (Preston, 2001). The fact of induced resistance or acquired diseased resistance of plants has predictable rising consideration as a novel approach to the integrated protection of crop plants (Ahl, 1984; Kuc, 1982) Induced resistance is the dedicated resistance in plant towards pathogen and as a result of preceding treatment with a pathogen or chemical that is not it self a pesticides. This system is based not on direct pathogen restraint but is incentive of natural defense mechanism in plants. Resistance is the capability of an organism to prohibit or overcome, entirely or in some extent, the cause of a pathogen or other detrimental aspect. Induced resistance is the phenomenon that a plant, formerly duly provoked, exhibits an enhanced resistance leading 'challenge' inoculation through pathogen.

Induced resistance or SAR is an activated phenomenon (Ryal *et al.*, 1996) and the time desired for improvement depends on inducing agent ranges one week to few weeks established it could last over a few weeks (Hammerschimidt *et al.*, 1995). Since Ross (1961a,b) has made first been organized enquiry about induced resistance, which could make awareness among pathologist (Hammerschmidt and Kuc, 1995). In the event of induced resistance plant exhibits an increased level of resistance to infection by a pathogen after proper stimulation this resistance response is first characterized by Ross (1961a, 1961b), is expressed systemically through out the plant growth and effective against a broad spectrum of viral, bacterial, and fungal pathogens (Hammerschimidt and Kuc 1995). In general, induced resistance can be triggered in three ways (1) by a prompt infection with a necrotizing pathogen (Ross, 1961a, 1961b; Kuc, 1982) (2) by management with assured chemicals, such as salicylic acid (SA; White, 1979) or 2, 6-dichloroisonicotinic acid (INA and Jasmonic acid (JA) or (3) by colonization of the rhizosphere with ideal plant growth-promoting rhizobacteria (PGPR; Alstrom, 1991; van Peer *et al.*, 1991; Wei *et al.*, 1991). Systemic resistance induced by the first two types of inducers is termed systemic acquired resistance (SAR; Ross, 1961b), whereas PGPR-mediated defense is usually referred to as induced systemic resistance (ISR; Kloepper *et al.*, 1992). Pathogen-induced SAR has been studied most comprehensively in tobacco and cucumber (KuC, 1982; Ward *et al.*, 1991) and was recently established in *Arabidopsis* as well (Uknes *et al.*, 1993; Cameron *et al.*, 1994). The resistant situation is characterized by an enhancing in endogenously synthesized SA at the beginning of SAR (Malamy *et al.*, 1990; Metraux *et al.*, 1990). Accumulation of SA appears to be critical for the induction of the SAR-signaling pathway because transgenic plants incapable to accumulate SA are incapable of developing SAR (Gaffney *et al.*, 1993). Furthermore, SAR is linked with bringing together expression of a set of so-called SAR genes (Ward *et al.*, 1991). These SAR genes incorporate genes encoding pathogenesis-related (PR) proteins of which some exhibit limited antifungal activity *in vitro* (Vigers *et al.*, 1991; Woloshuk *et al.*, 1991) as well as in vivo (Broglie *et al.*, 1991; Alexander *et al.*, 1993).

Systemic acquired resistance (SAR) is a pathogen-inducible defense mechanism in plants. The resistant state is dependent on endogenous accumulation of salicylic acid (SA) and is characterized by the activation of genes encoding pathogenesis-related (PR) proteins. Recently, preferred nonpathogenic, root-colonizing biocontrol bacteria have been shown to trigger a systemic resistance response as well.

Initially the, debate was over the serological basis of what was then referred to as required physiological immunity (Chester, 1933a,). From results of experiment using Tobacco mosaic virus to sensitize tobacco (*Nicotiana tabacum* L.) against subsequent "challenge" inoculations of Tobacco mosaic virus on infected leaves or on distal uninfected leaves, would clearly articulate the concepts of localized acquired resistance and SAR, respectively Cruikshank and Mandryk (1960) extended this concept to anon viral pathogen in field-grown tobacco plants using stem injection of *Peronospora tabacina* (D.B. Adam, also referred to as *P. hyoscyami* de Bary f. sp. *tabacina*) to trigger SAR against subsequent inoculations of the same pathogen, the causative agent of blue mold of tobacco. These landmark studies led to the development of the classic SAR models during the 1980s in other plants, such as cucumber (*Cucumis sativus* L.), common bean (*Phaseolus vulgaris* L.), rice (*Oryza sativa* L.), and Arabidopsis thaiana(L.) Heynh, demonstrating that SAR was conserved across diverse plant families and was effective against abroad range of viral, bacterial, and fungal pathogens (Sticher *et al.*, 1997). Additional interest in the biological control soil born diseases of plants led to the serendipitous discovery of another form of induced resistance associated with the colonization of plants root by certain plant growth promoting rhizobacteria (PGPR) referred to as induced systemic resistance (ISR) (van Loon. *et al.*, 1998). ISR is distinct from SAR in several keys physiological and biological phenotypes that are best divined in *A. thaliana* (Piterse *et al.*, 1996, Ton *et al.*, 1999, 2001; van Wees *et al.*, 1997). Result of lab and field studies now that, like SAR; ISR is effective agaiust a broad range of disease by viruses, bacteria, and fungi (Nand kumar *et al.*, 2001 Niranjan Raj *et al.*, 2003; Raupach and Kloepper, 1998,2000; van loon *et al.*, 1998; Wei *et al.*, 1996; Zehnder *et al.*, 2001).

Over of last preceding 20 years, research on SAR and ISR using model system has additional deliberation of the molecular basis of synthetic elicitors and PGPR formulation for use in agriculture (Benhamou and Nicole, 1999; Kessler and Baldwin, 2002; Kessmann *et al.*, 1994; Leroux 1996; Lucas, 1999; Lyyon *et al.*, 1995; Sticher *et al.*, 1997; van Loon *et al.*, 1998, Walling, 2000). The future use of SAR and ISR to control crop pests in conventional agriculture seems promising. Since synthetic elicitors and PGPR strains, in general, do not exhibit any direct antimicrobial activity, unlike traditional pesticides, they provide a way to control disease without asserting direct selective pressure on pathogen population.

Role of Different Variables in Disease Control and Plant Growth

Chemical Inducers

Rational chemicals are right now warned out in crop protection, and innovative agrochemicals are frequently being produced and evaluated. These chemicals can be used to control vectors for disease. However, any new chemicals must assure definite criteria prior to its can be commercial use. It must valuable and should not harm to the plant and other component of ecological unit nor to the consumers. A new generation of chemical compounds, produced either by chemical synthesis or by biotechnological processes, consists of molecules, which do not exhibit in vitro antimicrobial activity against the pathogens but instead. elicit plant defense responses. Several of these compounds are currently on the market in some countries and include derivatives of acetylsalicylic acid, *e.g.* benzotiadiazol (BTH) (Agrawal, 1999). In the future of plant control, currently provided mainly by chemically synthesized fungicides and bactericides, will soon be complemented or substituted by new disease-control technologies emerging from the basic knowledge of plant-microbe interactions, and consisting of new biotechnological approach. SA or INA has been reported to take off pathogen-induced SAR because they induce resistance to the same spectrum of pathogens and concurrently turn on the expression of SAR genes (Ward *et al.*, 1991; Ukries *et al.*, 1993). Although these observations are vastly stinking of a causal relationship between accumulation of PR proteins and SAR, eventual evidence that they are accountable for broad-spectrum resistance is still lacking. During the preceding several years, extensive studies have established that salicylic acid (SA) plays a crucial role in SAR by applications of a variety of pathogens and pathogen elicitors. Naturally, SA is phenolic compound and concerned in signal tranduction as well as local and systemic resistance to pathogens. The relevance of SA was exposed to mimic certain aspects of a pathogens infection, resulting in both SAR gene expression and induction of exogenous application of the chemicals. Salicylic acid (SA) is an important signal compound in plants and key signal in the induction of (SAR). In *Arabidopsis* salicylic acid (SA), jasmonic acid (JA), and ethylene (ET) function as key signaling molecules in which disease resistance is regulated by multiple signal tranduction pathways. In plants, resistance to pathogen infection is achieving through protective physical barriers and 0. a diverse array of antimicrobial chemicals and proteins. Many of these antimicrobial compounds are part of an active defense response, and their rapid induction is contingent on the plant's ability to recognize and respond to an invading pathogen (Baker *et al.*, 1997). SA is a necessary and sufficient signal for SAR because removing SA through the ectopic expression of salicylate hydroxylase (encoded by the bacterial *nahG* gene) blocks the onset of SAR (Gaffney *et al.*, 1993), whereas increasing SA concentrations by endogenous synthesis or exogenous application induces SAR (White, 1979; Malamy *et al.*, 1990, Metraux *et al.*, 1990; Rasmussen *et al.*, 1991; Yalpani *et al.*, 1991). Synthetic SA analogs such as 2,6-dichloroisonicotinic acid (INA) and benzothiodiazole are also effective inducers of SAR (Metraux *et al.*, 1991; Gorlach *et al.*, 1996). Through the phenomenon of SAR, SA in contact with pathogenic and nonpathogenic microorganisms triggers a wide range of defense mechanisms in plants that protect them against invasion. SAR, commonly triggered by local infection, can provide long-term resistance throughout the plant to subsequent infection by different pathogens. The SAR response correlates with the activation of pathogenesis-related (PR) genes. This process generally requires the involvement of the signal molecule salicylic acid (SA), which has been shown to increase in both infected and uninfected tissues (Sticher, 1997). Colonization of roots by certain rhizospheric bacteria confers another form of disease resistance termed induced systemic resistance (ISR). Several studies have shown that ISR is not an SA-dependent phenomenon but rather requires components of the jasmonic acid (JA) signaling pathway followed by the ethylene-signaling pathway (Pieterse, 2000, van Loon, 1998.). The first enzyme in the biosynthesis pathway of JA is lipoxygenase (LOX). Several components of the ethylene signal transduction pathway have been identified and include *ETRl* and *CTRl* (Clark, 1998). The *ETRl* gene is related to the super-family of catalytic receptors in prokaryotes referred to as two-compound regulators. The *ETRl* protein is an ethylene receptor with histidine protein kinase activity. The *CTRl* gene acts downstream of the ethylene receptor and encodes a protein most similar to the *Raf* family of serine/threonine protein kinases that initiate the mitogen-activated protein kinases cascades. Both *ETRl* and *CTRl* are negative regulators of ethylene signal (Huang, 2003.).

The gene encoding for phenylalanine ammonia-lyase (PAL) is believe to be activated by the JA/Ethylene signaling pathway in context of induced plant defenses (Diallinas, 1994). PAL is the first enzyme in the phenylpropanoid biosynthesis pathway, which provides precursors for lignin and phenols, as well as for SA. Other enzymes of the phenylpropanoid pathway, including peroxidases, are also induced in resistant reactions (Yedidia, 2000.) Peroxidases are also known for their role in the production of phytoalexins, reactive oxygen species, and formation of structural barriers PR proteins such as β-1, 3-glucanase (*PR2*) and Chitinase (*PR3*), known to disrupt the fungal mycelial wall, can be induced by SA as well as by pathogenic attack. However, a study has indicated that JA instead of SA induced the some chitinase classes, such as class III chitinases (*PR8*), (Samac, 1991). A number of reports have claimed resistance inducing activity of certain chemicals or plant extracts, but have failed to demonstrate that the compounds in question fulfills the criteria established for IR activation. fosethyl-Al, metalaxyle, triazides, 2,2-dichloro-3, 3-dimethyl cyclopropane (DCP), corboxylic acid and its analogs and probenezole are a few to name.

A variety of natural and artificial substances are inducers of IR the criteria practical to a plant defensive agent to be eligible as IR inducers by neither the agent nor metabolites derived from it has direct antimicrobial activity in vitro, or in plant or the agent modifies the plant pathogen interaction in such way that it related mechanism prior to or after challenge inoculation. Some molecules and synthetic products known to have resistance activity have been discussed by Kessmann *et al.* (1994). Transgenic plants expressing a high level of SA have been developed which show increased resistance to various pathogens. The synthetic inducers of plant defense mechanism like INA and BTH may act through the synthesis of Jasmonic acid.Several studies have provided evidence for existence of multiple signals in signal transduction pathway (Vidhyasekaran, 1998). Enhanced synthesis and release of these signals may activate defense gene and confer resistance. Kovats *et al.* (1991) also observed SAR against late blight disease in tomato. Localized infection of one part of a plant can result in the systemic expression of resistance in other part; therefore, it has been hypothesized that a signal is generated and mobilized from the initial infection site. Salicylic acid (SA), Jasmonic acid (JA), and ethylene apparently are involved in signaling pathways. The importance of SA in SAR signal transduction was documented using transgenic tobacco engineered to express the SA hydroxylase gene (*nahG*) (Gaffney, 1993). Jasmonate signaling during SAR has been studied in other plant-pathogen systems (Schweizer, 1993). Some evidence has shown that treatment of the plants with methyl jasmonate induced the accumulation of an antifungal defensins, but not of pathogenesis-related (PR) protein PR-I. In contrast, application of SA or iso-nicotinic acid (INA) caused the accumulation of PR-I, but not of defensins. This finding suggested the presence of two different pathways, both inducible by pathogens. One pathway led to PR-protein expression via SA. The other pathway, via jasmonate, led to the accumulation of defensins (Pennincky, 1997.) Bowling *et al.*, 1997 placed JA and SA in parallel pathways and proposed that the definition of SAR in *Arabidopsis* spp. should include both pathways. Ethylene is a volatile plant hormone that is involved in numerous physiological processes, including induced resistance (Boller, 1991). In particular, ethylene has played a role in the induction of SAR against tomato late blight (Spanu, 1989).

Plant Growth Promoting Rhizobacteria (PGPR)

Growing awareness of the environmental damage caused by the use of chemical substances for plant disease control in agriculture has raised the need to study biological alternatives, such as activating the defense response of plant crops by inducers that are not toxic to the environment. Certain strains of rhizosphere bacteria can suppress plant diseases by inducing resistance responses in the plant through induced systemic resistance (ISR) process (Alstrom, 1991; Van Peer, *et al.*, 1991, Wei, *et al.*, 1991). The mechanisms involved in ISR are just beginning to be elucidated (Buysens, S. *et al.*, 1996, De Mayer 1997, Lee man Metal 1995, Maurhofer, *et al.*, 1994, Press *et al.*, 1997), and appear to vary among bacterial strains and pathosystems. As biological agents, plant growth-promoting rhizobacteria (PGPR) is one of the potential alternatives for plant disease management. Research over the past two decades has demonstrated that in the phenomenon of SAR or ISR plants have latent defense mechanisms against pathogens, which can be systemically activated by exposure of plants to stress or infection by pathogens, which operates through the activation of defense genes and the accumulation of defense compounds at a site distant from the point of pathogen attack (Kessman *et al.*, 1994; van Loon *et al.*, 1998.) first reported that certain root-colonizing bacteria, or PGPR, could promote plant growth, and subsequent studies with PGPR demonstrated control of soil borne pathogens. Recently, research on mechanisms of biological control by PGPR revealed that some PGPR strains protect plants against pathogen infection through induction of systemic resistance, without provoking any symptoms themselves. Alstrom (1991) More recently, certain PGPR strains have been shown to protect plants through mechanisms associated with induced systemic resistance against pathogens that cause foliar disease symptoms (Alstrom, 1991; van Peer *et al.*, 1991; Wei *et al.*, 1991). Subsequent research with field-grown cucumber has demonstrated PGPR-induced systemic resistance (ISR) against foliar diseases caused by fungal and bacterial pathogens (Liu *et al.*, 1995a,b, Zehnder *et al.*, 2000) demonstrated that treatment of cucumber with PGPR reduced the incidence of wilt symptoms resulting from infection by the bacterial wilt pathogen *Erwinia tracheiphila.* In these studies, PGPR treatment was also associated with a reduction in numbers of cucumber beetles that are vectors of the bacterial wilt pathogen. The aforesaid studies were conducted using standard vegetable production practices, which include pre plant soil fumigation with methyl bromide. These diseases inhibit early plant growth and may delay harvest. As methyl bromide will be unavailable to US vegetable producers after 2005, it is important to determine if PGPR-mediated induced resistance can also occur without methyl bromide fumigation.

Table 9.1: Some Strains of Plant Growth Promoting Rhizobacteria (PGPR) to Control Plant Disease

Crop	Pathogen/Disease	PGPR Strain	Comments	Source
Arabidopsis	Fusarium oxisporum.f.sp. raphini	Pseudonomas FluforescencsWCS417	Vascular wilt	Piterse, et al., 1996
Pearl millet	Sclerospora glaucum/ downy mildew	Bacillus pumilus T4	Disease incidence	Niranjan Raj et al., 2003
Tomato	Non host pathogen	Pseudomonas Type-III	Suppresses PCD	Kang et al., 2004
Tomato	Leaf fungal	Bacillus cereus	Dialysates applied to roots protected	R.S. Romero et al., 2005
	Pseudomonas syringae pv. Lachrymans/ Angular leaf spot	INR+GB03	Plant Disease severity Vascular wilt	Raupach and Kloepper, 2000
	Fusarium oxysporum f. sp. lycopersici	Pseudomonas fluorescence WCS 417		Duijff et al., 1997
	Cucumber mosaic virus	Bacillus pumilus SE34	Viral titer	Zehnder et al., 2001
Munga bean	Pythium aphanidermatum/ root infecting fungi	Bacillus subtillis	Partial 16srDNA comparison	LiBin et al., 2005
Cucumber	Erwinia tracheiphilal/ Bacterial wilt	Bacillus pumilis INR7	% Vines wilts from a single field trial	Zehnder et al., 2001
	Pseudomonas syringae pv. Lachrymans/Angular leaf spot	Bacillus pumilis INR7	Disease severity and yield averaged across methyl bromide treated and non-treated field trials	Ranpach and Kloepper 2000
	Collectotricum Orbiculare	Peudomonas aureofaciens	Anthroenose	Wei, G. Kloepper, et al., 1991
Pepper	Meloidogyne incognita/ root knot nematode	GBO3 +IN937a	Disease severity	Kokilis-Baurell, et al., 2002
Pea nut	Cercosporidium	Bacillus cereus C_1	Disease Severity based on defolimion	Zhang, et al., 2001
Bean	Botrytis cinerea	Pseudomonas aeruginosa 7NSK2	Gray mold	Demeyer et al., 1997
Carnation	Fusarium oxysporum f. sp. dianthi	Pseudomonas fluorescenceWCS417	Vascular wilt	Van Peer, et al., 1991
Radish	Alternaria brassicicola	Pseudomonas fluorescence WCS41	Necrotic lesions	Hoffland et al., 1996
Rice	Rhizoctonia sonani/ sheath blight	Pseudomonas fluorescence PF1	Foliar symptoms	Nand Kumar et al., 2001

Selected PGPR, mainly fluorescent *Pseudomonas* sp. have been demonstrated to control plant diseases effectively by suppressing pathogens and deleterious microorganisms through siderophore-mediated competition for iron, or antibiosis (Thomashow and Weller, 1995). Maurhofer *et al.* (1994) showed that ISR induced by strain CHAO of *P. fluorescens* in tobacco against tobacco necrosis virus was accompanied by an increase in PR protein accumulation, suggesting that PGPR-mediated ISR and pathogen induced SAR are manifestations of a similar defense mechanism. However, Hoffland *et al.* (1995) were unable to establish an accumulation of PR proteins in radish displaying substantial ISR against *Fusarium oxysporum* when plants were treated with strain *WCS417r* of *P. fluorescence.* Therefore, it is unclear whether PGPR-mediated ISR and pathogen-induced SAR share a common signal transudation pathway.

Systemic acquired resistance (SAR) against late blight was previously demonstrated in tomato by inoculating *P. infestans* (Heller, 1986) or Tobacco necrosis virus (TNV) (Anfoka, 1997) or by applying chemicals onto tomato (Cohen, 1994) prior to challenge with the pathogen. Studies have shown that plant growth-promoting rhizobacteria (PGPR) can be applied on a wide range of plants for the purpose of disease control and growth enhancement (Kloepper, 1996; Van Loon, 1998). Induced systemic resistance (ISR) has been reported as one of the mechanisms by which PGPR reduce plant disease through the manipulation of the host plant's physical and biochemical properties (Kloepper, 1996; Van Loon, 1998). PGPR-elicited ISR has been demonstrated in many plant species, including *Arabidopsis* spp., bean, carnation, cucumber, radish, tobacco, and tomato (Van Loon, 1998). There are no reports of PGPR elicited ISR against late blight on tomato.

Siderophores

The vast majority of microorganisms produce siderophores (Barton *et al.*, 1993. Winkelmann, 1991), a diverse class of high-affinity iron chelating compounds that are produced in response to iron limitation, and exported from the cell where they chelate ferric iron. Ferric siderophore complexes are recognized and bound by specific outer- membrane receptor proteins, and iron is transported into the cell where it becomes available for metabolic functions. Among the siderophores produced by rhizosphere bacteria, only the pyoverdines (also called pseudobactins) produced by the *fluorescent pseudomonades* have been implicated in ISR (Duijff, *et al.*, 1994, Maurhofer, *et al.*, 1994). Bacterial determinants of ISR that have been identified as lipopolysaccharides (LPS), iron-regulated compounds, and certain antibiotics. LPS of *Pseudomonas fluorescens* strain *WCS417* has been implicated in triggering ISR against *Pseudomonas syringae* pv. tomato in *Arabidopsis* (Van Wees *et al.*, 1997), and to *Fusarium* wilt in carnation (Van Peer and Schippers, 1992), radish (Leeman *et al.*, 1995) and tomato (Duijff *et al.*, 1997). In radish, the O-antigenic side chains of the LPS of both *P. fluorescens WCS417* and *P. fluorescens WCS374* were major determinants of ISR against *Fusarium* wilt under iron-replete conditions, but not under iron-limiting conditions. Unknown iron-regulated factors in *WCS374* and *WCS417* caused ISR under low-iron conditions. Likewise, in *Pseudomonas putida BTP1*, an unknown iron-regulated metabolite appeared to be responsible for ISR in bean against *Botrytis cinerea* (Ongena *et al.*, 2002). De Meyer and Hofte (1997) found that the SA-producing PGPR *Pseudomonas aeruginosa 7NSK2* triggered ISR against *B. cinerea* in bean when the bacterium was cultivated on iron-poor media but not on iron-rich media, and that SA-negative mutants of this strain had lost their ability to induce systemic resistance. SA production by *P. aeruginosa 7NSK2* appeared to be essential for ISR also to tobacco mosaic virus (TMV) in tobacco (De Meyer *et al.*, 1999) and to *Colletotrichum lindemuthianum* in bean (Bigirimana and Hofte, 2002). Likewise, expression of SA-biosynthetic genes in *Pseudomonas fluorescens* P3 improved induction of resistance to tobacco necrosis virus (TNV) in tobacco (Maurhofer *et al.*, 1998). However, in *Pseudomonas fluorescens CHAO* the siderophore pyoverdine, rather than SA, appeared to be implicated in induced resistance against TNV (Maurhofer *et al.*, 1994). Similarly, in the SA-producing PGPR *Serratia marcescens* 90-166, an unidentified catechol-type siderophore rather than SA is involved in ISR (Press *et al.*, 1997). Antibiotics can also be involved in ISR. In *Arabidopsis*, Iavicoli *et al.* (2003) recently demonstrated the importance of 2,4-diacetylphloroglucinol (DAPG) produced by *P. fluorescens* CHAO for the induction of resistance against *Peronospora parasitica*. In tomato, Audenaert *et al.* (2002.b) showed that the phenazine antibiotic pyocyanin in combination with SA or the SA-containing siderophore pyochelin produced by *P. aeruginosa 7NSK2* act as determinants for induced resistance against *B. cinerea.*

Pseudomonas putida WCS358 is a PGPR originally isolated from the rhizosphere of potato (Geels and Schippers, 1983,a, b). It can suppress soil-borne plant diseases by siderophore-mediated competition for iron (Bakker *et al.*, 1986, 1993). *WCS358* can also induce systemic resistance in *Arabidopsis thaliana* against *Pseudomonas syringae* pv. *tomato* and *Fusarium oxysporum* f.sp. *raphani* (Van Wees *et al.*, 1997), but the strain is unable to induce resistance against Fusarium wilt in carnation (Duijff *et al.*, 1993) or radish (Leeman *et al.*, 1995).

Arabidopsis, bean and tomato model systems were compared to explore further the potential of *P. putida WCS358* to induce ISR and to identify bacterial determinants involved in this process.

PR-Proteins

PR-proteins were first observed as new proteins components induced by TMV in hypersensitive interaction between a plant and a pathogen results in the induction of numerous host-specific biochemical responses, so that the plant could with stand attack from pathogens. Chitin and glucan are cell-wall components of many pathogenic fungi and de polymerization of cell wall by the combined activity of chitinases and glucanases could kill fungi *in vitro*. Chitin is a naturally abundant polymer of β-1, 4-linked *N*-acetylglucosamine (GlcNac). Chitin hydrolysate can be used as a carbon and nitrogen source in the production of single-cell proteins. Production of lytic enzymes, chitinase and β-1,3-g1ucanase by several PGPR strains is considered as a major antagonistic property of these strains. It has been reported that chitinase and β-1,3-g1ucanase can function in defense against many fungal pathogens. Inhibition of growth of several fungi requires the presence of combinations of Chitinase and β-1,3-glucanase activities. Many plants to be known to induced by these enzymes activities in response to infection with fungal pathogens and also correlated with induced resistance. Moreover, these two enzymes act synergistically in the partial degradation of fungal cell walls. It was also suggested that combinations of the two enzymes could strongly inhibit growth of many fungi, including those that could not be inhibited by Chitinase or β-1,3-glucanase alone (Vogeli, 1988). It is also essential to analyse and distinguish different isozymes to elucidate their roles in host-pathogen interaction. Accumulation of PR-proteins is related with SAR in plants Chitinase and β-1,3-glucanase have not only the potential to hydrolyses cell components like chitin and β-1,3-glucan respectively, but they also release elicitors from the walls of fungi, which in turn stimulate various defense responses in plants (Ren, 1992).

The notion that the enhanced resistance apparent upon challenge inoculation depends on the same defense mechanisms as expressed after primary infection led to the identification of common. metabolic alterations induced systemically in response to local infection. Whereas induction of phytoalexins and cell wall stiffness are local reactions, accumulation of pathogenesis-related proteins (PRs) extends into non-inoculated plant parts that, upon challenge, exhibit acquired resistance (Rya.ls *et al.*, 1996). The proteins themselves are not transported from the primary inoculated leaves, as demonstrated elegantly by Gianinazzi and Ahl (1983) through the analysis of reciprocal grafts of

Nicotiana species expressing electrophoretically different proteins. While a link between PRs and acquired resistance in virus-infected tobacco was immediately hypothesized (Fraser (1982) pointed out that PRs became apparent in non-inoculated leaves distinctly later than acquired resistance appeared manifest. However, in tissues already primed to express PRs, challenge inoculation might lead to their earlier and faster accumulation. Moreover, a hybrid between *N. glutinosa* and *N. debneyi* constitutively expressed PRs and was highly resistant against TNV (Gianinazzi *et al.*, 1983). Enhanced defensive responsiveness throughout a plant resulting from local infection with a pathogen inducing necrotic lesions, such as in a HR. The term is appropriate and emphasizes the historic link with the pioneering work of Ross (1961a,b). Whereas Kloepper *et al.* (1992). This view has been reinforced by the characterization in *Arabidopsis* of mutants that either are compromised in both the production of PRs and the induction of SAR or are constitutive expressers of PR genes as well as SAR (*cpr1;* Bowling *et al.*, 1994). PRs have been defined as plant proteins that 756 are induced in pathological or related situations. Although they are implicated in plant defense, they have not been identified because of their anti-pathogenic action, but solely because of their accumulation in infected plants. Eleven families of PRs have now been officially recognized, but additional pathogen- induced proteins with potential anti-pathogenic action keep being described. PRs have been identified in at least nine plant families, with those in tobacco and tomato characterized best. It is now known that they comprise four families of chitinases (PR-3, -4, -8 and -11), one of β-1, 3-glucanases (PR-2), one of proteinase inhibitors (PR-6), and one specific peroxidase (PR-9), as well as the PR-l family with unknown biochemical properties, the thaumatin-like PR-5 family, and the birch allergen *Betv1-related* PR-10 family. Not all families are represented in any plant species, but each family may comprise several members. Together the PRs form a set of pathogen-induced proteins that may be considered as stress proteins. In the past decades it has become evident that plants, when exposed to various environmental stresses, respond by synthesizing sets of specific proteins. Well known are the heat-shock proteins, that appear to be common to all living organisms, and are transiently induced when ambient temperature exceeds some critical limit (Vierling, 1991). Different sets of proteins are induced by *e.g.* drought stress or freezing temperatures. For instance, during cold acclimation hardy cultivars of alfalfa synthesize a number of proteins that supposedly function in reducing the deleterious effects of low temperature on plant membranes.

In response to, particularly necrotizing, infections by viruses, viroids, fungi and bacteria, and thought to function in the acquired resistance against further infection. However, in contrast most other types of stress proteins, they accumulate in some of the PRs possess potential anti pathogenic activities. Chitinase, together with Glucanase, could be directed against fungal cell walls and, perhaps, insects. Insects are likely to be affected by proteinase inhibitors. The PR-8 family of chitinases, which also possesses lysozyme activity, may inhibit bacteria. The PR-9 peroxidase is of the lignin-forming type and could be involved in the strengthening of cell walls. The PR-10 family has sequence similarity to ribonucleases and is the only family consisting of cytoplasmic proteins, but there is no evidence that PR-10 proteins are active against *viz.* viruses. The PR-1 and PR-5 proteins are often strongly induced and seem to affect membranes, but their precise actions have not been elucidated. The inducible PRs are mostly acidic proteins that are secreted into the intercellular space of the leaf. Both through cDNA sequence homologies and on the basis of similar enzymatic activities, additional basic counterparts have been identified. These basic PRs occur at relatively low levels in the vacuole and, besides being induced upon infection, are expressed in a tissue-specific and developmentally controlled manner in leaves, roots and floral parts. Specific activities vary greatly, *i.e* from 5 nKat/mg with laminarin as a substrate for the inducible, acidic glucanase PR-2a to 23 and 1300 nKat/mg for -2b and -2c, respectively, and 1100 nKat/mg for the developmentally-controlled basic glucanase PR-2e. It has been suggested that in induced plants the accumulated intercellular proteins form the first line of defense to a challenging pathogen and, if this fails and the tissue is disrupted, the release of the vacuolar PRs functions as a second line, engulfing the pathogen with lytic enzymes. constitutive expression of several of the basic proteins in older leaves, roots and developing flowers could be similarly considered as a protective mechanism against pathogen invasion, possibly contributing to the often observed increase in resistance with plant age. However, the organ-specific expression of specific PR genes suggests that the proteins also play roles in normal developmental processes

Table 9.2: Pathogenesis-related (PR) Proteins in Plants

Pathogen–Related Family	Activity	Target Site
PR-1 (Antifungal)	?	Membrane
PR-2	β-1, 3-glucanase	Cell wall Glucans
PR-3	Endochitinase	Cell wall chitin
PR-4	Endochitinase	Cell wall chitin
PR-5	Osmotin	Membrane
Similar to thaumatin		
PR-6	Proteinase inhibitor	Proteinase
PR-7	Proteinase	?
PR-8	Endochitinase	Cell wall chitin
PR-9	Peroxidase	Indirectly through reinforcement of physical barrier with phenolics
PR-10	Rnase	?
PR-11	Endochitinase	Cell wall chitin
Unclassified	α-amylase Polygalacturonase inhibitor (PGI)	Cell wall glucan Polygalacturonase

Phytoalexins

Phytoalexin are one of the best and longest-studied defense responses of plants to infection are the induced accumulation of low molecular weight lipophilic antimicrobial secondary metabolites. While the phytoalexin assumption was first proposed in 1940, a role for these compounds in defense has been revealed through several experimental approaches. The idea that defenses can be activated after infection was crystallized by the phytoalexin hypothesis of Muller and Borger and the study of phytoalexin has been component of the basics of plant defense research still Phytoalexin have received much attention over the past 60 years, and much of this work has provided new insight into the regulation of gene activation, phytochemical variety, and the chemistry and biochemistry of secondary metabolites. The topic of phytoalexin and closely related defense responses has been reviewed numerous times (Dixon *et al.*, 1985; Dixon *et al.*,1990; Cruick Shank, 1963; Kuc J., 1997). Phytoalexins are compounds that accumulate rapidly around sites of incompatible pathogen infection and in response to an extensive array of biotic and aboitic elicitors. The degree of resistance of host has been correlated with the rate of production, accumulation of phytoalexins by the pathogens. Although, in few cases correlations not found between higher concentration of phytoalexin and greater resistance of hosts *viz.*, Bean-*Colletotricum lindemuthianum* and Potato-*Phytophthora infestants* (Agrios, 1997). Now ever in most cases positive correlation has been established between the two.

Table 9.3(a): Phenolic Compounds to Control Plant Diseases

Plant	Plant Parts	Phenol	Parasite
Allium cepa	Bulb, seedling	Catechol, Tannic acid	B. allii
Cocos nucifera	Plant	Caffeic acid, Coumarin, Gallic acid	Pythium sp.
Lycopersican esculentum	Plant	Catechol	Fusarium lycoperiscii
Nicotiana tobaccum	Root	Coumarin	TMV
Solanum melongena	Seedling	Catechol	Fusarium melongena
Solanum tuberosum	Tuber	Quinones, Polyhydroxy Phenol, Chlorogenic acid	Streptomyces scabies

Table 9.3(b): Some Phytoalexins which has been Reported so far

Phytoalexins	Hosts	Fungi
Ipomeamororane	Sweet potato	Ceratocystis fibriata
Orchinal	Orchid	Rhizoctonia
Isocoumerin	Carrot root	Ceratocystis fibriata
Pisatin	Pea (Pod)	M. fructicola
Phaseolin	French bean	M. fructicola
Rhishitin	Potato, Tomato	Phytophthora
Wyerone	Vicia faba	Botrytis fabae
Capsidal	Capsicum sp.	Fusarium sp.

Table 9.3(c): Occurrence of Some Different Phytoalexins in a Single Host Species

Host Species	Phytoalexins
Phaseolus vulgaris	Phaseolin, Phaseolliolin, Phaseollin iso flavone, 6-a-hydroxyp haseolin, Kietene, 2-methoxy phaseollin isoflavons
Pisum sativum	Pisatin, Maackiain, 3-hydroxy-2, 9-dimethoxy pterocarpon
Glycene max	Hydroxy phaseolin, Glyceollin, Wightione
Medicago sativa	Medicarpin, Salivan, Isosatiasnae, Coumestral, Medicalgal
Vicia faba	Wyerene acid, Wyerone, Wyerone epoxide, Medicarpin
Cicer arietinum	Medicarpin, Maackiain
Cajanus cajan	Cajanin, Cajanol, Pinostrobin chalcone
Solanum tuberosum	Rishitin Rishitinol, Lubimin, Phytotuberin
S. melongena	Lubimin, 9-oxonoberolidal
Capsicum frutescens	Capsidiol, Capsenone
Nicotiana tobacum	Capsidiol, Luimin, Rishitin, phytuberin
Beta vulgaris	Betavulgarin, Betagarin
Gossypium hirsutum	Gossypol, Vergosin
Daucus carota	Isocoumarin, Eugenin, 6-hydroxymellein, 6-methoxymellein
Orchis millitaris	Orchinol
Oryza sativa	Momilactone A and Momilactone B

Table 9.3(2d): Occurrence of Similar Phytoalexins in Different Host Species

Phytoalexins	Host Species
Pisatin	Pisum sativum, Lathyrus spp., Thermopsis spp., Trifolium pratense
Phaseolin	Phaseoles vulgaris, Vigna sinensis
Medicarpin	Medicogo sativa, Vicia faba, Trigonella spp., Melitotus spp., Canavalia eusiformis, Vigna spp.
Rishitin	Solanum tuberosum, N. tobacum, Lycopersicum esculentum
Kievitone	Lablab niger, Phaseoles lunatus, P. vulgaris, Vigna sinensis
Capsidiol	Capsicum frutescens, Datura stramonium, N. tobacum
Wyerons and Wyerone acid	Vicia faba, V. galilea

Phytoalexin biosynthesis occurs after a diversion of primary metabolic pathways. This diversion often arises from the de novo induction of enzymes, such as PAL, that control key branch point in the biosynthesis Pathways. However because the biosynthesis of most phytoalexin requires the activity of numerous enzymes, highly coordinated signaling events must be required in the attacked cells to establish successfully for this type of defense responses. Although the phytoalexin have an undeniable antimicrobial activity in vitro, the extent of their role in R-gene dependent responses in plants remain to be determined.

Signal Transudation and Possible Mechanism of Systemic Resistance

Two type of resistance components are found against pathogen, non-host and host or race/cultivars-specific. When pathogen attack does not result in growth and multiplication, a hypersensitive reaction/response characterized by localized cells and/or tissue death at the site of infection, is observed (Van loon, 1997) besides gene for gene recognition (R and *avr* genes mediated), non-host resistance operates by the recognition of specific pathogen or plant cell wall derived signal molecules, termed exogenous or endogenous elicitors, respectively. These elicitors may be glycoproteins, peptides, or oligosachharides, which activates multiple intracellular defiance signaling pathways. In short, this multi-component signal cascade response of plants to pathogens or inducing agent in host and non-host resistance is triggered by ligand/receptor interaction. The *avr* gene and pathogen inducing agent or plant surface generated elicitors serve as a ligand for membrane cytosolic receptor. The liganed receptor interaction changes the plasma membrane permeability resulting in Ca^{2+} and H^+ influx and K^+ and Cl^- efflux (Mc Dowell and Dangel, 2000). As a consequence, ion fluxes, reactive oxygen intermediates such bas superoxide (O^{-2}), hydrogen peroxide (H_2O_2) and hydroxyl free radical (OH^-) will be deduced by the activity of plasma membrane located NADPH oxidase and/or apoplastic-localized peroxidases. Change in ion influx, delay the hyper sensitive response by preventing the increase in cytosolic Ca^{2+} concentration through calcium channel inhibitors. Calcium channels pass the signal received from heterotrimeric GTP-binding proteins by kinase/phosphate activity, for down stream regulation. The active oxygen molecules and nitric oxide generated by ion flux and disturbances may act as a second messenger for HR-reaction and defense gene expression. Simultaneously, lipid bound unsaturated fatty acids in membrane release linolenic acid by induction of phospholipasese. Linolenic acid acts as an upstream substrate of jasmonic acid and its derivatives via an enzyme cascade. Reactive oxygen molecules generates during the oxidative burst to act as the central component of defense machinery in cell death and hypersensitive response in plants H_2O_2 production at plant cell surface rapidly stimulates peroxide mediated oxidative cross-linking of structural proteins in the cell wall to create stronger physical hindrances for the pathogen. The H_2O_2 is generated by superoxide (O^{2-}) usually; jasmonate, ethylene, and salicylic acid is the free major regulatory molecule in induction of defense related genes (Sticher *et al.*, 1997; Van Loon, 1997; Knoster *et al.*, 2001).

The exact role of ethylene in the defense response is not clear but this molecule is reported to preferentially induce basic PR-proteins. However, in combination with jasmonate, and ethylene regulates expression of several other genes. During SAR, accumulation of SA has been observe through out the plant and correlates well with the activation of PR-genes (Uknes *et al.*, 1993; Camron *et al.*, 1994.) Transgenic mutants (*NahG*) encoding salicylic acid is responsible for SAR signaling pathways (Lawton *et al.*, 1995).

Several reports support that both SA-dependent and SA-independent ISR responses are operative in host-pathogen interaction. In case of *P. aeruginosa 7NSK2* and SA-over production by genetically modified *P. fluoresecens P3* strain, triggering the SA-dependent SAR pathway has been reported (De Meyer *et al.*, 1999 a, b; Maurhofer *et al.*, 1998). Biological control strain *Serratia marcescens* 90-168 induces SA-dependent systemic resistance in both wild type and *NahG* mutant tobacco against *Pseudomonas syringae pv. tabaci* (Press et al., 1997). *Pseudomonas fluorescens strain WCS 417r* induces SA-independent ISR in *Arabidopsis* and showed normal level of ISR against *Pst DC3000* after root colonization (Pieterse *et al.*, 1996; Van Wees *et al.*, 1997). SA-independent systemic defense response by *WCS 417r* was confirmed in *sid1-I* and *sid2-I* mutant deficient in SA induction.

Mutant *jar*1-l and *etrl-1* deficient in jasmonic acid and ethylene responses and *npr*1-1 mutant did not exhibit ISR responses confirming that *WCS417r* mediated ISR requires both JA and ethylene and is dependent on *NPRI* (Pieterse *et al.*, 1998). The role of jasmonate and ethylene has also been confirmed by use of methyl jasmonate and ethylene precursor, l-aminocyclopropane-l-carboxylate (*ACC*), which effectively induced ISR responses in *NaHG* plants against Pst DC 3000. But when *jar*1-, *etr*1-1 and *nprl*-1 plants were tested; MeJA was not able to induce resistance in all the mutants whereas *ACC* did not induce it in *etrl*-1 and *nprl*-1 mutants. This suggested that ethylene mediated response acts downstream to jasmonic acid and *ntr* 1-1 down stream to ethylene. Downstream of *NPRI*, PR-genes are activated in SAR but not in ISR; this is dependent on the pathway activated up stream of it (Cao *et al.*, 1994; Pieterse *et al.*, 1998).

Future Perspective

Future research should address the time, location, and attentiveness of defense molecules as well as their control relationship to other putative defense response. With current technology the breeding of increased levels of induced resistance in to crop plants has made it possible to reduce the need for certain costly and potential harm full pesticides. The disease on leaves, stems and on fruits under the favorable condition result in de-foliation, drying-off of twig and pre-mature fruit drop and thus causing losses rising in fruit yields. With growing environmental consciousness through out the globe and persistent demand for eco-friendly products, there is need to produce fruits of good quality both for export and domestic consumption because most crops with great significance is widely grown in India. From many years, several reports indicated that viruses have been used in the green house and field to protect tomato against virulent strains of the viruses. In addition to viruses bacteria and fungi can be used as inducers for the control of disease. However, this may not be as economical in some cases, especially in area with highly developed agriculture and using pathogenic organism may cause additional problems. Biological control offers an eco-friendly approach to manage plant diseases and can be incorporated into limited chemicals usage for an effective integrated disease management. The natural protections of plants against pathogens are partly based on a variety of constitutive barriers present in the plant before the actual attack; current concern about the environment indicates a need to limit application of chemicals for plant disease control. Abroad use of pesticides is causing the environmental problems because they are ecologically harmful, and toxic to many organisms, and may lead to a development of pesticides resistance in pathogen. To produce pesticides free produce, and improve the environment, one needs to utilize an alternative pollution-free method for disease control. One of the most encouraging techniques for plant protection is the method of induced resistance and biological control. Induced resistance in plants can be a natural, safe effective, persistence alternative to pesticides in controlling plant diseases, it has not yet been utilized on a large scale, there are reports of the practical application of plants immunization.

The motivation for using induced resistance and biological control in my work is due to the use of chemical pesticides to protect and maintain yield in crops, which replaced the natural defense of plants.

ISR was described using plant growth promoting rhizobacteria (PGPR) colonizing the rhizosphere as biological agent. PGPR applied to the soil remain localized at the root surface and can induce resistance in the leaves and or the stem. Evidence from several experimental systems indicates that PGPR can protect plants systemically against various pathogens without causing any side effects.

New technology in all areas has improved agricultural production, and its sustainability. Today's agriculture is using best management practices (BMP's), by targating many of its applications, not broadcasting as was done in the past. New disease hybrides, biological pest control, reduced fungicides use; cultural practices that reduce the incidence of diseases and better placement and reduced amounts of fertilizers are all being employed Several reports indicate that non-pathogenic organism can be used successfully to induce the resistance in plants. Development of utilizes same of this organism to provide the best protection is essential. Seed treatment with selected PGPR strains (Wei et al.1991), or root treatment with same PGPR may reduce the cost of application.

As described pathogenesis related proteins (PRs) expression is correlated with increased resistance in several plants, but the exact function of PRs in the defense reaction against fungi remain to be defined more exactly.

A potential principle for chemically mediated disease control could be based on compound that would induce the IR response. The concepts of chemical, which have no direct anti-microbial activity for disease control, provide an exiting prospect. Salicylic acid (SA) plays a role in signal transudation after pathogen infection and has long been known as exogenous inducers of PR proteins accumulation and resistance.

Strains of genus *Pseudomonas* has recently gained considerable importance against a range of diseases and protection has been observed in a split root system.

The control of disease is achieved mainly with the fungicides/chemicals, alone or integrated with other strategies. No intensive efforts have been made to explore the possibility of controlling this disease with inducible host defenses response. The role of siderophore and inducers alone or with PGPR in induction of resistance against plant disease should be evaluated in proper manner because more work on these strategies is still remaining.

Chemical inducers may provide better means of applications possibilities for induction of resistance providing that they are easily accessible and not harm full, while found that, benzothiadiazole (BTH) alone sufficiently controlled bacterial spot and speck tomato; in some trials it was still necessary to add specific fungicides to control late and early blight. While one would predict that the combined use of SAR elicitors with standard pesticides should extended the effectiveness of the pesticides, experiment conducted by the Bousset and Pons-Kuhnemann (2003) suggest just the opposite's may actually on intensify the selective pressure exerted by certain pesticides on pathogen and pest populations. Clearly this is concern, especially when neither SAR nor ISR render the plant impervious to the pathogen, but rather seem to increase the plants overall tolerance.

In order to take advantage of plant immunization a plant's natural meaning protection against disease, we need more applied research to stimulate interest of industries in commercialization of current methodology and participating in the development of new technology. We also need to change your current thinking about the plant protection. Plant can be immunized and perhaps we are harming our selves by the application of pesticides to stop fungal activity in early stage of plant development when the induction process is the most effective. If wish of reduce pollution and strive for a healthy environment, we must consider alternatives to pesticides for the control of plant pests and diseases. It appears logical to take advantage of mechanisms, which helped plants with stands, the test of time and contribution to the survival of plants through out evolution. Although many exciting insights have emerged from recent research on plant defense signaling, our overall understanding of the process is still fragmentary. For example, we still know very little about the structural basis of pathogen recognition. In fact, we are less sure than before about what PR proteins actually recognize (*Avr* proteins, modified guardees, or complexes that include both?). Furthermore, many gaps remain in our models of the defense signal transudations network and these must be bridged before we can design truly rational strategies to activate the network. Nevertheless, useful applications are already being developed from our relatively limited knowledge base and others will undoubtedly follow as our level of basic understanding grows. In the short term, we expect that additional useful R genes will be cloned and that models of resistance signaling developed in *Arabidopsis* will continue to be evaluated for applicability in crops. We also anticipate the application of functional genomic tools to disease resistance, which will greatly accelerate the pace of discovery and provide new insights into interactions between defense signaling and other plant processes. In the longer term, a detailed understanding of the structural basis of recognition will enable us. to approach the holy grail of designing R proteins that recognize essential virulence factors (or important guardees). Given the assortment of strategy that pathogens use their ability to rapidly adapt, and rash to predict the development of magic shot for long-lasting, broad-spectrum resistance. However, it is rational to expect a forth-coming array of refined weapons that will provide effective protection in certain contexts, when thoughtfully incorporated with other control measures.

The productions of crops are partial due to its vulnerability to diseases. Strategies for increasing and stabilizing its production have been to expand resistant to diseases. The advantages and limitations of these techniques along with directions for future research are under discussion

Current studies documented that the specificity of phytoimmunity is at the beginning stages of genic associations between a pathogen and its plant host. Though elicitors, the products of pathogen a virulence genes (*avr-genes*), have been already investigated in aspect, there are few studies on the products of plant resistance genes (*R-genes*) recognizing these elicitors. The molecular-genetic studies of the relations between the pathogens and their plant host's appendage the already accessible armory for plant protection with new technologies, such as the genetically engineered plant resistance and the resistance induced by biogenic elicitors.

Acknowledgement

Authors are thankful to H.O.D., Department of Botany, BHU, Varanasi for providing facilities and one of the author (Anuj Kumar) extended his thanks to UGC New Delhi for financial support.

References

Agrawal, A.A., Tuzun, S., Bent, E. (eds.) (1999) Inducible plant defenses against pathogens and herbivores. In: *Biochemistry, Ecology, and Agriculture.* American Phytopathological Society, St. Paul, Minnesota.

Agrios, G.N. (1997). *Plant Pathology,* 4th edn. Academic Press, New York.

Ahl, P. and Gianinazzi, S. (1982b). Protein as a constitutive component in highly (TMV) resistant interspecific hybrids of *Nicotiana glutinosa* x *Nicotiana debneyi. Plant Science Letters,* 26: 173–181.

Ahl, P. (1984). Lutte cente les microorganisms pathogens des vegetaux: les in ducteur de resistance. *Phytopathol. Z.* 109: 45–64

Alenxander, D., Goodman, R.M., Gut Rella, M., Glascock, C., Weymann, K. (1993) Increased tolerance to two Oomycete pathogens in transgenic tobacco expressing pathogenesis related proteins *1a. Proc. Natl. Acd. Sci. USA,* 90: 7327–7331.

Alstrom, S. (1991). Induction of disease resistance in common beansusceptible to halo blight bacterial pathogen after seed bacterization with rhizosphere *pseudomonads. J. Gen. Appl. Microbiol.,* 37: 495–501.

Altman, A. (ed.) (1998). *Agricultural Biotechnology.* Marcel Dekker, New York.

Anfoka, G. and Buchenauer, H. (1997). Systemic acquired resistance in tomato against *Phytophthora infestans* by pre-inoculation with tobacco necrosis virus. *Physiol. Mol. Plant Pathol.,* 50: 85–101.

Audenaert, K., Pattery, T., Cornelis, P. and Hofte, M. (2002b). Induction of systemic resistance to *Botrytis cinerea* in tomato by *Pseudomonas aeruginosa 7NSK2:* role of salicylic acid, pyochelin and pyocyanin. *Mol. Plant-Microbe Interact.,* 15: 1147–1156.

Baker, B., Zambryski, P., Staskawicz, B. and Dinesh Kumar, S.P. (1997). Signalling in plant-microbe interactions. *Science,* 276:726–733.

Bakker, P.A.H.M., Lamers, J.G., Bakker, A.W., Marugg, J.D., Weisbeek, P.J. and Schippers, B. (1986). The role of siderophores in potato tuber yield increase by *Pseudomonas putida* in a short rotation of potato. *Neth. J. Plant Pathol.,* 92: 249–256.

Bakker, P.A.H.M., Raaijmakers, J.M. and Schippers, B. (1993) Role of iron in the suppression of bacterial plant pathogens by *fluorescent pseudomonads.* In: *Iron Chelation in Plants and Soil Microorganisms,* (Eds.) Barton, L.L. and Hemming, B.C. Academic Press, San Diego, pp. 269–278.

Barton, L.L. and Hemming, B.C. (1993). *Iron Chelation in Plants and Soil Microorganisms.* Academic Press, San Diego, CA.

Benhamou, N. and M. Nicol. (1999). Cell biology of plant immunization against microbial infection: The potential induced resistance in controlling plant disease. *Plant Physiol. Biochem.,* 37: 703–719.

Bigirimana, J. and Hofte, M. (2002). Induction of systemic resistance to *Colletotrichum lindemuthianum* in bean by benzothiadiazole derivative and rhizobacteria. *Phytoparasitica,* 30: 159–168.

Boller, T., Mattoo A. and Suttle, J. (eds.) (1991). Ethylene in pathogenesis and disease resistance. In: *The Plant Hormone Ethylene.* CRC Press, N.Y., pp. 293–314.

Bousset, L. and J. Pons-Kuhnemann. (2003). Effect of acibenzolar-S-methyl and ethrimol on the composition of laboratory population of barley powdery mildew. *Phytopathology,* 82: 304–315.

Bowling, S.A., Clarke, J. D., Liu, Y., Klessig, D. F. and Dong, X. (1997) The *cpr5* mutant of *Arabidopsis* expresses both *NPR1*-dependent and *NPR1*-independent resistance. *Plant Cell*, 9: 1573–1584.

Broglie, K., Chet, I., Holliday M., Cressman, R. and Biddle, P. (1991). Transgenic plants with enhanced resistance to fungal pathogen *Rhizoctonia solani. Science*, 254: 1194–1197.

Buysens, S., Heungens, K., Poppe, J. and Hofte, M. (1996). Involvement of pyochelin and pyoverdin in suppression of *Pythium*-induced damping-off of tomato by *Pseudomonas aeruginosa 7NSK2. Appl. Environ. Microbiol.*, 62: 865–871.

Cameron, R.K, Dixon, R. and Lamb, C. (1994). Biologically Induced systemic acquired resistance in *Arabidopsis thaliana. Plant J.* (In Press).

Cao, H., Bowling, S.A., Gordon, A.S. and Dong, X. (1994). Characterization of an *Arabidopsis. Cell*, 6: 1583–1592.

Chang, C., Kwok, S.F., Bleecker, A.B. and Meyerowitz, E.M. (1993). *Arabidopsis* ethylene-response gene *ETRI:* Similarity of product to two-component regulators. *Science*, 262: 539–544.

Chester, K. (1933). The problem of acquired physiological immunity in plants. *Quart. Rev. Biol.*, 8: 275–324.

Clark, K.L., Larsen, P.R., Wang, X. and Chang, C. (1998). Association of the *Arabidopsis CTRI* Raf-like kinase with the *ETRI* and *ERS* ethylene receptors. *Proc. Natl. Acad. Sci. USA*, 95: 5401–5406.

Cohen, Y. (1994). Local and systemic control of *Phytophthora infestans* in tomato plants by DL-3-amino-n-butanoic acids. *Phytopathology*, 84: 5559.

Cruikshank, I.A.M. and Mandryk, M. (1960). The effect of site infection of tobacco with *Pernospora tabacina adam.* on foliage reaction to blue molds. *J. Aust. Inst. Agric. Sci.*, 26: 369–372.

Dalisay, R.F. and Kuc, J.A. (1995). Persistence of reduced *penetration* by *Colletotrichum lagenarium* into cucumber leaves with induced systemic resistance and its relation to enhanced peroxidase and chitinase activities. *Physiol. Mol. Plant Pathol.*, 47: 329–338.

De Meyer, G. and Höfte, M. (1997). Salicylic acid produced by the rhizobacterium *Pseudomonas aeruginosa 7NSK2* induces resistance to leaf infection by *Botrytis cinerea* in bean. *Phytopathology*, 87: 588–593.

De Meyer, G., Audenaert, K. and Hofte, M. (1999). *Pseudomonas aeruginosa 7NSK2*-induced systemic resistance in tobacco depends on in planta salicylic acid but is not associated with PR1 a gene expression. *Eur. J Plant Pathol.*, 105: 513–517.

Delaney, T.P.L., Friedrich and J. Ryals. (1995). *Arabidopsis* signal transduction mutants defective in chemically and Biologically induced disease resistance. *Proc. Natl. Acad. USA*, 92: 6602–6606.

Delaney, T.P.L., Uknes, S., Vernooji, B., Friedrich and Weymann. (1994). A central role of salicylic acid in plant disease resistance. *Science*, 266: 1247–1249.

De Wit, P.J.G.M. (1997). Pathogen a virulence and plant resistance: A key role for recognition. *Trends Plant Sci.*, 2: 452–458.

Diallinas, G. and Kanellis, A.K. (1994). A phenylalanine ammonia-lyase gene from melon fruit: cDNA cloning, sequence and expression in response to development and wounding. *Plant Mol. Biol.*, 26: 473–479.

Dixon, R.A. and Paiva, N.L. (1995). Stress-induced phenyl propanoid metabolism. *Pl. Cell.*, 7: 1085–1095.

Duijff, B.J., Bakker, P.A.H.M. and Schipper, B. (1994). Suppression of *Fusarium* wilt of carnation by *Pseudomonas putida WCS358* at different levels of disease incidence and iron availability. *Biocontrol Sci. Technol.*, 4: 279–288.

Duijff, B.J., Gianinazzi-Pearson, V. and Lemanceau, P. (1997). Involvement of the outer-membrane lipopolysaccharides in the endophytic colonization of tomato roots by biocontrol *Pseudomonas fluorescens* WCS417r. *New Phytol.*, 135: 325–334.

Farmer, E.E., Weber, H. and Vollenweider, S. (1998). Fatty acid signaling in Arabidopsis. *Planta*, 206: 167–174.

Fraser, R.S.S. (1982). Are 'pathogenesis-related' proteins involved in acquired systemic resistance of tobacco plants to tobacco mosaic virus? *Journal of General Virology*, 58: 305–313.

Gaffney, T., Friedrich, L., Vernooij, B., Negrotto, D., Nye, G., Uknes, S., Ward, E., Kessmann, H. and Ryals, J. (1993). Requirement of salicylic acid for the induction of systematic acquired resistance. *Science*, 261: 754–756.

Geels, F.P. and Schippers, B. (1983a). Selection of antagonistic fluorescent *Pseudomonas* spp. and their root colonization and persistence following treatment of seed potatoes. *J. Phytopathol.*, 108: 193–206.

Geels, F.P. and Schippers, B. (1983b). Reduction of yield depressions in high frequency potato cropping soil after seed tuber treatments with antagonistic fluorescent *Pseudomonas* spp. *J. Phytopathol.*, 108: 207–214.

Gianinazzi, S. and Ahl, P. (1983) The genetic and molecular basis of β-proteins in the genus *Nicotiana. Netherlands Journal of Plant Pathology*, 89: 275–281.

Gorlach, J.S., Volroth, F., Knauf-Beiter, G., Hengy, U., Beckhove, K.H., Kogel, M., Oostendorf, T., Stsub, E., Ward, H., Kessmann and Rayls. (1996). Benzothiadiazole, a novel class of inducers of systemic resistance, activates gene expression and disease resistance in wheat. *Plant Cell*, 8: 629–643.

Hammerschmidt, R. and Smith-Becker, J.A. (2000). The role of salicylic acid in disease resistance. In: *Mechanism of Resistance to Plant Diseases*, (Eds.) Slusarenko, A., Fraser, R.S.S. and Van Loon, L.C. Kluwer Academic Publisher, pp. 37–53.

Hammerschmidt, R. and KuC, J. (1995). Induced resistance to disease in for the induction of systemic acquired resistance. *Science*, 261: 754–756.

Hammerschmidt, R., Nuckles, F. and Kuc, J. (1982). Association of enhance peroxidase activity with induced systemic resistance of cucumber to *Colletotrichum lagenarium. Physiol. Plant Pathol.*, 20: 73–82.

Heller, W.E. and Gessler, C. (1986). Induced systemic resistance in tomato plants against *Phytophthora infestans. J. Phytopathol.*, 116: 323–328.

Hoffland, E., Bakker, P.A.H.M. and Van Loon, L.C. (1997). Multiple disease protection by rhizobacteria that induce systemic resistance-reply. *Phytopathology*, 87: 138.

Hoffland, E., Pieterse, C.M.J., Bik, L. and Van Pelt, J.A. (1995). Induced systemic resistance in radish is not associated with accumulation of pathogenesis-related proteins. *Physiol. Mol. Plant Pathol.*, 46: 309–320.

Hua, J. and Meyerowitz, E. M. (1998) Ethylene responses are negatively regulated by a receptor gene family in *Arabidopsis thaliana. Cell*, 94: 261–271.

Huang, Y., Li, H., Hutchison, C.E., Laskeg, J. and Kieber, J.J. (2003). Biochemical and functional analysis of *CTRI*, a protein kinase that negatively regulates ethylene signaling in *Arabidopsis*. *Plant J.*, 33: 221–233.

Iavicoli, A., Boutet, E., Buchala, A. and Metraux, J.P. (2003). Induced systemic resistance in *Arabidopsis thaliana* in response to root inoculation with *Pseudomonas fluorescens* CHAO. *Mol. Plant-Microbe Interact.* 16: 851–858.

Kassanis, B., Gianinazzi, S. and White, R.F. (1974). A possible explanation of the resistance of virus-infected tobacco to second infection. *Journal of General Virology*, 23: 11–16.

Kessler, A. and I.T. Baldwin (2002). Plant responses to insect herbivory: The emerging molecular analysis. *Ann. Rev. Plant. Biol.*, 53: 299–328.

Kessmann, H., staub, C., Hofmann, T., Maetzke, and J. Herzog. (1994). Induction of systemic acquired disease resistance in plants by chemicals. *Ann. Rev. Phytopathol.*, 32: 493–459.

Kloepper, J.W. and Tuzun, S. (1996). Induced systemic to diseases and increased plant growth by growth-promoting rhizobacteria under field conditions. *Phytopathology*, 81: 1508–1516.

Kloepper, J.W. (1991). Plant growth-promoting rhizobacteria as biological control agents of soil borne diseases. In: *The Biological Control of Plant Diseases*, (Ed.) Bay-Petersen, J. FFTC Book Series No. 42. Food and Fertilizer Technology Center, Taipei, Taiwan.

Kloepper, J.W., Tuzun, S. and KuC, J.A. (1992). Proposed definitions related to induced resistance. *Biocontrol Sci. Technol.*, 2: 349–351.

Knoester, M., Linthorst, H.J.M., Bol, J.F. and Van Loon, L.C. (2001). Involvement of ethylene in lesion development and systemic acquired resistance in tobacco during the hypersensitive reaction to tobacco mosaic virus. *Physiol. Mol. Plant Pathol.*, 59: 45–57.

Kovats, K., Binder, A. and Hohl, H.R. (1991) Cytology of induced systemic resistance of tomato to *Phytophthora infestans. Planta*, 183: 491–496.

KuC, J. (1982). Induced immunity to plant disease. *BioScience*, 32: 854–860.

KuC, J. (1997). Plant immunization and its applicability for disease control. In: *Innovative Approaches to Plant Disease Control*, (Ed.) Chet. John Wiley and Sons, New York, pp. 225–274.

Lawton, K., Ukenes, S. and Ryals, J. (1994). The role of ethylene in systemic acquired resistance. *Plant Cell* (In Press).

Leeman, M., den Ouden, F.M., van Pelt, J.A., Dirkx, F.P.M., Steijl, H., Bakker, P.A.H.M. and Schippers, B. (1996). Iron availability affects induction of systemic resistance to *fusarium* wilt of radish by *Pseudomonas fluorescens. Phytopathology*, 86: 149–155.

Leeman, M., Van Pelt, J.A., Den Ouden, F.M., Heinsbroek, M., Bakker, P.A.H.M. and Schippers, B. (1995). Induction of systemic resistance against *fusarium* wilt of radish by lipopolysaccharides of *Pseudomonas fluorescens. Phytopathology*, 85: 1021–1027.

Leroux, P. (1996). Recent development in the mode of action of fungicides. *Pestic. Sci.*, 47: 171–197.

Liu, L., Kloepper, J.W. and Tuzun, S. (1995). Induction of systemic resistance in cucumber against bacterial angular leaf spot by plant growth promoting rhizobacteria. *Phytopathol.*, 85: 843–847.

Lucas, J.A. (1999). Plant immunization: From myth to SAR. *Pestic. Sci.*, 47: 191–197.

Lyon, G.D., T., Reglinski, and A.C. Newton. (1995). Novel disease control compounds: The potential 'immunize' plants against infection. *Plant Pathol.*, 44: 407–427.

Malamy, J., Carr, J.P., Klessig, D.F. and Raskin, I. (1990). Salicylic acid a likely endogenous signal in the resistance response of tobacco to viral infection. *Science*, 250: 1002–1004.

Maurhofer, M., Hase, C., Meuwly, P., Metraux, J.P. and Defago, G. (1994) Induction of systemic resistance of tobacco to tobacco necrosis virus by the root-colonizing *Pseudomonas fluorescens* strain CHAO: Influence of the *gac A* gene and of pyoverdine production. *Phytopathology*, 84: 139–146.

Maurhofer, M., Reimmann, C., Schmidli-Sacherer, P., Heeb, S., Haas, D. and Défago, G. (1998). Salicylic acid biosynthetic genes expressed in *Pseudomonas fluorescens* strain *P* 3 improve the induction of systemic resistance in tobacco against tobacco necrosis virus. *Phytopathology*, 88, 678–684.

McDowell, J.M. and J.L. Dangl. (2000). Signal transduction in the plant immune response. *Trends Biol. Sci.*, 25: 79–82

Metraux, J.P., Signer, H., Ryals, J., Ward, E., Wyss-benz, M., Gaudin, J., Raschorf, K., Schmid, E. and Blum, W. (1990). Inverardi B. Increase in salicylic acid at the one set of systemic acquired resistance in cucumber. *Science*, 250: 1004–1006.

Murphy, J.F., G.W., Zender, D.J., Sticher, E.J., Sikora, J.E., Polston, and J.W. Kloepper. (2000). Plant growth promoting rhizobactria mediated protection in tomato against *Tomato mottle virus. Plant Dis.*, 84: 779–784.

Nand Kumar, R.S., Babu, R., Viswanathan, T., Raguchander, and R. Samiyappan. (2001). Induction of systemic resistance in rice against sheath blight disease by *Pseudomonas flurescence. Soil. Biol. Biochem.*, 33: 603–612.

Niranjan, Raj, S.G., Vhaluaraju, K.N., Amruthesh, H.S., Shetty, M.S., Reddy and J.W. Kloepper. (2003). Induction of growth promotion and resistance against downy mildew on pearl millet (*Pennisetum glaucum*) by rhizobacteria. *Plant Dis.*, 87: 380–384.

Ongena, M., Giger, A., Jacques, P., Dommes, J. and Thonart, P. (2002). Study of bacterial determinants involved in the induction of systemic resistance in bean by *Pseudomonas putida BTP1. Eur. J. Plant Pathol.*, 108: 187–196.

Penninckx, I.A.M.A., Eggermont, K., Terras, F.R.G., Thomma, B.P.H.J., Samblanx, G.W., Buchala, A., Metraux, J.P., Manners, J.M. and Broekaert, W.F. (1997). Pathogen-induced systemic activation of a plant defensin gene in *Arabidopsis* follows a salicylic acid-independent pathway involving components of the ethylene and jasmonic acid responses. *Plant Cell*, 8: 2309–2315.

Pieterse, C.M.J. and Van Loon, L.C. (2004). NPR1: The spider in the web of induced resistance signaling pathways. *Current Opinion in Plant Biology*, 7: 456–464.

Pieterse, C.M.J., Van Pelt, J.A., Ton, J., Parchmann, S., Mueller, M. J., Buchala, A. J., Metraux, J.-P. and van Loon, L.C. (2000). Rhizobacteria-mediated induced systemic resistance (ISR) in *Arabidopsis* requires sensitivity to jasmonate and ethylene but is not accompanied by an increase in their production. *Physiol. Mol. Plant Pathol.*, 57: 123–134.

Pieterse, C.M.J., Van Pelt, J.A., Verhagen, B.W.M., Ton, J., van Wees, S.C.M., Leon-Kloosterziel, K.M. and van Loon, L.C. (2003). Induced systemic resistance by plant growth promoting rhizobacteria. *Symbiosis*, 35: 39–54.

Pieterse, C.M.J., Van Wees, S.C.M., Hoffland, E. and Van Pelt, J.A. (1996). Systemic resistance in *Arabidopsis* induced by biocontrol bacteria is independent of salicylic acid accumulation and pathogenesis-related gene expression. *Plant Cell*, 8: 1225–1237.

Pieterse, C.M.J., Van Wees, S.C.M., Van Pelt, J.A., Knoester, M., Laan, R., Gerrits, H., Weisbeek, J.P. and van Loon, L.C. (1998). A novel signaling pathway controlling induced systemic resistance in *Arabidopsis*. *Plant Cell*, 10: 1571–1580.

Press, C.M., Wilson, M., Tuzun, S. and Kloepper, J.W. (1997). Salicylic acid produced by *Serratia marcescens* 90–166 is not the primary determinant of induced systemic resistance in cucumber or tobacco. *Mol. Plant-Microbe Interact.*, 10: 761–768.

Press, C.M. and Kloepper, J.W. (1997). Role of bacterial cell constituents in PGPR-mediated induced systemic resistance of cucumber. In: *Plant Growth: Promoting Rhizobacteria, Present Status and Future Prospects*, (Eds.) A. Ogoshi, K. Kobayashi, Y. Homma, F. Kodama, N. Kondo, and S. Akino. *Proc. Int. Workshop Plant Growth–Promoting Rhizobacteria, 4th Japan–OECD Joint Workshop*. Nakanishi Printing, Sapporo, Japan, pp. 260–264.

Preston, G.M., Bertrand, N. and Rainey, P.B. (2001). Type III secretion in plant growth-promoting *Pseudomonas fluorescens SBW25*. *Mol. Microbiol.*, 41: 999–1014.

Rasmussen, J.B., Hammerschmidt, R. and Zook, M.N. 1991. Systemic induction of salicylic acid accumulation in cucumber after inoculation with *Pseudomonas syringaepv. syrigae*. *Plant Physiol.*, 97: 1342–1347.

Raupachh, G.S. and J.W. Kloepper. (2000). Biocontrol of cucumber disease in field by plant growth promoting rhizobacteria with and without methyl bromide fumigationpathogens. *Phytopathology*, 84: 1073–1075.

Raupachh, G.S. and J.W. Kloepper. (1998). Mixture of plant growth promoting rhizobacteria enhanced biological control of multiple cucumber pathogens. *Phytopathology*, 55: 1158–1580.

Ren, Y.Y. and West, C.A. (1992). Elicitation of diterpenes biosynthesis in rice (*Oryza sativa* L.) by chitin. *Plant Physiol.*, 99: 1169–1178.

Romeiro, R.S., Lanna Filho, Vieira Junior, J.R., Silva, H.S.A., Baracat-ereira, M.C. and Carvalho, M.G. (2005). Macromolecules released by a plant growth-promoting *Rhizobacterium* as elicitors of systemic resistance in tomato to bacterial and fungal. *Pathogens*, 153: 2–120.

Ross, A.F. (1961). Systemic acquired resistance induced by localized virus infections in plants. *Virology*, 14: 340–358.

Ross, A.F. (1961). Localized acquired resistance to plant virus infection in hypersensitive hosts. *Virology*, 14: 329–339.

Ryals, J.A., Neuenschwander U.H., Willits, M.G., Molina, A., Steiner, H.Y. and Hunt, M.D. (1996). Systemic acquired resistance. *The Plant Cell*, 8: 1809–1819.

Samac, D.A. and Shah, D.M. (1991) Developmental and pathogen induced activation of the *Arabidopsis* acidic chitinase promoter. *Plant Cell*, 3: 1063–1072.

Schweizer, P., Gees, R.E.M. and M. Singer, E. (1993). Effect of jasmonic acid on the interaction of barley (*Hordeum vulgare* L.) with the powdery mildew *Erysiphe graminis f.sp. hordei*. *Plant Physiol.*, 102: 503–511.

Simpson, A.J.G., Reinach, F.C., Arruda, P., Abreu, F.A., Acencio, M., Alvarenga, R., Alves, L.M.C., Araya, J.E., Baia, G.S., Baptista, C.S., Barros, M.H., Bonaccorsi, E.D., Bordin, S., Bove, J.M. and Birones, M.R.S., *et al.* (2000). The genome sequence of the plant pathogen *Xylellafastidiosa. Nature,* 406: 151–157.

Spanu, P. and Boiler, T. (1989) Ethylene biosynthesis in tomato plants infected with *Phytophthora infestans. Plant Physiol.,* 134: 533–537.

Sticher, L., Mauch-Mani, B. and Metraux, J.P. (1997). Systemic acquired resistance. *Ann. Rev. Phytopath.,* 35: 235–270.

Thomashow, L.S., Weller, D.M., Bonsall, R.E. and Pierson, I.S. (1990). Production of the antibiotic phenazin-l-carboxylic acid by fluorescent *Pseudomonas* species in the rhizosphere of wheat. *Appl. Environ. Microbiol.,* 56: 908–912.

Ton, J. (2001). Rhizobacteria-mediated induced systemic resistance in Arabidopsis: molecular-genetic basis of induced resistance in relation to basal resistance. *Ph.D. Thesis,* Utrecht University, pp. 1–136.

Ton, J., Pieterse, C.M.J. and Van Loon, L.C. (1999). Identification of a locus in *Arabidopsis* controlling both the expression of rhizobacteria-mediated induced systemic resistance (ISR) and basal resistance against *Pseudomonas syringae* pv. *tomato. Mol. Plant-Microbe Interact.,* 12: 911–918.

Uken, S., Lawton, Ward, E., Gaffeney, T. and Friedrick, L. (1993). The molecular biology of systemic acquired resistance. In: *Plant Responses to the Environmental,* (Ed.) P. Gresshoff. CRC Press, Boca Raton, 2: 1–10.

Van Loon, L.C. (2000). Systemic induced resistance. In: *Mechanisms of Resistance to Plant Diseases.* (Eds.) Slusarenko, A., Fraser, R.S.S. and Van Loon, L.C. Kluwer Academic Publisher, pp. 521–574.

Van Loon, L.C. and Antoniw, J.F. (1982). Comparison of the effects of salicylic acid and ethephon with virus-induced hypersensitivity and acquired resistance in tobacco. *Neth. J. Plant Pathol.,* 88: 237–256.

Van Loon, L.C., Bakker, P.A.H.M. and Pieterse, C.M.J. (1998). Systemic resistance induced by Rhizobacteria. *Ann. Rev. Phytopathol.,* 36: 453–483.

Van Loon, L.C. (1975). Polyacrylamide disc electrophoresis of the 'SamsunNN' IV. Similarity of qualitative changes of specific proteins after infection with different viruses and their relationship to acquired resistance. *Virology,* 67: 566–575.

Van Loon, L.C. (1997). Induced resistance in plants and the role of pathogenesis-related proteins. *European Journal of Plant Pathology,* 103: 753–765.

Van Peer, R. and Schippers, B. (1992) Lipopolysaccharides of plant-growth promoting *Pseudomonas* sp. strain *WCS417r* induce resistance in carnation to *Fusarium* wilt. *Neth. J. Plant Pathol.,* 98: 129–139.

Van Peer, R., Niemann, G.J. and Schippers, B. (1991) Induced resistance and phytoalexin accumulation in biological control of Fusarium wilt of carnation by *Pseudomonas* sp. strain *WCS417r. Phytopathology,* 81: 728–734.

Van Wees, S.C.M., Pieterse, C.M.J., Trijssenaar, A., Van't Westende, Y., Hartog, F. and Van Loon, L.C. (1997). Differential induction of systemic resistance in Arabidopsis by biocontrol bacteria. *Mol. Plant-Microbe Interact.,* 10: 716–724.

Vidhyasekhran, P. (1998). *Physiology of Disease Resistance in Plants.* Vol. II. CRC Press, Inc., Boca Raton, Florida, pp 128.

Vigers, A.J., Roberts, W.K. and Selitrennikoff, C.P. (1991). A new family of antifungal proteins. *Mol. Plant. Microbe. Int.,* 4: 315–323.

Vogeli, U., Meins, F. Jr. and Boiler, B.C. (1989). Coordinated regulation of chitinase and (–1, 3–glucanase in bean leaves. *Planta,* 194: 364–372.

Waffa, M. Haggag. (2002). Sustainable agriculture management of plant disease. *Online Journal of Biological Sciences,* 2(4): 280–284.

Walling, L.L. (2000). The myriad plant response to herbivores. *J. Plant Growth Regual.,* 19: 195–216.

Ward, E.B.W. (1984). Suppression of Metaxylic activity by glyphosphate: Evidence that host defence mechanism contribute to metalaxyl inhibition of *Phytophthora megaspermaf sp.* glycines in sobeans. *Physiol. Plant Pathol.,* 25: 381–386.

Ward, E.R., Uknes, S.J, Williams, S.C., Dincher, S.S. and Wiederhold, D.L. (1991). Coordinate gene activity in response to agents that induce systemic acquired resistance. *Plant Cell,* 3: 1085–1094.

Wei, G., Kloepper, J. W. and Tuzun, S. (1991). Induction of systemic resistance of cucumber to *Colletotrichum orbiculare* by select strains of plant growth-promoting rhizobacteria. *Phytopathology,* 81: 1508–1512.

Wei, Z.M., Wei, Z., Laby, R.J., Zumoff, C.H., Bauer, B.W., He, S.Y., Collmer, A. and Beer, S.V. *et al.* (1992). Harpin, elicitor of the hypetsensitive response produced by the plant pathogen *Erwinia amylovora. Science,* 257: 85–88.

White, R.F. (1979). Acetylsalicylic acid (aspirin) induced resistance to tobacco masaic virus in tobacco. *Virology,* 99: 410–412.

Winkelmann, G. (1991). *CRC Handbook of Microbial Iron Chelates.* CRC Press, Boca Raton, FL.

Woloshank, C.P., Meulenhoff, J.S., Sela-Buurlage, M., van den Elzen, P.J.M. and Comelissen, B.J.C. (1991). Pathogen-induced protein with, inhibitory activity towards *Phytophthora Infestant Cell,* 3: 619–628.

Yalpani, N., Silverman, P., Wilson, T.M.A., Kleier, D.A. and Raskin, I. (1991). Salicylic acid is a systemic signal and an inducer of Pathogenesis-related proteins in virus infected tobacco. *Plant Cell,* 3: 809–818.

Yedidia, I., Benhamou, N., Kapulnik, Y. and Chet, I. (2000) Induction and accumulation of PR proteins activity during early stages of root colonization by the mycoparasite *Trichoderma harzianum* strain *T-203. Plant Physiol. Biochem.,* 38: 863–873.

Zehnder, G.W., J.F. Murphy, Ej. Sikora, and J.W. Kloepper. (2001). Application of rhizobacteria for induced resistance. *Eur. J. Plant Pathol.,* 107: 39–50.

©R.C. Ram & Asha Sinha (*Editors*)
Microbial Diversity: Modern Trends, **2007** **152–160**
Daya Publishing House, New Delhi

Chapter 10

Micronutrients: Key for Disease Management

Raghavendra Singh,* Jai Prakash Rai** and Alok Kumar Singh**

***Department of Agronomy, **Department of Plant Pathology,**
Faculty of Agriculture, Udai Pratap Autonomous College, Varanasi – 221 002, U.P.

Introduction

Among the twenty essential mineral elements or mineral nutrients, eleven are classified as micronutrients. These micronutrients are manganese (Mn), iron (Fe), zinc (Zn); copper (Cu), boron (B), molybdenum (Mo), chlorine (Cl), silicon (Si), vanadium (Va), nickel (Ni) and chromium (Cr). Micronutrients, as the name indicates, are the nutrients that are required by plants in lesser amounts (less than 50 ppm) for proper growth and development (Marshner, 1988). If a plant is suffering from the deficiencies of any of these micronutrients, it will be weakened and will become more susceptible for the onslaughts of diseases. The correction of the deficiency will reduce the disease severity. It is important to note that overcoming nutrient deficiency of various nutrients will have various effects on resistance to diseases.

The effect of nutrient deficiencies on susceptibility of plants to diseases has been known for more than 75 years (Rosen and Ellio, 1923), the mechanistic explanations of the phenomena are still lacking. However, considerable work has been done on trying to unravel how micronutrients contribute to increase the resistance level of the plants to diseases (Graham, 1983; Huber and Wilhelm, 1988; Graham and Webb, 1991; Rengel *et al.*, 1994; Huber and Graham, 1999).

A number of micronutrients are known for their fungicidal effects (*e.g.* Cu, Zn, Mn) and therefore, are used in commercial preparations manufactured for utilization in decreasing incidence of diseases. The situation where plant nutrition by micronutrients is below optimum and, as a consequence of the metabolic changes caused by such apparent deficiencies, plants become more susceptible to certain specific disease. In such a situation alleviating nutrient deficiency will increase plant vigour as well as specific effect on increasing plant resistance to a particular disease.

Micronutrients are not only required for higher plants but also for the growth of pathogenic micro flora and microfauna. Therefore, the soil may be deficient in available micronutrients not only for growth of plants but also for the growth of pathogenic micro flora as well. In such conditions, supplying micronutrients to severely deficient soils may initially improve pathogenic growth and virulence and thus, increase disease severity in comparison with non-amended micronutrient deficient soil, [*e.g.* in case of Mn and the fungus *Gaeiimannomyces graminis* var. *tritici.*, as the causal agent of take all disease of wheat (Wilhelm *et al.*, 1990; Rengel *et al.*, 1994) also Zn and the fungus *Sclerotium rolfsii* causing stem rot disease of soybean (Deb and Dutta, 1992)]. Further addition of micronutrient may start, benefiting plant more than the pathogenic loss and will, therefore, result in greater resistance of plant to the pathogenic infections.

Micronutrients are equally important for proper growth and reproduction of fungi and other pathogenic micro flora. However the limits of micronutrient concentration in soil for proper growth and developments of plants and microorganisms may vary to significant extents. Consequently, the soil with the level of micronutrients too low to be of use to plants may though, be deficient in micronutrients for proper crop cultivation but it may not necessarily be deficient in micronutrients for a given microorganism and can support and promote its growth and reproduction. Supplementation of micronutrients in a soil that is deficient in micronutrients for both plants and pathogens, will improve pathogenic growth and virulence initially and in consequence, disease severity will be increased. This had been proven true in case of manganese and take-all disease of wheat (caused by *Gaeiimannomyces graminis* var. *tritici.* Ggt) and zinc with *Sclerotium rolfsii.* This could be reckoned as increased micronutrient level to such an extent that the micronutrient deficiency for pathogens had been removed but the level of micronutrient is still lower for the proper growth of crop plants. Further addition of micronutrients raises their level in soil that becomes useful to the plants and simultaneously toxic to the pathogens.

The chapter will more concentrate towards the role of micronutrients in various diseases of plants (including field and horticultural crops).

Diseases Managed by Improving Micronutrient Levels

Take-all disease of wheat is a classical example that has been managed effectively by application or enhanced uptake of Manganese (Mn) by wheat plants. The pathogen *Gaeiimannomyces graminis* var. *tritici* is a soil borne ascomycete fungus. Its hyphae penetrate the root cortex followed by invasion of vascular system of the infected plant. A blocked phloem results in cessation of transport of assimilates (photosynthates) to the growing roots and thus, the root system of affected plant is destroyed. A plant deprived with its functional rot system is no more able to survive, its productivity and yield omitted.

It has been observed that well-nourished plants often suffer less damage due to take-all. This could be due to the fact that the extensive root system of well-nourished plants 'escape' the damaging effect of the disease than the roots being resistant to the disease. In more recent research involving measurement of root infection, it has been shown that supplying wheat crop with NH_4–N, Mg, P, K and various other micronutrients reduced the incidence and severity of take-all under controlled conditions and in the field (Lucas *et al.*, 1997; Huber and Graham, 1999).

With take-all disease of wheat it has been a general observation that the disease symptoms are followed by and not preceded by the symptoms of micronutrient deficiency and it is shown by Graham and Webb (1991) that correcting deficiency has a positive correlation with increase in resistance to the disease. However, this may not hold true in case of Zn where the symptoms of the disease appear before Zn deficiency is apparent. This speaks of importance of Zn compartmentation and/or a form of Zn at the critical sites of penetration than the total Zn concentration in root cells in determining tolerance level to Ggt in wheat (Rengel, 2001).

Manganese (Mn) availability is inversely correlated with severity of take-all and therefore, all the measures including Mn fertilization that increase Mn availability decrease the disease (Graham and Webb, 1991; Huber and Graham, 1999). Simultaneously, increasing soil pH that decreases Mn availability has been shown to increase.take-all incidence (Mason *et al.*, 1994). Copper (Cu) as well as boron (B) fertilization can stimulate biocontrol activity by the fungus *Trichoderma koningii* against take-all resulting in effective decrease in disease severity (Duffy *et al.*, 1997). Manganese activates a wide variety of enzymes which regulate cellular oxidation-reduction (oxido-reductase) processes; hydrolysis and joining together two molecules coupled with breakdown of phosphate bonds in ATP and also acts as regulator of many physiological defense systems of the plant to disease.

Although many micronutrients when applied in suitable amounts have been shown to decrease the losses by Ggt, the most striking effects among these have been shown by application of Mn. Graham and Rovira (1984) have put forward three hypotheses regarding mode of action of this important micronutrient:

1. Direct toxicity of Mn to the Ggt fungus if available in amounts enough to produce deposits of MnO_2 on roots.
2. Role of Mn in enhancing production of defense materials since enzymes like PAL (phenylalanine ammonia lyase) and peroxidase important for phenolic compounds and lignin biosynthesis, respectively, are stimulated by or dependent on Mn^{2+} ions (Engelsma, 1972; Gross *et al.*, 1977).
3. Role of micro flora in vicinity of roots (rhizosphere microflora) is also important in affecting availability and therefore determining the action of micronutrients. For example, the ratio between Mn-oxidizing and Mn-reducing micro flora in rhizosphere has a pronounced effect on Mn availability to the wheat plants since oxidized form (Mn^{4+}) is not available and reduced form (Mn^{2+}) is available to the wheat plants. Therefore the less above stated ratio will be, the more Mn will be available to the plants and consequently increased tolerance to Ggt. The Mn-reducing micro flora also place Ggt, which are Mn-oxidizers, in competition. Reduction in take-all incidence due to continuous monoculture of wheat is attributed to the decrease in the Mn oxidizing capacity of Ggt isolates observed in fields where take-all decline appeared by wheat monoculture.

Another disease in which rhizosphere micro flora have an important role to play is the cavity spot disease of carrots caused by *Pythium coloratum.* If the soil pH is raised by application of lime, population of a number of non-streptomycete actinomycetes is increased. These prokaryotic micro flora are antagonistic to the pathogenic fungus *P. coloratum* (El-Tarabily *et al.*, 1996).

'Bare patch' disease of wheat is caused in Australia by *Rhizoctonia solani.* The disease is more prominent in Zn-deficient soils. This is hypothesized that plants in such soils exude more photosiderophores, sugars, phenolics, amino acids and other organic compounds due to membrane leakiness (since such plants suffer from high levels of toxic free oxygen radicals which lead to peroxidation of membrane lipids, loss of membrane integrity and increased membrane permeability).

This increased root exudation may serve as a chemoattractant to R *solani* resulting in accelerated infection process.

Zn application in *Medicago truncatula* infected by *R. solani* (AG 8) is not inhibitory to the pathogen by direct toxicity of the element, but yield loss is observed to be reducing. This, however, may be due to increased root growth by micronutrient that increases the tolerance to the root pruning effect by the fungus (Streeter *et al.*, 1999).

R. solani is known to produce a Zn-chelating agent, swainsonine that deprives the plants of available Zn; thus making the membranes of their root cells leaky, increased root exudation and consequently increased pathogenic infection.

Elmer (1997) reported alleviation of Rhizoctonia root and crown rot in *Beta vulgaris* caused by *R. solani* (AG 2-2) following application of various Chlorine (Cl)–containing salts. It was found that the treatments increased Cl and Mn contents in plant tissue.

Deficiency of zinc is also known to aggravate crown rot disease of wheat caused by *Fusarium graminearum* (*Gibberella zeae*) (Sparrow and Graham, 1988). Pathogenic infection at the basal part of the culm is significantly decreased following Zn application (Grewal *et al.*, 1996). It has been observed that cultivars tolerant to Zn-deficiency are also tolerant to the disease. This is attributed to the Zn extraction capacity of such cultivars from soils with low Zn levels.

Zn fertilization of soybean decreased severity of stem rot disease (Deb and Dutta, 1992). Externally supplied Zn protected canola (*Brassica napus*) plants from blackleg disease caused by *Leptosphaeria maculans* by suppressing sirodesmin PL toxicity (Rouxel *et al.*, 1990).

In rice, it is widely accepted fact that decreased silicon content in plants (brought about either by soils deficient in silicon or by reduced uptake of this micronutrient by plants) makes them vulnerable to the pathogenic attacks of blast, brown spot, husk discolouration and many other diseases. Infection of rice plants by *Pyricularia oryzae* (Pershin *et al.*, 1995) and *Pyricularia grisea* (*Magnaporthe grisea*) (Filippi and Prabhu, 1998) have been shown to decrease by application of silicon (Si) coupled with Zn. Heavy nitrogen fertilization in rice field is known to cause reduced silicon uptake by plants and thus, increased severity of blast disease.

A range of multiple fungi and bacteria cause rice husk discolouration disease which causes losses in the form of reduced grain quality. It has been shown that the disease is much lower in japonica ecotypes compared to indica types. It has been found out to be due to more accumulation of Si in japonica types than in indica types that have considerably lower husk Si concentration.

The halo region surrounding the penetration sites of *Erysiphe graminis* in the barley leaf epidermis has been found to contain accumulated Si and Mn. Si is also known to accumulate in the cell walls at infection sites of a number of pathogens (Kunoh, 1990). This deposition of Si is considered to be the physical barrier posed against the pathogen by the host plant. Thus, role of Si in host defense system is quite important.

Kostandi *et al.* (1997) reported at least 10 per cent decrease in severity of smut disease (*Ustilago maydis* syn. *U. zeae*) n *Zea mays* following fertilization of plants with soil applied Zn. This decrease in disease severity was correlated with increased Zn content of leaves and grain yield.

Downy mildew of pearl millet (*Pennisetum glaucum*) caused by *Sclerospora graminicola* was effectively controlled by soil applications of boron (B), Zn or iron (Fe) (Gupta and Singh, 1995).

Decreased content of Zn, Mn and Cu in the leaves of *Lathyrus sativus* was accompanied by increased severity of downy mildew caused by *Peronospora lathyri-palustris* (Prasad *et al.*, 1997).

The proportion of plants dead due to root rotting disease in cotton (*Gossypium hirsutum*) [caused by *Phymatotrichum omnivorum* (*Phymatotrichopsis omnivora*)] was decreased from 61 per cent to 36 per cent following stem drenching or soil application of Fe chelate or spraying leaves with nickel (Ni) (Matocha and Vacek, 1997).

Micronutrients contribute to a great extent in the biosynthesis of chemical tools of plant defense system. Soybean (*Glycine max*) seeds, when treated with 10-100 per cent M concentrations of Zn, Mn or Cu, an increase in total phenol and *o*-dehydroxyphenol content and peroxidase activity was observed with simultaneous decrease in pectolytic enzyme activity (Bag and Sinha, 1997). Such biochemical changes enabled the plants to resist *Sclerotium rolfsii* (*Corlicium rolfsii*) infections. Even young plants infected with the pathogen showed decreased mortality. The pathogenic fungus causes stem rot disease. It is important to note that micronutrient concentration used for seed treatment did not inhibit sclerotial germination rejecting any possibility of direct toxicity of the chemicals to the fungus. Thus, the notion of increased tolerance to the disease through activating and improving plant defense system by micronutrient application is strengthened.

Micronutrients are equally important in alleviating diseases of horticultural crops. Spraying leaves of *Cucumis sativus* with micronutrients (B, Cu and Mn) in low doses can prevent development of powdery mildew (*Sphaerotheca fuliginea*) (Reuveni *et al.*, 1997). These micronutrients are supposedly involved in various enzymes like peroxidase, glucanases etc., that constitute plant defense mechanism. Burgess *et al.* (1992) found least occurrence of powdery scab in potato caused by *Spongospora subterranea* in the fields with the highest levels of native soil-Zn. Dik *et al.* (1998) reported increased resistance in *Cucumis sativus* genotypes to powdery mildew caused by *Sporothrix fuliginea* when Si was added in the nutrient solution. Similarly, severity of gerbera (*Gerbera jamesonii*) root rot caused by *Phytophthora cryptogea* was reduced if Cu was added in the nutrient solution in which the plant was grown (Toppe and Thinggaard, 1998).

Obtaining better productivity from healthy plants in micronutrient deficient soils through correction of deficiency by adding or applying micronutrients in suitable quantity and quality is not so easy. Most of the time it is found that a particular micronutrient is right there in the soil but it is in unavailable form, or as soon as micronutrient is added in soil, it is readily converted into the forms usually unavailable to the crop plants under the actions of rhizosphere micro flora. Alternatively, manipulation of rhizosphere micro flora to make the forms of micronutrients already present in soil available to the crop plants can ease the task with greater success and efficiency. For example, increasing the population of Mn-reducing micro flora such as *Pseudomonas* spp. and *Serratia* spp. helps increase Mn availability to the asparagus roots and thereby increased resistance of asparagus to Fusarium crown and root rot disease (Elmer, 1995). Similarly addition of green manure in soil has been found to support the growth and population build-up of Mn-reducing rhizosphere micro flora that increase Mn availability to potato plants conferring resistance to potato scab disease (Rogers, 1969).

Micronutrients also have a definite role in determination and regulation of interactions between plant pathogens and their biocontrol agents. The biocontrol agent *Pseudomonas fluoroscens* strain CHAO produces an antibiotic 2,4-diacetylphloroglucinol that limits the growth of pathogenic fungus, *Fusarium oxysporum* f. sp. *radcis-lycopersici* (Duffy and Defago, 1997). Zn reduces production of fusaric acid, the phytotoxin responsible for pathogenicity of the fungus *Fusarium oxysporum* f. sp. *radicis-lycopersici.* This inhibition of phytotoxin is a condition ideal for increased antibiotic production by *Pseudomonas fluoroscens* strain CHA 0. This, in total, results in reduction of crown and root rot disease of tomato (*Lycopersicon esculentum*). Copper also shows similar effects as Zn (Duffy and Defago, 1997).

Boron (B) is another micronutrient important in decreasing incidence and severity of many diseases, if found in sufficient amounts. Its deficiency may aggravate diseases like, brown heart (water core) in radish, heart rot of beets, brown heart rot of cauliflower, internal brown rot of sweet potato and cracked stem of celery (Gupta, 1993). Sufficient boron in soil reduces incidence of clubfoot in swede and other crucifers, fusaria in beans, tomato and cotton, rhizoctonias in mungbean, pea and cowpea, TMV in bean and tomato and yellow leaf curl virus in tomato (Graham and Webb, 1991).

Antimicrobial property of chlorine is known worldwide, but as a micronutrient, it can help decrease a number of diseases if present in sufficient amounts in soil. Diseases particularly controlled by Cl as a micronutrient are stalk rot and northern leaf blight of maize and take-all of wheat, downy mildew of millet and root rot of barley (Graham and Webb, 1991; Heckman, 1998). Engel *et al.* (1994) controlled powdery mildew and leaf rust of winter wheat with application of Cl at seven out of nine experimental locations.

Copper has been utilized as and antifungal element since long. Barring its antifungal property, it can help reduce incidence of many diseases when applied as a micronutrient. Soil application of Cu decreases many fungal and bacterial diseases including downy mildew of wheat and ergot of rye and barley (Graham and Webb, 1991).

Although holding up of iron by various bio-chemicals known as siderophores deprived pathogenic micro flora of their iron nutrition, its supplementation as a micronutrient can also cause reduction in various other diseases such as rust and smut infections in wheat and *Colletotrichum musae* infections on banana. Topical application of iron as spray enhances resistance in apple and pears to *S. phaeriopsis malorum* and tolerance in cabbage to *Olpidium brassicae* (Graham and Webb, 1991).

Although, it is a matter of research how molybdenum (Mo) protects plants from diseases since its specific role in such a case is still unclear, it is known to suppress verticillium wilt in tomato (Graham and Webb, 1991). Also interesting is the case with take-all of wheat and the opposite effects of Mn and Mo on it. Mn, Cu, Cl, Fe and Zn are known to reduce take-all severity whereas Mo increases disease severity of take-all in wheat. The reasons and modes of action are still to be found out.

Epilogue

In modem agriculture where intensive cropping and maximum land utilization have become a necessity to feed everyday bulging human population and increase earnings from crop cultivation, people have also become aware of micronutrient depletion from their soils and the consequent effects in the form of increased disease severity and reduced crop yields. Also high demand of modem crop cultivars for micronutrients and the concept of sustainable agriculture (to maintain the status quo of the soil at least, if not improve its nutritional status) has compelled the contemporary agriculturists and bioscientists to understand the importance and role of micro nutrients in successful crop cultivation. Compared to macronutrients, micronutrients are required in lower amounts and serve mainly as constituents of prosthetic groups in metalloproteins and/or as activators of enzyme reactions. Rates of application of micronutrients may range from 0.2 to 100 kg ha^{-1} depending upon the micronutrient, crop requirement and method of application.

Further studies are required to understand the exact mode of actions of micronutrients in alleviating plant diseases. Once the mechanism is found out, the strategy of crop protection may be manipulated and our dependence on hazardous pesticides for disease management may be reduced and who knows agriculture may become totally independent of such Frankenstein's monster in the future.

References

Bag, T.K. and Sinha, A.K. (1997). Control of stem rot of soybean incited by *Sclerotium rolfsii* through metal salts. *Indian Phytopathol.*, 50: 350–357.

Burgess, P.J., Burnett, F.J., Brereton, P.S., Wale, S.J. and Sinclair, A.H. (1992). An overview of the influence of zinc on the severity of powdery scab in potatoes. *Aspects Appl. Biol.*, 33:143–150.

Deb, P.R. and Dutta, B.K. (1992). Effect of inorganic fertilizers on the development of foot rot disease of soybean caused by *Sclerotium rolfsii. Indian Phytopathol.*, 45: 89–97.

Dik, A.J., Verhaar, M.A. and Belanger, R.R. (1998). Comparison of three biological control agents against cucumber powdery mildew (*Sphaerotheca fuliginea*) in semi– commercial scale glasshouse trials. *Eur. J. Plant Pathol.*, 104: 413–423.

Duffy, B.K. and Defago, G. (1997). Zinc improves biocontrol of fusarium crown and root rot of tomato by *Pseudomonas fluorescens* and represses the production of pathogen metabolites inhibitory to bacterial; antibiotic biosynthesis. *Phytopathology*, 87: 1250–1257.

Duffy, B.K., Ownley, B.H. and Weller, D.M. (1997). Soil chemical and physical properties associated with suppression of take-all of wheat by *Trichoderma koningii. Phytopathology*, 87: 1118–1124.

Elmer, W.H. (1995). Association between Mn-reducing root bacteria and NaCl applications in suppression of fusarium crown and root rot of asparagus. *Phytopathology*, 85: 1461–1467.

Elmer, W.H. (1997). Influence of chloride and nitrogen form on rhizoctonia root and crown rot of table beets. *Plant Disease*, 81: 635–640.

El-Tarabily, K.A., Hardy, G.E.S.J. Sivasithamparam, K. and Kurtböke, I.D. (1996). Microbiological differences between limed and unlimed soils and their relationship with cavity spot disease of carrots (*Daucus carota* L.) caused by *Pythium coloratum* in Western Australia. *Plant Soil*, 183: 279–290.

Engel, R.E., Eckhoff, J. and Berg, R.K. (1994). Grain yield, kernel weight and disease responses of winter wheat cultivars to chloride fertilization. *Agron. J.*, 86: 891–896.

Engelsma, G. (1972). A possible role of divalent manganese in the photoinduction of phenylalanine ammonia-lyase. *Plant Physiol.*, 50: 599–602.

Filippi, M.C. and Prabhu, A.S. (1998). Relationship between panicle blast severity and mineral nutrient content of plant tissue in upland rice. *J. Plant Nutr.*, 21: 1577–1587.

Graham, R.D. (1983). Effects of nutrient stress on susceptibility of plants to disease with particular reference to the trace elements. *Adv. Bot. Res.*, 10: 221–276.

Graham, R.D. and Rovira, A.D. (1984). A role for manganese in the resistance of wheat plants to take-all. *Plant Soil*, 78: 441–444.

Graham, RD. and Webb, M.J. (1991). Micronutrients and disease resistance and tolerance in plants. In: *Micronutrients in Agriculture*, 2nd Edition, (Eds.) Mortvedt, J.J., Cox, F.R., Shuman, L.N., Welch, R.M. Soil Science Society of America, Inc., Madison, W.I., pp. 329–370

Grewal, H.S., Graham, R.D. and Rengel, Z. (1996). Genotypic variation in zinc efficiency and resistance to crown rot disease (*Fusarium graminearum* Schw. Group. I) in wheat. *Plant Soil*, 186: 219–226.

Gross, G.G., Janse, C. and Elstner, E.F. (1977). Involvement of malate, monophenols, and the super oxide radical in hydrogen peroxide formation by isolated cell walls from horse radish (*Armoracia lapathifolia* Gilib). *Planta*, 136: 271–276.

Gupta, G.K. and Singh, D. (1995). Downy mildew control in pearl millet through nutrient application. *Indian Phytopath.*, 48: 439–443.

Gupta, U.C. (1993). Factors affecting boron uptake by plants. In: *Boron and its Role in Crop Production*, (Ed.) Gupta, U.C. CRC Press, Boca Raton, FL, pp. 87–123.

Heckman, J.R. (1998). Corn stalk rot suppression and grain yield response to chloride. *J. Plant Nutr.*, 21: 149–155.

Huber, D.M. and Graham, R.D. (1999). The role of nutrition in crop resistance and tolerance to diseases. In: *Mineral Nutrition of Crops: Fundamental Mechanisms and Implications*, (Ed.) Rengel, Z. The Haworth Press, New York, pp. 169–204.

Huber, D.M. and Wilhelm, N.S. (1988). The role of manganese in resistance to plant diseases In: *Manganese in Soils and Plants*, (Eds.) Graham, R.D., Hannam, R.J., and Uren, M.C. Kluuwer Academic Publishers, Dordrecht, The Netherlands, pp. 155–173.

Kostandi, S., Soliman, M. and Ghaly, A. (1997). Smut disease and yield performance in corn (*Zea mqys* L.) as influenced by nitrapyrin, urea and zinc applications in coarse-textured soils. *J. Agronomy and Crop Sci.*, 179: 219–226.

Kunoh, H. (1990). Ultrastructure and mobilization of ions near infection sites. *Ann. Rev. Phytopathol.*, 28: 93–111.

Lucas, P., Jeuffry, M.H., Schoeny, A. and Sarniguet, A. (1997). Basis for nitrogen fertilization management of winter wheat crops infected with take-all. *Aspects Appl. Biol.*, 50: 255–262.

Marshner, H. (1988). Mechanisms of manganese acquisition by roots from soil. In: *Manganese in Soils and Plants*, (Eds.) Graham, R.D., Hannam, R.J. and Uren, N.C. Kluwer Academic Publishers, Dordrecht, The Netherlands, pp. 191–204.

Mason, M.G., Porter, W.M. and Cox, W.J. (1994). Effect of an acidifying nitrogen fertilizer and lime on soil pH and wheat yields. 2. Plant response. *Aust. J. Exp. Agric.*, 34: 247–253.

Matocha, J.E. and Vacek, S.G. (1997). Efficacy of fungicidal and nutritional treatments on cotton root rot suppression. In: *Proceedings of Beltwide Cotton Conferences*, New Orleans, L.A., U.S.A. January 06–10, Vol. 1, National Cotton, Council, Memphis, U.S.A. pp. 135–137.

Pershin, B.M., Pershina, A.N. and Egorina, L.M. (1995). Silicori and rice production in the Primorsaii region. *Agrokhimiya* 10: 68– 74.

Prasad, B.K., Shambhu, D., Sinha, M.P., Singh, S.P., Kumar, S., Prasad, R.L. and Dayal, S. (1997). Change in mineral content in *Lathyrus sativus* infected with *Peronolpora lathyri-palustris*. *Indian Phytopath.*, 50: 65–67.

Rengel, Z. (2001). The role of micronutrient fertilization in managing plant diseases. In: *Perspectives on the Micronutrient Nutrition of Crops*, (Eds.) Singh, K., Mori, M. and Welch, R.M. Scientific Publishers (India), Jodhpur, pp. 53–69.

Rengel, Z., Pedler, J.F. and Graham, R.D. (1994). Control of Mn status in plants and rhizosphere: Genetic aspects of host and pathogen effects in the wheat take-all interaction. In: *Biochemistry of Metal: Micronutrients in the Rhizosphere*, (Eds.) Manthey, J.A., Crowley, D.E., Luster, D.G. Lewis Publishers/CRC Press Boca Raton, FL, U.S.A., pp. 125–145.

Reuveni, M., Agapov, V. and Reuveni, R. (1997). A foliar spray of micronutrient solutions induces local and systemic protection against powdery mildew (*Sphaerotheca fuliginea*) in cucumber plants. *Eur. J. Plant Pathol.*, 103: 581–588.

Rogers, P. (1969). Organic manuring for potato scab control and its relations to soil manganese. *Ann. Appl. Biol.*, 63: 371–378.

Rosen, H.R. and Ellio, J.S. (1923). Pathogenicity of *Ophiobolus cariceti* to in its relationship to weakened plants. *J. Agric. Res.*, 25: 351–358.

Rouxel, T., Kollmann, A. and Bousquet, J.F. (1990). Zinc suppresses sirodesmin PL toxicity and protects *Brassica napus* plants against the blackleg disease caused by *Leptospheria maculans*. *Plant Sci.*, 68: 77–86.

Sparrow, D.H. and Graham, R.D. (1988). Susceptibility of zinc deficient wheat plants to colonization by *fusarium graminearum*. Schw. Group I. *Plant Soil*, 112: 261–266.

Streeter, T.C., Rengel, Z., Neate, S.M. and Graham, R.D. (1999). Zinc fertilization increases tolerance to *Rhizoctonia solani* (AG 8) in *Medicago truncatula* genotype Caliph. *Plant Soil* (submitted).

Toppe, B. and Thinggaard, K. (1998). Prevention of Phytophthora root rot in gerbera by increasing copper ion concentration in the nutrient solution. *Eur. J. Plant Pathol.*, 104: 359–366.

Wilhelm, N.S., Graham, R.D. and Rovira, A.D. (1990). Control of Mn status and infection rate by genotype of both, host and pathogen in the wheat take-all interaction. *Plant Soil*, 123: 267–275.

©R.C. Ram & Asha Sinha (*Editors*)
***Microbial Diversity: Modern Trends*, 2007**
Daya Publishing House, New Delhi
161–168

Chapter 11

Postharvest Diseases: Strategies and Management

R.C. Ram

Department of Mycology and Plant Pathology, Institute of Agricultural Sciences, Banaras Hindu University, Varanasi – 221 005 (U.P.)

ABSTRACT

Postharvest diseases develop on plant produces or plant products during harvesting, grading and packing, during transportation to the market and to the consumer, and while the produce is in the possession of the consumer until the movement of actual consumption or use. All types of fruits and vegetables or any plant products are susceptible to postharvest diseases. Postharvest diseases destroy 10 to 30 per cent of the total yield of crops and in some perishable crops, especially in developing like countries, they destroy more than 30 per cent of the crops yield. Postharvest diseases usually cause great losses of fresh fruits and vegetables by reducing their quality, quantity or both. Postharvest diseases of grains and legumes also result in the production by some infecting microorganisms of toxic substances known as mycotoxins. Mycotoxins are also produces by some fungi in infected fruits and vegetables but in these cases they are generally removed by different treatment or discarding before consumption.

Postharvest diseases are caused primarily by a relatively small number of Ascomycetes and Imperfect fungi and by few species of Oomyceties, Zygomyceties, Basidiomyceties and Bacteria. The control of postharvest deterioration and spoilage by fungi of grains, legumes, fodders and commercial feeds depends on certain precautions and conditions that must be mate before and during harvesting and then during storage. Investigations carried out by various researchers clearly indicate that the real causes of the spoilage of fruits in transit and storage are the high temperatures and injuries sustained by the fruits during the process of marketing.

Introduction

Different diseases arise after crops are harvested because the seeds, fruits, vegetables or other storage organs are essentially dormant structures and their cells are physiologically unlike those of the growing plant. In India, huge losses are incurred on account of different type of postharvest disease. The postharvest microbial spoilage of fruits, vegetables and grains has been grouped under Field or market Pathology. This specialized area of study deals with problems encountered during *picking, packing, transportation* and *storage* of produces. Thus it occupies special attention in field of Plant Pathology.

Economic losses caused by postharvest diseases are considerably more than realized because fruits and vegetables increase manifold in unit value while passing from the field at harvest to consumer.

A substantial portion of these losses is due to diseases caused by microorganisms. In India also, there is a considerable loss due to postharvest diseases. The data collected from some studies carried out India on postharvest diseases of fruits and vegetables, put the average loss at 20–30 per cent (Mehta, 1975; Parpia, 1976; Primental, 1983 and Submali *et al.*, 2001). The National Committee appointed by the Government of India has estimated food grain losses during the postharvest period to be 9.3 per cent and part of it can be due to fungal detoriation (Dharam Vir, 1974).

Therefore, area of research in postharvest disease management that require immediate attention are, to develop low cost eco-friendly less energy-consuming farm storage structure, standardization of preservatives for different crops, packing materials. Besides this chemical residue management and developing new products would add value to the produce.

In addition to the loss of food calories and nutritional elements, mycotoxins and toxic stress metabolites found in vegetable and fruits colonized by postharvest pathogens are emerging as a major dietary problem. The numerous diseases which occur in transit and storage result mainly from the activity of fungi and bacteria or non-parasitic origin. Stakman and Harrar (1957) divide the pathogenic storage disease in two categories:

1. Disease of dry bulk material, such as grains.
2. Disease of succulent storage organs, such as fruits, rhizomes, bulbs and tubers.

Management of Postharvest Diseases

Proper Harvesting and Handling Practices

Care must be exercised to handle the fruits and vegetables carefully at harvesting. The harvesting of way produce is critical because many pathogens are involved in storage diseases and enter through wounds. Harvesting of fruits and vegetables should be done at the proper stage and as well as keeping in mind the distance of the market. It should be done preferably in the early morning to avoid heat.

Packaging

In our country mostly gunny bags, bamboo baskets and wooden crates lined with news paper are used for packing. In foreign countries, at first vegetables and fruits are cleaned, trimmed, pretreated, pre cooled and packed in bags of transparent film, trays or cartons over wrapped with transparent film.

Transportation

In India, transport of commodities is done mainly by road or Railway wagons resulting in a lot of damage. It is necessary to modify the long distance transport vehicles by introducing more ventilation, reducing temperature, increasing relative humidity, improving loading and unloading systems.

Storage

It is one of the most important aspects of postharvest handling as the storage life of almost all the vegetables is short due to dehydration, losses of respiration, and rotting and so on. This causes deterioration in quality by wilting, shrinking and weight of loss. Refrigeration is one of the useful methods for long term storage of fresh vegetables. Various sprout inhibitions like maleic hydrazide (MH) and naphthalene acetic acid (NAA) have been found to check sprouting, rotting and weight loss.

Temperature and Relative Humidity

Temperature, Humidity and Ventilation determine to a very large extent the rate of disease development during transit and storage. Fruits and vegetables continue to respire after they are harvested and as a result heat is produce. This is known as vital heat. High temperature accelerates respiration and ageing; hence it must be cooled quickly to remove their field heat. This can be achieved in several ways:

1. By spraying or immersing in water
2. By vacuum treatment (induced rapid evaporation).

Near freezing temperature prevents most bacterial and fungal rots. The development of some non-parasitic diseases. Humidity during transit and storage should be generally high, other wise the produce dries and losses its prime market quality.

Tactics to Overcome the Pre-harvest Infection Management

The principal way to control pre-harvest infection is by the application of protectant or eradicant *i.e.*, fungicides to the crop in the fields. Some common chemicals which are commonly used in various postharvest diseases management are given below:

Name of the Disease	Chemical Used
Botrytis rot of strawberry	Thiram and Captan or Benomyl
Anthracnose of Banana	Benomyl
Anthracnose of Papaya	Maneb (every ten days interval)
Lentilcel rotting of Apples	Captan, Benomyl, Thiabendazole
Fruit rot summer Saqash	Aureofungin (Antibiotic)
Fruit rot of Guava	Aureofungin (Antibiotic)
Oospora Tomato fruit rot	Castor and Coconut oil
Blue mould rot of Apples	Carbendazim, Thiophanate methyl

Besides this, application of Calcium chloride Iprodione, Vinclozolin, Meclozolin, Anilazine, Carbendazim several times during the season have given good control of major postharvest diseases of various crops. Field spray of fungicides such as Carbendazim Benomyl Captan, Thiabendazole prevented the establishment of many diseases.

Thus, pre-harvest spray of systemic fungicides can eradicate quiescent infections as well as protect against their establishment before harvest and reduce disease pressure.

Strategies for Postharvest Disease Management

1. Orchard management
2. Disinfection and washing
3. Postharvest chemical treatment
   (*a*) Heated fungicide suspension
   (*b*) Sugar analogues as potential fungicide
   (*c*) Salt solution
   (*d*) Management of fungicide resistant-strain
4. Fumigation
5. Heat treatment
6. Fruits wraps, waxing, coating and packing
7. Low Temperature storage
8. Control atmosphere storage
9. Maintenance of host resistance
10. Moderation of disease symptoms
11. Preventing spread of pathogen
12. Chemical treatments
13. Inorganic sulphur compounds
14. Biological control

Orchard Management

1. Infected parts (dead woods, leaves, and other parts of fruit trees) should be removed from vicinity of the healthy plants.
2. Harvesting time can be changed.
3. Nutrient deficiency sometimes posses fruit resistant: *e.g.*, die back of mango (*Botrytiosphaeria ribis*).
4. Fruits should be cultivated in regions which are unfavorable for infection.
5. Resistant cultivars should be cultivated.
6. Proper handling during harvesting, packaging, should be made to minimize fruits disease.

Disinfection and Washing

These practices have some disadvantage of being corrosive to the box material. For avoiding such a problem dilute the chemicals with a hot borax solution before use. As for example:

Formaldehyde and nitrogen dichloride	Lemon storage boxes
Sulfur dioxide	Figs packing boxes
Sodium-o-phenyl phenate (cooled water)	Brown rot peach
Sec-butyl ammonium phosphate	Orange de-greening
Hypo chlorite or Sodium-o-phenyl phenate	Citrus and peach
Water	Fusarium spp.
3% formaldehyde solution	Penicillium spp.

Postharvest Chemical Treatment

Two major strategies are taken.

1. Cover the surface of the produce with the fungicide before harvest.
2. Treat the produce after harvest with inhibitory quantity of fungicide (Fungistatic).

Sodium-o-phenyl phenate; Sec-butyl amine; Bi phenyl and Thiabendazole: Sodium carbonate	Used under usual condition
Sodium hypo chloride	Used in in vitro
Growth regulator like IAA and MH	Against Aspergillus and Rhizopus rot of Papaya

Following chemicals are also used against this practices

Heated fungicide suspension	Benomyl + Dichloran	Rhizopus rot of Prunus
Sugar analogues as potential fungicides	1% solution of 2-deoxi-D-glucose	Penicillium expansum Botrytis cinerea
Salt solutions	Sodium lignin sulphonate + Sodium-o-phenyl phenate	Penicillium expansum Botrytis cinerea and Mucor piriformis
Fungicide resistant strains	Benzimidazoles + other fungicides	Penicillium expansum

Fumigation

Some volatile fumigants are ideal for particularly the lenticel rots.

SO_2 fumigation on grapes	Botrytis cinerea
SO_2 + $NaHSo_4$ on grapes	Botrytis cinerea and Aspergillus niger
SO_2 on apple and pear	Trichothecicea roseum, Glomerella cingulate, Penicillium expansum and Monilinia spp.

Heat Treatments

Heat treatments have the advantage of low cost, relatively simple equipment and no chemical residues on the treated product.

Following recommended temperature and time duration are essentials for heat treatment of particular organisms.

Phytophthora spp.	60°C for 1.5 minutes
Trichothecicea roseum, Glomerella cingulate, Penicillium expansum and Monilinia spp., Rhizopus stolonifer	50°C for 5 minutes
Decay of peaches and other stone fruits	46°C for 5 minutes

Following fruits are recommended for heat treatment against particular organism.

Mango	Penicillium spp. and Diplodia spp., Colletotrichum spp.
Papaya	Colletotrichum spp.
Banana	Crown rot
Cantaloups	Stem scar and rind mould

Fruits Wraps, Waxing, Coating, and Packing

Various antifungal chemicals have been used for fruits wrappers and box liners.

Tactics Follows	Chemical Used	Pathogen
Wrapping grapes in tissue paper	o-phenyl butyrate and sodium meta bisulphate	Aspergillus niger P. canescens
Wraping apple	Potassium iodide	Glicladium roseum
Individual fruit wrapping	Polyethylene	Prevent spread of the disease
Wrapping with news paper	Carbendazin and Thiobendazole	Blue mould

Fruit waxes reduced the growth of *Rhizopus stolonifer, Botrytis cinerea, Alternaria alternata in vitro* condition.

Skin coating of fruits with oils, waxes and colloidal solution can improve the quality of fruits.

Low Temperature Storage

In certain cases, postharvest decay of fruits and vegetables can be delayed by storing at low temperatures. In storage, at 0°C most of the pathogen become inactive. The low storage temperature is most effective for apples and grapes which tolerate near freezing temperature. Greater the reduction in respiration and general metabolism, longer the storage life of fruits just above its freezing point. Pathogen such as *Botrytis, Penicillium, Sclerotium, centrospora* almost inactive near freezing temperature. Low temperature delay is the development of disease by:

1. Delay ripening of fruits, so prolong the resistant to pathogen.
2. Direct inhibition of pathogens at low temperature

Control Atmosphere Storage

Fruits stored in conditions 10 win oxygen, high in CO_2, or both. It also suppress the respiration rate are known as controlled atmosphere storage. For example, growth of *Botrytis cinerea* in apple was checked only if oxygen level is reduced 1 per cent.

Maintenance of Host Resistance

Among the postharvest disease management, maintainer of host resistance is an important practice. Application of gibberllic acid reduces *Penicillium* decay of oranges by delaying senescence of the skin of the fruit.

Moderation of Disease Symptoms

The value of fresh fruits and vegetables is reduced significantly by superficial fungal growths or by visible contamination with spore or other debris arising from an adjacent disease. As for example, Sooty moulds of citrus fruits or on other surfaces of crowns of bananas at the cut surfaces, the fungicidal treatment after harvesting may overcome these conditions.

Preventing Spread of Pathogen

The spread of disease by contact frequently accounts for the major losses caused by the pathogen. Fungicide-treated wraps check the secondary spread of pathogen. Although this does not control the disease in the fruits already infected. Some common practices are:

Botrytis on pears can be prevented by oil paper wraps containing Copper sulphate.

Rhizopus on tomato may check by impregnated with organic compounds.

Grape in storage are fumigated with sulphur dioxide to prevent the spread of *Botrytis.*

Chemical Treatments

The first treatment employed to control postharvest decays of citrus fruits was with an alkaline solution of sodium tetraborate (borex) and sodium carbonate. The most widely used treatment has been a combination of 4 per cent borex and 2 per cent boric acid solution at a temperature of 43.5°C. At present day borex is replaced by o-phenyl-phenate which more effective.

6–8 per cent of borax effective against Penicillium moulds and stem end rots due to *Diplodia natalensis* and *Phomopsis citri* of orange.

Inorganic Sulphur Compound

Protective fungicide like sulphur dust commonly used for postharvest treatment of peaches to protect against *Monilinia fructicola* during marketing.

Biological Control

Application of microorganism or microbial products for the control of fruit pathogens are relevant method at present.

Biological Agents	Against Pathogenic Strains
Trichoderma spp.	Number of fruit rots
Bacillus subtilis	Brown rot of stone fruit
Yeast (Pichia qoilliermondi strain US-7) and Hanseniaspora uvarum strain-138	Brown rot of citrus, stone fruits, pome fruits

Conclusions

Most of researches have been carried out on stored grains pests than on deteriorating pathogens possibly because of two reasons. Firstly, the insects are obvious enemies and secondly, the majority of the plant pathologists are attracted to work and are overburdened with fungal diseases that affect plant in the field.

Most of the postharvest fungi and bacteria can macerated very quickly causing deterioration very fast as a in the soft rot by *Ervinia, Rhizopus, Geotrichum, Sclerotinia* etc. and may spread by contact to adjoining units within containers as in *Monilia fructicola, Gleosporium* spp. on apple *Botryodiplodia theobromae* on different fruits, and *Phomopsis* citri on citrus etc. Many postharvest pathogens may complete infection by direct penetration in the field, as in *Colletotrichum gloeosporiodes* on banana, citrus, mango, papaya, etc. remain latent in mature fruits. Further progresses of this disease take place in the postharvest stage. Mechanical injuries are impossible to avoid during harvesting and postharvest operation. Coupled with injury, the predisposing factors such as low and high temperatures oxygen deficiency, abundance of inoculum in containers and handling atmosphere, water and equipments as for certain virus diseases, are also responsible for postharvest damages.

The chief approaches towards postharvest chemical and physical treatments are the following:

1. Prevention on infection (by reducing the inoculum level, by interrupting the infection process by preharvest care and treatment, by proper storage);

2. Eradication or attenuation of the established pathogens (pre- or postharvest chemical treatment, heat therapy, ionizing radiation and refrizeration);
3. Moderation of disease symptoms (preventing superficial growth and sporulation of pathogens and preventing spread of the pathogen from diseased to healthy fruits and vegetables in the package by keeping the fruits physically separated in smaller packets or in compartments within larger container, etc. and
4. Spraying growth regulators to retard senescence. The postharvest tolerances of most commercially important fruits and vegetables to low O_2 and elevated CO_2 and the optimum atmosphere combination to achieve maximum product stombility have been established.

Most research on atmosphere effect on plant metabolism has concentrated on metabolism of growing plants often of roots under anoxia (Geigerberger, 2003). In fruits and vegetables the focus has been on fermentation metabolism under low O_2 partial pressure. (Imahori *et al.*, 2002). Modified atmosphere packing (MAP) technology has been successfully used to control both discolouration and decay. It is essential to maintain the optimum balance of O_2 and CO_2 in the package to minimized and decay. It has been shown in recent years that certain agents such as hydrated sodium calcium aluminosilicated, binds to mycotoxins, and if such agents are added to moldy corn before it is fed to swine the effect of many mycotoxins present are reduced considerably.

References

Dharam Vir. (1974). Study of some problems associated with postharvest fungal spoilage of seeds and grains. In: *Current Trends in Plant Pathology*, (Eds.) S.P. Raychaudhari and J.P. Verma. Botany Department, Lucknow University, pp. 221–226.

Geigerberger, D. (2003). Response of plant metabolosm to too little oxygen. *Current Opinion Plant Biol.*, 124: 696–701.

Imahori, Y.M., Kota, H., Farukawa, Y. and K. Chachin. (2002). Ethenolic fermentation enzymes, their products and transcription of alcohl dehydrogenase from bell pepper fruit held under various low oxygen atmosphere of postharvest. *Soil. Technol.*, 25: 159–167.

Mehta, P.R. (1975). Fungicidal umbrella. In: *Advances in Mycology and Plant Pathology*, (Eds.) S.P. Raychaudhari *et al.* Paranassus Publishers and Printers, New Delhi, pp. 330–341.

Parpia, H.A.B. (1976). Postharvest losses: Impact of their prevention on food supplies nutrition and development. (Eds.) Nsemshaw, N.H. and Behar, M. pp. 195–206.

Primental, D. (1983). Environmental aspect of post management. In: *Chemistry and World Supplies: The New Frontier Chemirawan II*, (Ed.) L.W. Scheimitt. Pergaman Press Oxford.

Stakman, E.C. and J.G. Harrar. (1957). *Principles of Plant Pathology*. Ronald Press, New York.

Submali, *et al.* (2000). *Diseases of Fruits and Vegetables and their Management*, T.S. Thind (ed.) Kalyani Publishers, pp. 474.

©R.C. Ram & Asha Sinha (*Editors*)
***Microbial Diversity: Modern Trends*, 2007** **169–183**
Daya Publishing House, New Delhi

Chapter 12

Rhizoctonia solani in Reference to Banded Leaf and Sheath Blight of Maize

Alok Kumar Singh**, Jai Prakash Rai* and Satish Chand Saxena**

***Department of Plant Pathology, Udai Pratap Autonomous College, Varanasi - 221 002**
****Centre of Advanced Studies in Plant Pathology, College of Agriculture, G.B. Pant University of Agriculture and Technology, Pantnagar- 263 145**

Introduction

Maize (*Zea mays* L.) is the third most important cereal crop in the world agricultural economy as food for man and food for livestock. By virtue of its immense potentiality, it is known as 'Queen of Cereals.' It is grown throughout the temperate, subtropical and tropical zones. It belongs to the grass family (Gramineae) and is a cross pollinated, monoecious plant.

The total area under maize cultivation in the world is about 127.38 mha with a total production of 470.5 million tones with the average yield of 3694 kg/ha (FAO Report, 1993). In India alone area under cultivation of this crop is about 6 mha with a total production of 9.4 million tones giving the yield 1570 kg/ha. The major maize growing states are Andhra Pradesh, Bihar, Haryana, Himachal Pradesh, Karnataka, Madhya Pradesh, Punjab, Rajasthan and Uttar Pradesh.

About 112 diseases of maize have been reported so far from different parts of world, out of these, 65 are known to occur in India. The major diseases in different agro-climatic regions are, seed rots and seedling blights, leaf spot and blights, downy mildews, stalk rots, banded leaf and sheath blight, brown stripe downy mildew, Erwinia stalk rot, smuts and rusts leading to about 15–20 per cent yield loss annually. Pantnagar has been the hot spot for many diseases of maize. At present, it has been the main centre for maize pathology research for at least three major maize diseases, *viz.*, brown stripe downy mildew, Erwinia stalk rot and banded leaf and sheath blight.

Banded leaf and sheath blight was first reported in India from the *tarai* region of Uttar Pradesh (Payak and Renfro, 1966). The importance of the disease was realized only in 1972 when an epidemic occurred in warm and humid foothills in Mandi district of Himachal Pradesh. Banded leaf and sheath blight of maize is known under many names, *viz.*, sclerotial disease, banded sheath rot, banded sclerotial disease, sharp eye spot, oriental leaf and blight, Rhizoctonia ear rot, leaf and sheath blight, sheath blight, sheath rot and corn sheth blight.

The disease is caused by *Rhizoctonia solani* f. sp. *sasakii* Kühn, perfect stage being in Basidiomycotina as *Thanatephorus cucumeris* (Frank) Donk. It is a destructive plant pathogen with an almost unlimited host range. *R. solani* is both a pathogen and a saprophyte; it aggressively colonizes organic debris and is one of the fastest growing fungi. The cells of the fungus are multinucleate and cytoplasm is interconnected through a septal pore (dolipore) that is the characteristic of Basidiomycotina. No asexual spores are formed; only sclerotia are formed as soil borne propagules. The sclerotia are knots of undifferentiated intertwined, pigmented, monilioid cells.

The symptoms of the disease are observed on all aerial parts of the plant except tassel. The disease manifests itself on leaf, leaf sheaths stalk and ears as leaf and sheath blight. Under natural conditions disease appears at pre-flowering stage on 30–40 days old plants but infection can also occur on young plants (Baruah, 1979). The disease appears on basal leaf sheaths as water soaked, straw coloured, irregular to roundish spots on both the surfaces. Profuse mycelial growth is seen on the affected areas of leaf sheaths and in between the leaf sheath and stem. Later on, as the infection becomes older, numerous sclerotial bodies develop on, the affected part. The disease results ill direct losses as premature death, stalk breakage, destruction of leaves, leaf sheaths and ear rot. Losses to the extent of 10–97 per cent have been reported on different cultivars of maize (Singh and Sharma, 1976; Butchaiah, 1977).

Since 1986, the disease is occurring in high proportions at Pantnagar in almost all the maize germplasms. So far, no material has shown resistance to the banded leaf and sheath blight disease in epiphytotic conditions. Because of its seriousness, there are increasing losses in the existing cultivars. Considering the above aspects, the Pantnagar centre has been recognized as the main centre for research on banded leaf and sheath blight disease at national level.

Banded leaf and sheath blight of maize, caused by *Rhizoctonia solani* f. sp. *sasakii* (Kühn) Exner [*Thanatephorus cucumeris* (Frank) Donk.] was described for the first time from Sri Lanka (Bertus, 1927) as sclerotial disease. Subsequently, the disease was reported from Sierra Leone (Deighton, 1932) and Philippines (Reyes, 1941). The disease has been reported from different areas under many names, *viz.*, banded sheath rot from Ivory Coast.) and Wiltshire (1956) also named the pathogen as *Corticium solani* (Prill and Delacr.) Costan and Dufar, Sharp eye spot (Knight and Burrill, 1964) from Great Britain, Sheath Blight from Japan caused by *Pellicularia sasakii* (Shirai) Ito (Kujwara, 1968). Then, the same disease was reported from China (Zhu, 1982), as sheath blight from USA (Hirrel *et al.*, 1988), and Korea (Lee *et al.*, 1989). In India, the disease was reported as banded leaf and sheath blight of maize caused by *Hypochnus sasakii* by Ullstrup in 1960 from the *tarai* region of Uttar Pradesh (Payak and Renfro, 1966). The disease was also reported as ear rot caused by *Corticium sasakii* on maize in 1977 from Meghalaya state (Maiti, 1978). Since then the disease has been reported from Himachal Pradesh, Rajasthan and West Bengal (Ahuja and Payak, 1988).

Symptomatology

Symptoms of the disease have been observed on all the aerial plant parts of maize except the tassel, with its frequent occurrence on lower leaf sheaths and ears. The disease was also found on drooping leaves showing partial or complete blotching. The disease was also recorded from leaves, leaf sheaths ear husk and ears. Wiltshire (1956) found it on lower leaf sheaths only. Knight and Burrill (1964) reported sharp eye spot on leaves both on lodged and unlodged plants. However, rarely observed symptoms of *R. solani* f. sp. *sasakii* include stalk lesions (rind spotting), stalk breakage, clumping and cracking of styles, horse shoe shaped lesions with banding on caryopses. Under natural conditions, disease appears at the pre-flowering stage on 30 to 40 day old plants but infection can also occur on young plants (Baruah, 1979) which may subsequently result in severe blighting and death of apical region of growing plants.

Leaves

Under natural conditions, dropping blades especially the distal halves of leaves proximate to soil surface are affected. Infection spreads from leaf sheaths to the basal portion of leaves involving mid ribs. Lesions appear as in irregular patches, similar in colour but larger in size and spread more rapidly than on leaf sheath, covering greater areas with alternating dark bands. Many blotches interspersed with dark, 1–3 cm broad bands are produced perpendicularly to the long axis of leaves and sheathes become thin and papery, loose and can be torn easily. A slightly raised and spreading type of mycelial growth is apparent on both the surfaces of the infected leaf.

Leaf Sheaths

The symptoms are more common on sheaths than on leaves. The disease appears on basal leaf sheaths as water soaked, straw coloured, irregular to roundish spots on both the surfaces. A sort of wave pattern of disease advancement can be seen not only on leaves but also on sheaths and husk leaves. In early stages, marginal chlorosis and rotting of lamnae proceed Inwardly. Later as the Infection becomes older, numerous sclerotial bodies are also seen.

Stalk

The pathogen also causes elongated dark brown to black spots or lesions on the rind of the stalk under the affected sheaths. These spots coalesce together extending the lesions and covering almost an internode. Individual lesion range from 2–10 × 3–15 mm in size to those which cover the whole internodes. Sometimes these lesions are transformed into cankers. Generally, lesions develop lengthwise but a few of them girdle near the nodes. Upon artificial inoculation entire rind is sometimes affected, the stalk thereby weakened and breaks easily.

Ear

The disease is observed first of all, on basal part of the outermost husk leaves forwarding to sheath from which the ear emerges. The same types of lesions are found on ear but the bands are fairly prominent, giving a blackened appearance. The affected ears become brown and numerous sclerotia are observed on husk, lightly attached to the cob. Whitish mycelium and sclerotia are also seen frequently on silks between and on kernel rows and glumes.

The grain shows light grayish to dark brown discolouration. They are drastically reduced in size and wrinkled, and under severe conditions, the grain becomes chaffy in appearance and light in weight.

Causal Organism

The causal organism of banded leaf and sheath blight disease has been first reported more than 100 years ago, the German scientist Julius Kühn observed a fungus on diseased potato tubers and named it *Rhizoctonia solani* (Kühn, 1858). Since then this fungus has gained the notoriety of being a widespread, destructive and versatile plant pathogen. It occurs in all parts of the world and is probably indigenous to uncultivated areas.

The genus *Rhizoctonia* was erected by de Candolle (1815) to accommodate the non-sporulating root pathogen *Rhizoctonia crocorum* D.C. ex. Fr. The basic characters of the genus, as set forth by de Candolle were production of sclerotia of uniform texture with hyphal threads emanating from them and the association of the mycelium with roots of living plants.

R. solani is capable of attacking a tremendous range of host plants including maize causing seed decay, damping-off, stem canker, root rots, fruit decay and foliage diseases. The combination of competitive saprophytic ability with lethal pathogenic potential and almost unlimited host range makes *R. solani* an economically important and destructive pathogen.

The perfect stage of the fungus has been identified as *Thanatephorus cucumeris* (Frank) Donk., belonging to the Subdivision Basidiomycotina, on the basis of large size of basidia and spores, frequent occurrence of adventitious septa in protosterigmata and primitive type of sclerotia development.

Diagnosis of *R. solani* as a Species

Because of wide variation in morphology, pathogenicity and physiology, the taxonomy and nomenclature of *R. solani* has been subject of controversy. In spite of some commendable efforts our present state of knowledge does not permit the development of a system of taxonomy and nomenclature that removes all the major sources of error. Too few mycelia have been studied and even fewer perfect states have been induced. As a result, there is no assurance that all the mycelia fitting the description of *R. solani* are actually *R. solani.* A number of species of *Rhizoctonia* are very similar to *R. solani.* In addition, mycelia of some ascomycetous fungi may closely resemble *R. solani.* It is largely from the work of Duggar (1915) that provides a detailed discussion of the types of diseases associated with it, its morphology and growth habits and its probable relationship to other *Rhizoctonia* species that the present species concept has evolved. *Rhizoctonia solani* has been characterized as:

1. Pale to dark brown, rapidly growing mycelium of relatively large diameter with branching near the distal septum of hyphal cells, often at nearly right angles in older hyphae.
2. Constriction of branches of hyphae at the point of origin.
3. Formation of a septum in the branch near the point of origin.
4. Production of moniliod cells, often called barrel-shaped cells or chlamydospores, in chains or aggregates sometimes referred to as sporodochia.
5. Production of sclerotia of nearly uniform texture and varying in size and shape. From small, round sclerotia, often less than 1mm in diameter, to thin crusts several centimeters across.
6. Pathogenicity to a wide range of hosts, resulting in a variety of symptoms including damping-off, rotting of roots and other underground plant parts, blighting of hypocotyls, stem, leaves and decay of fruits and seeds.
7. Possession of a basidiomycetous perfect state variously referred to as *Hypochnus, Corticium, Pellicularia, Ceratobasidium* or *Thanatephorus.*
8. Possession of a prominent septal pore apparatus.
9. Possession of multinucleate cells in actively growing hyphae.

Some of the above mentioned characters vary widely whereas others appear to be uniform and are more reliable diagnostic features.

Hyphal Characters

Pigmentation of Hyphae

Pigmentation appears to be restricted to various shades of brown. Young colonies on the media may be nearly white but all the older colonies have some shade of brown. For these reasons, the brown pigment appears to be a stable diagnostic character. Any mycelia remaining permanently white or showing pigmentation other than any shade of brown are not considered as *R. solani.*

Hyphal Diameter

Hyphal diameter varies widely both within and among the isolates. Measurements by Thomas (1925), Richter and Schneider (1953) indicated that moat isolates fall within 5–14μ range, the diameter of hyphae within a colony varies widely according to age and position and also according to composition of the mycelium and temperature (Dahl, 1933). Within broad limits, hyphal diameters have litde diagnostic value. However, colonies in which hyphae consistently are less than 5–61μ wide are not likely to be *R. solani.*

Angle of Branching

Right angle branching is sometimes cited as a diagnostic feature but it is not considered reliable. Other features of branching are diagnostically more important. Branching almost occurs near the distal septum of a cell in yoring and advancing hyphae. In older hyphae branching may occur at any place along the cell. Constriction of branch near the point of origin appears to be stable and reliable characteristic of *R. solani* (Parmeter, 1970).

Septal Pore Apparatus

The dolipore apparatus is a prominent feature in *R. solani.* This apparatus is also common in many other basidiomycetes and its presence in a non-sporulating mycelium is not itself indicative of *R. solani.*

Nucleus

Number of nuclei per cell also provides valuable diagnostic data. Studies indicate that the cells of young, actively growing hyphae are multinucleate, though there is great variation in the number of nuclei per cell (Sanford and Skoropad, 1955; Flentie and Stretton, 1964). Tip cells generally contain 4–8 nuclei with a range of 2–25; young cells mostly 7–11 with a range of 2–25; and straight hyphae mostly 6–11 with a range of 3–19 (Sanford and Skoropad, 1955). These workers observed fewer nuclei in older hyphae and no nuclei in old yellowish hyphae. Saksena (1961) reported that the number of nuclei per vegetative cells of unspecified type as 4–25. All the isolates that produce the perfect stage *Thanatephorus cucumeris* (Frank) Donk. have multinucleate cells. This characteristic, therefore, is used to distinguish *R. solani* from similar fungi like *Ceratobasidium* with binucleate cells (Parmeter *et al.,* 1967).

Monilioid Cells

Monilioid cells may be hyaline or brown, barrel shaped, pyriform, irregular or lobate, numerous or rare, loosely grouped or tightly clampled. Saksena and Vaartaja (1961) have placed considerable emphasis on the use of monilioid cells in distinguishing species of *Rhizoctonia.* However, as reported by Saksena and Vaartaja (1961) the wide variation in dimensions of moniliod cells does not provide suitable criteria for distinguishing the species. In addition, many fungi, including both ascomycetes and basidiomycetes produce monilioid cells similar to those of *R. solani.*

Sclerotia

Sclerotial cells of *R. solani* are essentially similar and while the outer cells may be darker with thicker walls unlike sclerotium there is no obvious differentiation into a rind and a medulla.

Sclerotia of *R. solani* are quite variable. Studies by Papavizas (1965) on variation among single spore isolates showed that sclerotia of sibling isolates varied widely in size, shape, shade of brown, surface texture and distribution on culture medium. In addition, sclerotia may be absent in some isolates. The basic structure of the sclerotium may serve to exclude a mycelium from *R. solani* but the size or shape of the sclerotium within broad limits, does not.

Anastomosis

Anastomosis, while not a diagnostic feature itself, has been used as a criterion for the recognition of species and grouping of isolates of *R. solani.* Flentje and Stretton (1964) recognized varying degrees of anastomosis. Some isolates fused readily to produce mixing of the cytoplasm. Others showed cell wall fusion without mixing of the cytoplasm of the fused cells. Yet other isolates failed to fuse at all. It has been shown that cytoplasmic mixing often leads to the death of fusion cells and sometimes adjacent cells.

Kernkamp *et al.* (1952) observed that hyphae of single spore isolates from the same parent may fail to anastomose. In the light of these observations and our nearly complete ignorance of the mechanism underlying the process of anastomosis in *R. solani*, successful anastomosis, however, can be considered as evidence that mycelia represent the same species such as *R. solani.*

Pathogenicity

From taxonomic point of view three characteristics are significant:

1. Isolates may cause several types of diseases including aerial blights, damping-off, seed decay, fruit decay and root rot.
2. Isolates may vary from avirulent to aggressively virulent.
3. Host range among isolates may vary from limited to extremely wide.

Since *R. Solani* has a very wide host range it is difficult to assume that a given isolate is avirulent if it fails to attack a number of hosts. Similarly, host specificity is also difficult to determine. Variation in virulence was shown by some isolates when tested repeatedly and when tested in different soils. Prior nutrition may also influence virulence. Thus, while pathogenicity to a wide range of plants is characteristic of *R. solani*, avirulence and restricted host range do not exclude isolates from *R. solani.*

Perfect Stage

Generation of perfect stage, *Thanatephorus cucumeris* (Frank) Donk. is the only sure means to recognize *Rhizoctonia solani.* Lack of standard fruiting methods, however, makes the task difficult.

Colony Characters

The basic colour of the colony of *R. solani* is brown. Within the substrate the mycelium is generally hyaline and irregular in shape. Surface mycelium may be sparse or may consist of a thick stroma-like layer. The aerial portion on or above the substrate may consist of hyaline or brown runner hyphae, irregular large or small masses of monilioid cells and sclerotia. Houston (1945) and Richter and Schneider (1953) attempted to place large collections of culture from various hosts into three cultural groups. Type A produced, on the surface of the medium, a heavy stroma-like layer which was white at first and became pale grey later. On the surface of this layer were light cinnamon drab strands of hyphae; sclerotia were natal brown and formed as slightly raised areas radiating from the inoculum and aerial mycelium was sparse. Type B culture produced no stroma-like layer, few globose natal brown sclerotia with irregular surfaces and with little aerial mycelium. Type C cultures were characterized by a slight stroma-like layer which was at first buff but turned wood brown with age. Sclerotia were natal brown and globose with very irregular surface. The three groups also differed in their growth rates. Kim *et al.* (1988) suggested that isolates obtained from 12 locations could be divided into colony type Land D based on the cultural appearance on PDA. Colony type L was colourless to light brown in culture and produced variable number of sclerotia. Colony type D showed brown to dark brown discolouration of mycelia in culture media, and usually produced few sclerotia.

Growth Rate

Growth rate of isolates of *R. solani* may vary from a few hundredth of a milimetre per hour to at least 1 mm per hour. Usually isolates from aerial plant parts including the '*sasakii.*' type grow rapidly. Isolates from near the soil surface, including the *praticola* type, grow quite rapidly and isolates from subterranean plant parts grow more slowly (Sherwood, 1970). In a study of 86 clones, Durbin (1959) found that aerial clones had growth rates ranging from 20–37 mm per day with a mean of 28 mm; the surface clones grew 13–37 mm/day, with a mean of 22 mm and subterranean isolates grew 5–12 mm/day with a mean of 11 mm. Ruppel (1972) suggested that foliar and crown isolates of sugar beet were fast growing.

El Zarka (1965) found no relation of growth rate to cultural type of 15 isolates from potato. However, Galindo (1982) has confirmed the relationship between growth rate and cultural types.

Sherwood (1969) showed that relationship between anastomosis groups, growth rate and virulence. AG–1 and AG–4 were fast growing than AG–2 and AG–3; thus being different. Fast growing isolates were more virulent than slow growing isolates. Similarly, Ruppel (1972) also showed the relationship between growth rate and anastomosis group.

Growth rate of a given isolate at a given temperature can differ on different media. According to Le Clerg (1934) the growth rate of a particular isolate on PDA was not always correlated with that on malt extract agar; some isolates grew more rapidly on one than on the other, while the reverse was true for other isolates. Juliatti *et al.* (1+88) found the growth rate of potato isolates on PDA varied from 20–31 mm/day.

Growth Responses

Temperature

Despite the significant variation within the species, several investigators have achieved some success in correlating growth-temperature responses with other characters used to describe strains. Thomas (1925) observed that isolates originating in warm environments did not grow at 3–7°C, while those from temperate regions did. Richter and Schneider (1953) demonstrated that isolates from warm areas or from greenhouses tend to have higher minima and maxima than to the isolates from colder regions. Isolates having relatively low temperature optima may also have low minima and maxima, and those having high optima have high maxima and minima as well (Thomas, 1925; Richter and Schneider, 1953).

Houston (1945) found that optimum and maximum temperature of his type C isolates were 25°C and 33°C and for type A and B 28°C and 40°C, respectively. Richter and Schneider (1953) observed that temperature cardinals for the 'potato' and 'crucifer' groups were lower than those of the four other groups studied. The isolates of the *'sasakii.'* type grew best at 28–31 °C. The lower and upper limits for growth varied between 5–12 °C and 30–40 °C. The fungus did not grow below 8°C and above 35°C. Sherwood (1970) determined that optimum temperature for growth was 28°C. Temperature in the range of 24–28 °C is known best for mycelial growth in culture and 16–28 °C for sclerotial formation. Roy (1972) observed that isolates grow better at 27°C than at 20 and 13 °C on PDA. Hemmi and Yokogi (1972) noted that isolates did not grow between 5 and 8 °C. Optimum growth rate usually lies between 20–30 °C though higher and lower optima are occasionally reported. Studies by Sherwood (1970) indicate that minima may range from 0–10 °C and maxima from 29–38 °C.

Regarding effect of temperature on the colour of colony of maize isolate, Singh (1998) found that the colour of the colony varied with the temperature. The colony remained light grey at lower temperatures whereas at higher temperatures, it turned light brown to pale brown in colour.

Light

Light has an important effect on *in vitro* growth and sclerotial production of *R. solani.* Butler (1957) found that continuous fluorescent light of 150–400 f.c. suppressed vegetative growth. Endo (1940) stated that sunlight partially inhibited linear growth of *C. sasakii.* However, experiments did not seem to rule out temperature increases inside culture dishes exposed to sunlight as a possible cause of inhibition. Durbin (1959) compared darkness with fluorescent light of 200 f.c. and energy peaks at 580 and 480. Illumination significantly decreased mycelial dry weight of seven out of the nine clones tested, but did not influence the linear growth rate. The length and development of lateral hyphae were decreased by illumination, terminal hyphae being unaffected.

Lokesh and Somashekhar (1988) observed influence of light on growth pattern of *R. solani,* a maize isolate on PDA under different conditions of light. The test fungus was cultured under near ultraviolet (NUV) light ($\lambda = 360$ nm), visible light ($\lambda = 400–700$ nm) and complete darkness. They found variation in the growth potential of mycelium and sclerotium. The growth activity of sclerotia under visible light was greater with higher sclerotial production as compared to the cultures under NUV and darkness.

pH

Most isolates of *R. solani* preferred an acidic medium for growth. The optimum pH for *in vitro* growth of most isolates of *R. solani* range between 5.5 and 7.0 (Le Clerg, 1934). The data on various initial pH values must be interpreted cautiously, because most investigators used unbuffered media, and pH values probably became modified during the culture period. Only one isolate has been reported to grow at pH 2.0 (Richter and Schneider, 1953) Jackson (1940) buffered a liquid medium with sodium glycerophosphate and renewed the solution every second day. The two isolates tested did not grow at pH 2.5, grew slightly at pH 3.5, showed a progressive increase between 5.5 and 6.5, a decrease at 7.5, and very little growth at 8.5. Roth and Riker (1943) and Bateman (1962) obtained bimodal growth curves. Bateman (1962) found that maximum growth of isolates on buffered PDA occurred at 5.8; less growth at 6.8 than at 5.8 or 7.5. Roth and Riker (1943) found bimodal response with growth peaks at 4.2 and 6.2 for an isolate grown on PDA buffered with sodium phosphate.

Source of Primary Inoculum and Secondary Spread

The source of primary inoculum are the sclerotia in soil or in infected host debris, and the active mycelium on the other grass hosts that grow in the vicinity of maize plants in field.

In undisturbed soil, sclerotia are present up to a depth of 2.0 cm. Population of sclerotia in ploughed field is higher between 6 and 12 cm than in upper layers and the sclerotia buried deep in soil have better buoyance and germinability. Sclerotia which survive on plant debris often come up on the soil surface during field preparation and other operations. They come in contact with newly planted seedlings/plants and cause infection. The germinated sclerotia come in touch with lower leaf sheaths or drooping leaves, those which are in innate contact with soil. The infection progresses upwards in bands killing the tissues in advance for better growth of the pathogen. The primary inoculum becomes more active during the period of pre-flowering stage (30–40 days old plants) which falls in late July and early part of August. Secondary spread is due to contact of healthy plants with infected leaves or sheaths or *vice-versa.* The infection continues to mid-dough stage infecting the entire plant including the ear; the tassel remaining unaffected. The fungus begins forming sclerotia on the infected plant organs immediately when the conditions become no more conducive to growth. These sclerotia fall down separately or with infected host leaves/sheaths in soil to serve as a source of primary inoculum for infection on to other host crops in the field or over-winter for the next season.

Host Range

The rice pathogen has wide host range and infects plants belonging to over 32 families in 188 genera. *R. solani* f. sp. *sasakii* can be inoculated artificially in a number of crop plants belonging to families, Gramineae, Papilionaceae and Solanaceae: *Paspalum scrobiculatum, Pannisetum pupurium, Panicum miliaceum, Zea mexicana, Zea mays, Oryza sativa, Saccharum officinarum, Sorghum bicolor, Arachis hypogea, Clycine max, Pisum sativum, Vigna radiata* and *Lycopersicum esculentum* (Baruah and Lal, 1981; Ahuja and Payak, 1988). Maize has also been found to be infected by isolates from rice, sugarcane, arrowroot and some grasses. Cross inoculation tests indicated that rice and turmeric isolates did not differ either in pathogenicity or morphological characters (Lal *et al.,* 1980). Cotton isolates show a clean distinction from cereal isolates and within latter, wheat isolates are considered to be different from those of maize and rice isolate.

Estimation of Yield Losses

Several reports are available on the estimation of yield losses due to the banded leaf and sheath blight disease of maize. A direct correlation exists between disease incidence and grain yield. Singh and Sharma (1976) evaluated 10 different maize cultivars for grain yield and reported that grain yield losses ranged from 12–41 per cent with an average of 19 per cent. This reduction in grain yield showed highly significant correlation with disease severity. Butchaiah (1977) also observed a drastic reduction in yield of 97 per cent in composite D-745 and hybrid Ganga Safed-2. Baruah (1979) observed a direct correlation between the severity of banded leaf and sheath blight disease of maize caused by *R. solani* and other yield parameters. Lal *et al.* (1980) reported that the losses in grain yield due to infection were 24–32 per cent in 10 different cultivars depending on disease severity levels of 70–87 per cent. According to Summer and Minton (1989) grain yields averaged 6890 and 8760 kg/ha, with high and low inoculum levels, respectively compared with 9890 kg/ha for non-infected plots indicating that the yields were reduced to 47, 42 and 8 per cent in soils infected with high inoculum levels and 15, 19 and 1 per cent with the low inoculum levels from first to third year, respectively. Disease incidence was found to be positively correlated with yield loss by Jiang *et al.* (1991). Pandey (1992) also showed a significant negative correlation between the disease severity and grain yield per plot and cobs per plant.

As far as qualitative loss in seed health is concerned, early infection of *R. solani* on cobs results in production of infected seeds exhibiting horse shoe shaped lesions. Such grains are chaffy, poor and light in weight. If cob infection occurs at the late stage, the germination of seed is adversely affected resulting into seed rot and seedling blight.

Apart from *R. solani, Bipolaris maydis, Botryodiplodia theobromae, Colletotrichum graminicola* and *Fusarium moniliforme* have also been found responsible for reduction in germination and field emergence of discoloured seeds of maize.

Disease Management

Cultural Practices

Mitrasena and Adikary (1986) reported that wide row to row and plant to plant spacing reduced sheath blight in rice but the yield was not superior and close spacing could not compensate yield losses due to increased disease severity levels. In general, it has been observed that irrigation treatments did not influence *R. solani* severity in maize. When maize crop was grown after *Canavalia ensiformis* in rotation, maize growth was significantly better in spite of higher incidence of *R. solani. Rhizoctonia* population is greater in sandy loam during mid-season than early or late in the season. As far as fertilizer application is concerned, higher applications of phosphorus do not affect the disease levels of *R. solani.*

Sharma and Hembram (1990) suggested a simple and economical measure for controlling the disease caused by *R. solani* f. sp. *sasakii* consisting of stripping the lower 2 or 3 leaf sheaths, which arrest upward movement of the pathogen by severing its connection with the succeeding leaf sheath. However experiments with similar treatments conducted at the Pantnagar centre (Anonymous, 1994) which is a hot spot for banded leaf and sheath blight disease of maize showed that there was non-significant reduction in disease and the yield levels were at par or lower when compared with the check. Further, another observation in plants where lower leaf sheaths were removed an extensive lodging was observed before silk emergence due to poor support of stalk at the base.

Chemicals

Roy and Saikia (1976) showed excellent control of the disease by application of benomyl and mancozeb on rice, whereas and Verma and Menon (1977) reported that Hinosan and Kitazin granules reduced the disease severity effectively. Butchaiah (1977) tested 25 fungicides and 9 antibiotics against the pathogen *in vitro*. Thiabendazole, Du ter and Vitavax proved to be effective over others against the disease under field conditions. Baruah (1979) observed the effective control of the disease with 3 fungicides, *viz.*, carboxin, TPTH and thiabendazole out of 10 chemicals tested. Thiabendazole, followed by TPTH and carboxin significandy increased not only the grain yield but also 1000-kernel weight and the number of cobs per plant. Ahuja and Payak (1986) presented the effectiveness of validamycin-A, aureofungin, benodanil, carbendazim, dichlorine, thiabendazole and thiophanate-methyl on isolates of *R. solani* f. sp. *sasakii.* Jiang *et al.* (1991) recommended a specific duration of the crop for the application of chemicals on the diseased plants, *i.e.* when infection rate is less than 2.3 per cent and also reported that the antibiotic, Jinggangmycin gave the best control up to 73.7 per cent. Sinha (1992) reported most effective control of banded sclerotial disease with Bavistin 50 WP (carbendazim) resulting in 87 per cent disease control, followed by Benlate 50 WP (benomyl), Brestan 60 WP (fentin), Calixin 75 EC (triademorph) and Difolatan 80 WP (captafol) which controlled the disease by 82, 77, and 72 per cent, respectively. Singh (1998) reported application of Bavistin (0.05 per cent) and Tilt (0.1 per cent) gave maximum control of the disease when applied at 10 days interval startillg from 30th day after planting.

Biological Control

Various fungi such as *Trichoderma harzianum, T. viride, Gliocladium virens, Aspergillus* sp., *Neurospora crassa* and *Penicillium* sp. are known to inhibit the mycelial growth; suppression of formation and germination of sclerotia of *R. solani.* Isolates of *T. viride* are highly antagonistic to *R. solani* and the effectiveness can be increased by combination biocontrol agent with fungicidal treatments. *Trichoderma* sp. gives successful control of seed rot and early stem rot in maize caused by *R. solani.*

Baicu *et al.* (1989) reported that effectiveness of some *Trichoderma viride* mutants tolerant to fungicides such as thiram and thiophanate-methyl and the mutant Td-k was found most effective against *R. solani.* Herman *et al.* (1989) observed that the combination of *Trichoderma harzianum* strains and solid matrix priming can improve that biological seed treatment in soils not infested with *R. solani.* Hebber *et al.* (1992) reported that strains of *Pseudomonas cepacia* can inhibit *R. solani* of maize *in vitro.*

Baby and Manibhusan Rao (1993) used two fungal antagonists, *viz., Gliocladium virens* Miller and *Trichoderma longibrachiatum* Rifai agge. to control rice sheath blight and they found that these two antagonists significantly enhanced the growth of rice plants as well as grain yield in addition to the disease suppression. Sharma (1996) evaluated the biocontrol fungi and bacteria against *R. solani,* following dual culture technique in laboratory condition, the fungal bioagents, *viz., Trichoderma harzianum, T. viride, Gliocladium virens* and *Trichoderma* sp. (field isolate) were found highly effective in parasitizing the growth of *R. solani* within 10 days, whereas none of the fluorescent pseudomonads could inhibit or cover the growth of *R. solani.*

Integration of Control Measures for the Management of Banded Leaf and Sheath Blight of Maize

Scanty work has been done on the integrated management of this disease on maize; although it has been attempted in case of rice sheath blight. Krishnaiah *et al.* (1986) observed that in Krishna delta (Andhra Pradesh), the combination of moderately resistant varieties, conservation of natural enemies and need based use of pesticides reduced sheath blight and other pests and average yield increased by 36 per cent at farmers' field.

Integration of management practices such as seed coating by antagonists, seed treatment by Vitavax, foliar spray of mancozeb and removal of lower leaves was done by Pandey (1992) in different combinations. In most of the cases the treatments were reported effective when compared to check.

Sharma (1996) managed the disease through combining the chemicals (Savistin and Tilt), bio-control agent (*Trichoderma harzianum*) and cultural practices (removal of lower leaves). None of the treatments effectively reduced the disease but foliar sprays of *T. harzianum*+ Tilt followed by *T. harzianum* sprays, Savistin + Tilt + *T. harzianum,* Savistin + Tilt and savistin alone could exhibit some reduction in disease severity. Removal of lower leaves alone was not so effective.

Epilogue

With quite large host range, *Rhizoctonia solani* Kühn, the causal fungus of the disease, banded leaf and sheath blight in maize, becomes a threatening problem in almost all the areas of the maize cultivation. It survives forming sclerotial bodies in soils or the infected host debris to resurge and play havoc to the crop the very next year as soon as the conditions become favourable to its growth. Not only maize, the: same fungus may equally infect other crops of economic importance such as rice, mungbean, urdbean and soybean causing considerable losses to the farmers' economy. In India where farmers have little to gain because of their small land holdings and limited resources like water for irrigation and power to operate machinery, this amount of loss sometimes becomes a serious threat to their economy. Moreover, less awareness among farmers about this disease and the loss caused by it make the problem worse. Although much work has been done on the biology and pathology of this notorious pathogen, it has rendered the plant breeders little to do with development of resistant or tolerant crop varieties by virtue of its strong saprophytic ability and high growth rate. The best option, however, seems to integrate the management practices in such a form so as to loose less to the disease. Many of the investigators working on the management of the disease have strongly advocated adoption of the biological measures to integrate with chemicals for effective and durable management of this fungus. Cultural practices such as crop rotation, sanitation and selection of crop cultivars that may escape the virulent form of the fungus may provide added advantage. Lastly and most importantly, it remains still to get a simple and effective means to stop the loss caused by fungi like *Rhizoctonia solani.*

References

Ahuja, S.C and Payak, M.M. (1986). *In vitro* response of maize and rice isolate of *Rhizoctonia solani* to antibiotics and fungi toxicants. *Int. Rice Res. Newslett.*, 11: 16–17.

Ahuja, S.C. and Payak, M.M. (1988). Banded leaf and sheath blight of maize. In: *Perspectives in Mycology and Plant Pathology*, (Eds.) Agnihotri, V.P., Sarbhoy, A.K. and Kumar, D. Malhotra Publishing House, New Delhi, pp. 178–186.

Anonymous. (1994). Maize at Pantnagar, *Annual Report: Maize Pathology*, AICMIP, pp. 26–39.

Baby, U.I. and Manibhusan Rao, K. (1993). Control of rice sheath blight through the integration of fungal antagonists and organic amendments. *Tropical Agril.*, 70: 240–244.

Baicu, T., Sesan, T. and Gogoasa, C. (1989). Effectiveness of some *Trichoderma viride* mutants tolerant to fungicides in biological control of some phytopathogenic fungi. *Probleme-Protectia-Plantelor*, 17: 205–218.

Baruah, P. (1979). Studies on some aspects of banded sclerotial disease of maize caused by *Rhizoctonia solani* f. sp. *sasakii. M.Sc. (Ag.) Thesis*, G.B. Pant University of Agriculture and Technology, Pantnagar, 56 p.

Baruah, P. and Lal, S. (1981). Host range of *Rhizoctonia solani* f. sp. *sasaktii*, the incitant of banded sclerotial disease of maize. *Indian Phytopath.*, 34: 494–946.

Bateman, D.F. (1962). Relation of soil pH to development of poinsettia root rots *Phytopathology*, 52: 559–566.

Bertus, L.S. (1927). A sclerotial disease of maize due to *Rhizoctonia solani* Kühn. *Yearbook Department of Agriculture*, Ceylon, pp. 46–48.

Butchaiah, K. (1977). Studies on banded sclerotial disease of maize. *M.Sc. (Ag.) Thesis*, G.B. Pant University of Agriculture and Technology, Pantnagar, 65 p.

Butler, E. E. (1957). *Rhizoctonia solani* as a parasite of fungi. *Mycologia*, 49: 354–373.

Dahl, A.S. (1933). Effect of temperature on brown patch of turf (Abstr.). *Phytopathology*, 23: 8.

de Candolle, A.P. (1815). *Memorie sir les rhizoctones, nouveau genre de chamignons quiatta-que les recines, des planets et en parliculier celle dela Luzerne cultivee. Mem. Mus. Nat. Hist. Nat. Sar.*, 2: 209–216.

Deighton, F.C. (1932). Mycological work. *Annual Report Agric. Deptt. Sierra Leone*, pp. 20–25.

Dugger, B.M. (1915). *Rhizoctonia crocorum* [Pers. D.C. and *R. solani* Kühn (*Corticium vagum* B. and C.)] with notes on other species. *Ann. Missouri Botan. Garden*, 2: 403–458.

Durbin, R.D. (1959). Some effects of light on the growth and morphology of *Rhizoctonia solani*. *Phytopathology*, 49: 59–60.

El Zarka, A.K. (1965). Studies on *R. solani.* Kühn, the cause of the black scurf disease of potato. *Mededel. Landbouwhogeschoolen, Wageningen*, 217: 73.

Endo, R.M. (1940). Physiological studies on the causal fungi of sclerotium diseases of rice plants with special reference to some factors controlling the occurrence of the disease. *Bull. Mryazaki Coll. Agr. Forestry*, 11: 55–218.

FAO Yearbook (1993). *Production*, 47: 77–78.

Flentje, N.T. and Stretton, H.M. (1964). Mechanism of variation in *Thanatephorus cucumeris* and *T. parlicolus. Aust. J. Biol. Sci.*, 12: 686–704.

Galindo, J.J. (1982). Variability among isolates of *Rhizoctonia solani* associated with snapbean hypocotyls and soils in New York. *Plant Dis.*, 66: 390–394.

Hebber, K.P., Atkinson, D., Tucker, W. and Dart, P.J. (1992). Suppression of *Fusarium moniliforme* by maize root associated *Pseudomonas cepacia. Soil Biol. and Biochem.*, 24: 1009–1020.

Hemmi, T. and Yokogi, K. (1972). Studies on sclerotium disease of rice plant. *J. Agri. and Hort.*, 2: 955–1094.

Herman, G.E., Taylor, A.G. and Stasz, T.E. (1989). Combining effective strains of *Trichoderma harzianum* and solid matrix priming to improve biological seed treatments. *Plant Dis.*, 73: 631–637.

Hirrel, M.C., Lee, F.N., Dale, J.L. and Plunkett, D.E. (1988). First report of sheath blight (*Rhizoctonia solani*) on field corn in Arkansas. *Plant Dis.*, 72: 644.

Houston, B.R. (1945). Culture types and pathogenicity of isolates of *Corticium solani. Phytopathology*, 35: 371–393.

Jackson, L.W.R. (1940). Effects of H-ion and Al-ion concentrations on damping-off of conifer and certain causative fungi. *Phytopathology*, 30: 563–579.

Jiang, H.L., Ding, X. and Ma, H. (1991). The occurrence and chemical control of maize sheath blight (*Rhizoctonia solani*) in Rugao of Jiangsu. *Plant Prot.*, 6: 12.

Juliatti, F.C., Chaves, G.M. and Zambolirn, L. (1988). Growth rate, sclerotia production and pathogenicity of *Rhizoctonia solani* isolates from potato. *Fitopathologia–Brasileira*, 13: 243–246.

Kernkamp, M.F., Dezeeuw, D.J., Chen, S.M., Ortega, B.C., Tsiang, C.T. and Khan, A.M. (1952). Investigations on physiologic specialization and parasitism of *Rhizoctonia solani. Minnesota Agr. Expl. Sta. Bull.*, 200: 36.

Kim W.G., Kim, C.K. and Yu, S.H. (1988). Anastomosis grouping and cultural characteristics of isolates of *Rhizoctonia solani* Kühn from sclerotia overwintered in paddy fields. *Korean J. Pl. Patho.*, 4: 136–142.

Knight, B.C. and Burrin, W.D. (1964). Sharp eye spot of maize. *Plant Pathol.*, 13: 138.

Krishnaiah, K., Reddy, P.C. and Rao, C.S. (1986). Integrated pest control in rice in Krishna Delta of A.P. *Indian J. Plant Prot.*, 14: 1.

Kühn, J.G. (1858). *Die Krankheiten der Kulturegewächse, ihre ursachen und ihre Verhütung, Gustav Bosselmann, Berlin,* 312 p.

Kujwara, T. (1968). Important disease of maize in Japan. pp. 165–172. In: Proceedings of Symposium on maize production in South East Asia Present Situation and Future Problems. *Tropical Agric. Res. Series Agric. Fir Fish Res. Conc. Minist. Agric.,* Japan, 187 p.

Lal, S., Baruah, P and Butchaiah, K. (1980). Assessment of yield losses in maize cultivars due to banded sclerotial disease. *Indian Phytopath.,* 31: 274–278.

Le Clerg, E.L. (1934). Parasitism of *Rhizoctonia solani* on sugar beets. *J. Agril. Res.,* 49: 407–431

Lee, S.B., Kim, J.G., Han, M.S. and Han, H.J. (1989). Studies on the occurrence of corn sheath blight caused by *Rhizoctonia solani.* I. Influence of Growth season and cultural environment on the occurrence of sheath blight disease. *J. Korean Soc. Grassland Sci.,* 9: 174–178.

Lokesh, S. and Somashekhar, R.K. (1988). Influence of light on growth pattern of *Rhizoctonia solani*: A maize isolate. *Curr. Sci.,* 51: 414–415.

Maiti, S. (1978). Two new era rots of maize from India. *Pl. Dis. Reptr.,* 62: 1047–1076.

Mitrasena, Y.J.P.K. and Adikary, W.P. (1986). Effect of plant density on sheath blight incidence. *Int. Rice Res. Newslett.,* 11: 20.

Pandey, M.K. (1992). Studies on banded leaf and sheath blight of maize. *M. c. (Ag.) Thesis.* G.B. Pant University of Agri. and Tech., Pantnagar, 79 p.

Papavizas, G.C. (1965). Comparative studies of single basidiospore isolates of *Pelliculari filamentosa* and *P. praticola. Mycologia,* 57: 91–103.

Parmeter, J.R., Jr. Whitney, H.S. and Platt, W.D. (1967). Affinities of some *Rhizoctonia* Species the resemble mycelium of *Thanatephorus cucumeris. Phytopathology,* 57: 218–223.

Parrneter, J.R. Jr. (1970). *Rhizoctonia solani: Biology and Pathology.* University of California Press, Berkeley, 255 p.

Payak, M.M. and Renfro, B.L. (1966). Diseases of maize new to India. *Indian Phytopathol. Soc. Bull.,* IARI, New Delhi, 3: 14–18.

Reyes, G.M. (1941). Notes of disease affecting maize in the Philippines. *Philipp. J. Agric.,* 12: 61–69.

Richter, R. and Schneider, R. (1953). *Untersuchungen Zur Morphologische and Biolo-gischen Diffirenzierung Von Rhizoctonia solani* K *Phytophathol Z.,* 20: 167 –226.

Roth, L.F. and Riker, A.J. (1943). Influence of temperature, moisture and soil reaction on the damping-off of red pine seedlings by *Pythium* and *Rhizoctonia. J. Agr. Res.,* 67: 273–293.

Roy, A.K. and Saikia, U.N. (1976). Chemical control of sheath blight of rice. *Indian Phytopath.,* 24: 334–356.

Roy, A.K. (1972). Growth of *Rhizoctonia solani* in relation to temperature. *Phytopath. Z.,* 74: 28–36.

Ruppel, E.G. (1972). Correlation of cultural characters and source of isolates with pathogenicity of *R. solani* from sugar beet. *Phytopathology,* 62: 202–205.

Saksena, H.K. and Vaartaja, O. (1961). Taxonomy, morphology and pathogenicity of *Rhizoctonia* sp. from forest nurseries. *Can. J. Bot.,* 39: 627 –647.

Saksena, H.K. (1961). Nuclear phenomenon in the basidium of *Ceratobasidium praticolom* (Kotila) Oliv. *Can. J. Bot.*, 39: 717–725.

Sanford, G.B. and Skoropad, W.P. (1955). Distribution of nuclei in hyphal cells of *Rhizoctonia solani. Can. J. Microbiol.*, 1: 412–415.

Sharma, R.C. and Hembram, D. (1990). Leaf stripping, a new method to control banded leaf and sheath blight of maize. *Curr. Sci.*, 59: 745–746.

Sharma, G. (1996). Integrated management of banded leaf and sheath blight caused by *R. solani* Kühn. *M.Sc. (Ag.) Thesis*, G.B. Pant University of Agri. and Tech., Pantnagar, 90 p.

Sherwood, R.T. (1969). Morphology and Physiology in four anastomosis groups of *Thanatephorus cucumeris. Phytopathology,* 59: 1924–1929.

Sherwood, R.T. (1970). Physiology of *Rhizoctonia solani.* In: *Rhizoctonia solani: Biology and Pathology,* (Ed.) J.R. Parmeter Jr. University of California Press, Berkeley, pp. 69–92.

Singh, A.K. (1998). Studies on *Rhizoctonia solani* Kühn, the incitant of banded leaf and sheath blight of maize (*Zea mays* L.) with special reference to variability and integrated management *Ph.D. Thesis,* G.B. Pant University of Agriculture and Technology, Pantnagar, U.P., 167 p.

Singh D.M. and Sharma, Y.R. (1976). Evaluation of maize germplasm to banded disease and assessment of yield loss. *Indian Phytopath.*, 29.

Sinha, P.P. (1992). Chemical control of banded sclerotial disease of maize. *Orissa J. Agric. Res.*, 5: 54–57.

Summer, D.R. and Minton, N.A. (1989). Crop losses in corn induced by *Rhizoctonia solani* AG-2-2 and nematodes. *Phytopathology,* 79: 934–941.

Thomas, K.S. (1925). Onderzoekingen over *Rhizoctonia.* Electr. Drukkerij "De Industrie" *J. Van Druton Utrecht.* 98 p.

Verma, A.S. and Menon, M.R. (1977). Fungicidal trial on the control of sheath blight of rice. *Madras Agric. J.*, 64: 416–417.

Wiltshire, S.P. (1956). Plant disease in British colonial dependencies, a half yearly report. *FAO Plant Prot. Bull.*, 5: 66.

Zhu, H. (1982). A disease caused by *Rhizoctonia solani* (Kühn) on maize. *Acta Phytopath. Sinica,* 12: 61–62.

©R.C. Ram & Asha Sinha (*Editors*)
***Microbial Diversity: Modern Trends*, 2007** **184–189**
Daya Publishing House, New Delhi

Chapter 13

Rhizobium: A Potential Biofertilizer

Neerja Asthana* and Niranjan Kumar**

***Department of Botany, B.R.A. Bihar University, Muzaffarpur – 842 001, Bihar**
****Department of Mycology and Plant Pathology, Institute of Agricultural Sciences, B.H.U., Varanasi – 221 005, U.P.**

Introduction

Man began to cultivate land in an organized way for food grains around 8000 B.C. very soon he learned that the same land cannot support the growth of plants endlessly and this led him to think about ways to improve the fertility of soil. In the dawn of new millennium, challenges in the agriculture sector are quite different from those met in the previous decades. Population has been increasing at very fast rate. This accounts for the enormous pressure to produce more food from less land. It is high time to transform the traditional agricultural scenario to meet the food demand for the growing population and simultaneously conserve the environment.

For a sustainable agriculture system, it is imperative to utilize renewable inputs which can maximize the ecological benefits and minimize the environmental hazards. One possible way of achieving this is to decrease dependence on use of chemical nitrogen fertilizers by harvesting the atmospheric nitrogen through biological processes.

One of the major concerns in today's world is the pollution and contamination of soil. The use of chemical fertilizers and pesticides has caused tremendous harm to the environment. An answer to this is the biofertilizer, an environmentally friendly fertilizer now used in most countries. Biofertilizer are organisms that enrich the nutrient quality of soil. The main sources of biofertilizer are bacteria, fungi and cynobacteria (blue-green algae). The most striking relationship that these have with plants is symbiosis, in which the partners derive benefits from each other. Plants have a number of relationships with fungi, bacteria, and algae, the most common of which are with mycorrhiza, rhizobium, and cyanophyceae. These are known to deliver a number of benefits including plant nutrition, disease resistance, and tolerance to adverse soil and climatic conditions. These techniques have proved to be successful biofertilizers that form a health relationship with the roots. The biological soil fertility management is an ecological approach for sustainable development and is mainly concerned with the maintenance of yield, closely associated with desires to conserve natural resources, including the need to utilize all available resources within economic limits. Biofertilizers help to solve such problems as increased salinity of the soil and chemical run-offs from the agricultural fields. Thus, biofertilizers are important to ensure a healthy future for the generations to come.

During last two decades, agricultural production has increased in developed and certain developing countries due to the use of high yielding varieties and enhanced consumption of chemical fertilizers and water. Man made fertilizers, those containing N_2, P_2O_5 and K_2 increase the output of agricultural products. The increase in world fertilizer requirement by 2000 AD would be approximately three times the rate of current consumption (Verghese, 1977). Chemical fertilizers increase the consumption of non-renewable forms of energy. Therefore it is essential for us to evolve and adopt a strategy of integrated nutrient supply by using a judicious combination of chemical fertilizers, organic manures and biofertilizers. Biofertilizer harness atmosphere nitrogen with the help of specialized soil microorganisms.

Biofertilizers include microorganisms and their metabolites that are capable of enhancing soil fertility, crop growth, and/or yield. These include both indigenous microbes and microbial inoculants, that is, microorganisms that replace fertilizers or increase a crop's fertilizer use efficiency. Soil microorganisms such as bacteria, ectomycorrhiza, arbuscular mycorrhizal fungi, and soil algae, especially the N_2-fixing cyanobacteria have potential as biofertilizers. Nitrogen-fixing inoculants based on Rhizobium species were among the first biofertilizers introduced into agroecosystems in the nineteenth century.

Basically Biofertilizers can be divided into two broad categories *viz.*, Nitrogen fixers and Phosphorus solubisers. Further categorization is as following:

Nitrogen Fixers

1. *Rhizobium* spp.
2. *Azotobacter*
3. *Azospirillium*
4. Blue Green Algae (BGA)
5. *Azolla–Anabaena*

Phosphate Solubilisers

1. *Pseudomanas*
2. *Bacillus*
3. *Penicillium*
4. *Aspergillus*

Mycorrhiza

1. Endo-mycorrhiza
2. Ecto-mycorrhiza
3. Vescular Arbuscular Mycorrhiza (VAM)

Biofertilizers accelerate certain microbial processes in the soil which augment the extent of availability nutrients in a form easily assimilated by plants. Very often microorganisms are not as efficient in natural surroundings as one would expect them to be and therefore artificially multiplied cultures of efficient selected microorganisms play a vital role in accelerating the microbial processes in soil. Use of biofertilizers is one of the important components of integrated nutrient management, as they are cost effective and renewable source of plant nutrients to supplement the chemical fertilizers for sustainable agriculture. Several microorganisms and their association with crop plants are being exploited in the production of biofertilizers. They can be grouped in different ways based on their nature and function.

1. *N_2 Fixers*
   (*a*) Free living: *Azotobacter, Acetobacter, Beijerinckia, Clostridium, Klebsiella, Anabaena, Nostoc*
   (*b*) Symbiotic: *Rhizobium, Frankia, Anabaena azollae*
   (*c*) Associative symbiont: *Azospirillum*
2. *Phosphorus Solubilizers Bacteria*
   (*a*) *Bacillus megaterium var. phosphaticum, Bacillus subtilis, Bacillus circulans, Pseudomonas striata*
   (*b*) *Penicillium* sp. *Aspergillus awamori*
3. *Fungi P Mobilizers*
   (*a*) VAM fungi: *Glomus* sp., *Gigaspora* sp., *Acaulospora* sp., *Scutellospora* sp. and *Sclerocystis* sp.
   (*b*) Ectomycorrhizal: *Laccaria* sp., *Pisolithus* sp., *Boletus* fungi sp., *Amanita* sp.
   (*c*) Ericoid mycorrhizae: *Pezisella ericae*
   (*d*) Orchids: *Rhizoctonia solani* (usage restricted as it is pathogenic on other plants)
4. *Silicate and Zinc Solubilizers*: *Bacillus* sp.
5. *Plant Growth Promoting Rhizobacteria*: *Pseudomonas* spp., and many more

The term 'biofertilizers' or more appropriately 'microbial inoculants' can be generally defined as a preparations of containing live or latent cells of efficient strains of nitrogen fixing, phosphate solubilizing or cellulolytic microorganism used in maintaining soil fertility and increasing crop productivity. Thus biofertilizers are ecofriendly, low cost inputs playing a significant role in improving the quality of agricultural produce and sustaining the productivity over a longer period of time. There are a number of biofertlizers like *Rhizobium, Azotobacter, Azospirillum,* Blue green algae and *Azolla-Anabaena* symbionts. Except *Rhizobium* all are free-living, while it is symbiotic bacteria legume-*Rhizobium* symbiosis is the most promising as it supplies 80-90 per cent of the total nitrogen requirement of the legumes. Biological nitrogen fixation contributes to 69 per cent of the global nitrogen fixation.

Background

First commercial biofertilizer started as *Rhizobium* in 1895 by Nobbe and Hiltner with the product 'Nitragin' in USA. Nitragin was produced on gelatin and agar nutrient media. However, agar based inoculants were soon replaced by peat based ones. After, 1895 modest laboratory preparations in the mid-1930s in the United States, rhizobial inoculants (also known as legume inoculants) are become industrial propositions in the United States, Europe, Australia and India. In India use of *Rhizobium* inoculant was initiated in 1920, but systematic production began after 1950 with gradual introduction of other biofertilizers on commercial scale started only during late 1960s when yellow seeded soybean was introduced for the first time. Recognition of Indian peat as suitable carrier for production of biofertilizer in 1967 further augmented the growth of biofertilizer industry in India. The performance of Indian peat based biofertilizer at Indian Agricultural Research Institute, New Delhi, was found to be comparable to that obtained with imported 'Nitragin' biofertilizer from USA. Since then, the process of development of biofertilizer, especially of rhizobial biofertlizer for various crops in India has made a tremendous success.

Rhizobium inoculant–*Rhizobium* is a symbiotic nitrogen fixing bacteria.

1. 80 per cent of the total global nitrogen fixation, through biological means comes from legume-*Rhizobium* symbiosis alone. Scientists conclusively proved that nodules on legume roots are responsible for fixing atmospheric nitrogen
2. Legumes absorb the nitrogen fixed by *Rhizobium* and in turn supply the nutrients required by *Rhizobium.* Inoculation with *Rhizobium* helps to save nitrogen fertilizer and increase grain yield by about 10–20 per cent depending upon the crop and soil conditions. The root secretions from the leguminous plants and biological material released from the decomposed nodule enrich the fertility states of soil. Seed bacterization with *Rhizobium* is recommended for all leguminous crops.

Rhizobium in Root Nodules

Rhizobia enter the roots of legumes either through root hairs (as in clovers and lucerns) or directly at the point of emergence of lateral roots (as in ground nut).

A transverse section of a nodule reveals a central 'bacteriod zone' which is enveloped in a nodule cortex. The bacteroid zone is made up of host cells containing 'bacteroids' encased in membranous envelopes which are of host origin.

Function of the Nodule

The nodule is merely a protective structure since it is now well established that bacteroids are the seats of nitrogen fixation. Nitrogenase is the enzyme which mediates the reduction of N_2 to NH_3. Nitrogen fixation is essentially an anaerobic process and hence the nodule must have a mechanism to exclude O_2 from the bacteroid which is the site of fixation. This is accomplished by the presence of leghaemoglobin around bacteroids enclosed by membranous envelops of host origin. The amount of leghaemoglobin and the extent of bacteroid tissue in nodules have a direct bearing on the amount of nitrogen fixed by legumes (Bergersen and Briggs, 1958; Chopra and Subba Rao, 1967; Verma and Bal, 1976).

The functioning of the nodule may be influenced by changes in temperature, light intensity, photoperiods, the presence of combined nitrogen in soil, pH of soil, mineral nutrition (especially Cobalt and Molybedenum since the latter happens to be an integrate part of the enzyme nitrogenase) growth substances, and the presence of salt and antagonistic microorganisms in soil Nutman (1976), and Subba Rao (1977, 1979).

Cultivation and Mass Production

Rhizobia are usually maintained by sub-culturing at frequent intervals on yeast extract mannitol agar (YEMA) medium. The accepted medium for the growth of *Rhizobium* on a large scale is the YEM liquid medium containing sucrose or glucose.

Agar-based inoculants have become out dated except for small scale experiments broth cultures are in vogue in European days carrier (peat) based inoculants are more frequently used in United States, Australia and New Zealand. In India, peat like material available in the Nilgiri Valley to an estimated extends of 5.5 million tones has been found to be a good carrier (Iswaran *et al.*, 1969). Lignite is another carrier which is widely used (Kandasmy and Prasad, 1971) and an estimated quantity of 3 million tones is available annually from Neyveli lignite mines in South India.

Quality Control

The quality of cultures varies between individual manufacturers within a country and standards expected to be reached by inoculants producers depend on the extent of expertise and local conditions obtainable within a country.

Conclusion

Industrial production of nitrogenous fertilizers is a costly process dependant on energy derived from fossil fuel, which is getting depleted at a faster rate. On the contrary biofertilizers are eco-friendly low cost inputs playing a significant role in improving the quality of agricultural produce and sustaining the productivity over a longer period of time. Since use of biofertilizer is a living system and the influence is subject to environmental, biological and nutritional stress.

To conclude, it may be said that when the country is shifting its agricultural system gradually to organic farming there will be high demand of biofertilizer in future. Biofertilizer, for sustainable agriculture needs to be practical rather than theoretical. The farmers have to be made convinced of the facts that points to it and make them adopt latest technologies. Extension activities such as group meetings, trainings, field demonstrations, field visits and success stories motivate and encourage them to change their old farming practices.

References

Berersen, F.J., and Briggs, M.J. (1958). Studies on the bacterial component of soybean root nodules: Cytology and organization in the host tissue. *J. Gen. Microbiology*. 19: 482–490.

Chopra, C.L. and Subba Rao, N.S. (1967). Mutual relationship–among bacteroids leghaemoglobin and nitrogen content of Egyptian clover and gram. *Arch. Microbiol.*, 58: 71–76. (Arch Fur Mickbiol.)

Iswaran, V., Sundara Rao, W.V.B., Magu, S.P. and Jauhri, K.S. (1969). Indian peat as a carrier of *Rhizobium*. *Curr. Sci.*, 38: 468–469.

Jeewa Ram, K.L. Jain and P.C. Lodha (2004). Compatibility of seed protectant with *Rhizobium leguminosarum* and its biological nitrogen fixation ability in mungbean. *J. Mycol. Pl. Pathol.*, 34(2): 644–666.

Kandaswamy, R. and Prasad, N.N. (1971). Lignite as a carrier of rhizobia. *Curr. Sci.*, 40: 496.

Kurz, N.G.W. and La Rue, T.A. (1975). Nitrogenase activity in rhizobia in the absence of host plant. *Nature*, London, 256: 407–409.

Mc. Comb, T.A., Elliot, J. and Dilworn, J.H. (1975). Acetylene reduction by *Rhizobium* in pure culture. *Nature*, London, 256: 409–410.

Nutman, P.S. (Ed.) (1976). *Symbiotic Nitrogen Fixation in Plants*. International Biological Programme, 7, Cambridge Univ. Press, London.

Pandy, S., S. Nehra and P.C. Trivedi. (2004). Effect of Inoculation time of Heterodera cajani and *Rhizobium* on nodulation in cowpea. *J. Mycol. Pl. Pathol.* 35(2): 332.

Pangem, J.D., Child, J.J., Scowowft, W.R. and Gibson, A.H. (1975). Nitrogen fixation by *Rhizobium* cultured on a defined medium, *Nature,* London, 156,406–407.

Postgae, J.R. (1975). *Rhizobium* as a free-living nitrogen fixer. *Nature,* London, 256: 363.

Solanky, K.U., B.P. Mehta and A.N. Sabalpara. (2004). Rhizobial seed bacterzation and nitrogen dose on yield in pigeonpea in South Gujarat. *J. Mycol. Pl. Pathol.,* 34(2).

Subba Rao, N.S. (1976). Field response of legumes in India to inoculation and fertilizer application. In: *Symbiotic Nitrogen Fixation in Plants,* (Ed.) P.S. Nutman. International Biological Orogramme, Wl. Cambridge Uni. Press, London, pp. 255–268.

Subba Rao, N.S. (1977). *Soil Microrganisms and Plant Growth.* Oxford and IBH Publishing Co., New Delhi.

Subba Rao, N.S. (1979). Chemically and biologically fixed nitrogen: potentials and prospects. In: *Recent Advances in Biological Nitrogen Fixation,* (Ed.) N.S. Subba Rao. Oxford and IBH Publishing Co., Pvt. Ltd., New Delhi, pp. 1–7.

Subba Rao, N.S. (Ed.) (1979). *Recent Advances in Biological Nitrogen Fixation.* Oxford and IBH Publishing Co., New Delhi.

Subba Rao, N.S., Tilak, K.V.B.R., Laksmi Kumari, M. and Singh, C.S. (1996). *Azospirillum*: A new bacterial fertilizer for tropical crop. *Sci. Reporter,* C.S.I.R. (India), pp. 690–692.

Tiwari, R.K.S. and Ashok Singh. (2004). Efficacy of fungicides on *Rhizoctonia solani and Sclerotium rolfsii* and their effect on *Trichoderma harzianum* and *Rhizobium Ieguminousorum. J. Mycol. Pl. Pathol.,* 34(2).

Verghese, M.C. (1977). Issues facing the world fertilizers industry. In: *Proceeding of the FAI–IFDC Fertilizer Seminar, 1977, Trends in Consumption and Production.* The Fertilizer Association of India, New Delhi, pp. 1–41.

Verma, D.P.S. and Bat, A.K. (1976). Intracellular site of synthesis and locatization of leghaemoglobin in root nodules. *Proc. Nath. Acad. Sci.* USA, 73,3843–3847.

©R.C. Ram & Asha Sinha (*Editors*)
***Microbial Diversity: Modern Trends*, 2007** **190–202**
Daya Publishing House, New Delhi

Chapter 14

Participatory Production of *Trichoderma*

Ramesh Chand*, Chanda Kushwaha*, V.K. Chandola** and A.K. Joshi***

***Department of Mycology and Plant Pathology, **Department of Farm Engineering, ***Department of Genetics and Plant Breeding**
Institute of Agricultural Sciences, Banaras Hindu University, Varanasi – 221 005, U.P.

Introduction

Importance of *Trichoderma harzianum* as a fungal bio-agent for controlling of a number of soil borne and foliar plant pathogens is well documented (Papavizas, and Lumsden, 1980; Blakeman, and Fokema, 1982; Harnby, 1983; Timothy and Belanger, 2001). This bio-agent may be applied in different ways *viz.*, mixing the conidia and mycelium in nursery soil, dipping roots of seedlings, foliar sprays or by seed treatment. *Trichoderma* also serves as a decomposer of agricultural waste and can be used in the preparation of compost. Recently, *Trichoderma* has attracted the attention of a number of industries that are now marketing the bio-agent with different trade names. In one of our interactions with a group of farmers involved in participatory research, it was found that most of the *Trichoderma* products available in the market are not effective. After analysis of 50 random samples of such commercially available products it was found that *Trichoderma* population as Colony Forming Unit (cfu) g^{-1} in these formulations was far below the required level. This could be attributed due to various reasons *viz.*, poor quality of formulation, inappropriate storage and packing etc. It was also observed that poor quality of *Trichoderma* formulation was serving as an impediment in reckoning it popular among the farmers of this region.

Farmers of eastern India are mostly resource poor and often receive a new technology quite late. Consequently they are not benefited by the new technology to its full potential. Therefore, participation of the farmers in technology development and transfer processes is desirable in order to derive maximum benefits. The participation of the farmers is expected to ensure quick adoption and dissemination of the technology. It also provides rapid feedback for its further improvement.

In this article we have described the methodology for the mass production and quick adoption of *Trichoderma* at village community level.

Review of Literature

Trichoderma spp. is now regarded as an important bio-agent for the control of plant pathogenic fungi and for the decomposition of organic matter.

Backman *et al.* (1975) reported that a diatomaceous earth granule impregnated with 10 per cent molasses solution was suitable for growth and delivery of *Trichoderma harzianum* to peanut fields. *Trichoderma* spp. was grown on the sterile granules for 4 days and applied 70 and 100 days after planting.

Lewis *et al.* (1983) grew *Trichoderma* spp. and produced chladomydospores as well as conidia in molasses corn steep liquor (M-CSL), sucrose nitrate (SN) and glucose tartrate (GT) media. Isolates of *Trichoderma hamatum, Trichoderma viride,* and *Trichoderma harzianum,* formed 10.4, 5.9 and 1.1 × 10^8 chlamydospores g^{-1} dry weight of mycelium. *Trichoderma* spp. was abundantly produced on a variety of solid substrates moistened with liquid nutrient or water, Bran corn meal and peanut hull meal having a pH of 4, were better solid substrates for the production of chlamydospores and conidia.

Papavizas *et al.* (1984) reported that the large batches of biomass of *Gliocladium virens, T. hamatum, T. harzianum, T viride,* and *Talaromyces flavaus* were produced in liquid fermentation in 20 liters vessels. Simulating industrial condition by utilizing commercially available inexpensive ingredients (Molasses and brewer's), the maximum biomass yield of *T. harzianum, T. viride* and *G. virens* was obtained after 15 days of agitated incubation.

Windham *et al.* (1986) found increased growth rate of tomato, tobacco and radish by the inoculation of *Trichoderma.* This enhanced growth was due to hormones produced by the *Trichoderma.*

Kazachkova *et al.* (1992) reported a new, efficient method for the production of trichodermin from *T. lignorum* [*T. viride*] strain Harz 6 K by using different culture media. They suggested that this method can be used for commercial production.

Panicker and Jeyaranjan (1993) studied five substrates and found that farmyard manure was the best for multiplication of *T. viride* and *T. harzianum* followed by wheat and rice bran. Peat soil and rice straw were poor substrates.

Jin *et al.* (1996) reported that the Richard's medium plus V8 juice (RM 8) and glycerol to improved both biomass production and desiccation tolerance of *T. harzianum* and transfer the results from shake flask culture to a laboratory scale system fermenter.

Hari *et al.* (1998), evaluated sugarcane byproducts (bagasse, trash and pressmud) as substrates for mass multiplication of the bio-control agent *T. viride.* Best multiplication of *T. viride* (3.1 × 10^8 cfu / gram) was obtained on bagasses. The population on bagasses was comparatively higher than those on sorghum grains 5 × 10^6 and 1 × 10^6 respectively.

Zhao Lei (1999) described that *T. viride* has become an important biological control agent against plant pathogenic fungi. Two stage liquid/solid media, the liquid media were prepared by adding 5 gram of maize starch and 95 ml of water to flasks, which was then sterilized. One mm bit of the inoculum (a solution containing 1000000 conidia /ml) was added and the liquid media was maintained on oscillator at 28°C and 180 rotations/minute for 24 hours. The solid media were prepared using sufficient quantities of sterilized bran and added at 2 hours intervals to the liquid media, until culture with water contents of 50 per cent, 53 per cent, 56 per cent, 58 per cent, 60 per cent, 63 per cent, and 66 per cent had been formed. These cultures were then maintained at 28°C. The greatest production of conidia occurred on the 58 per cent medium (40–60 billion conidia /gram).

Chaube *et al.* (2000) published a list of growth media used for the production of *Trichoderma.* Diatomaceous earth granule impregnated with 10 per cent molasses solution, attapulgus clay granules, wheat bran, chopped straw, sorghum seed, low quality oat seeds, finger millet seeds, were recommended for the production of *Trichoderma harzianum* and *Trichoderma viride.*

Desai and Kulkarni (2002a) evaluated the efficacy of grains and seeds for multiplication of native isolates of *T. viride* and *T. harzianum.* Sterilized grains were separately inoculated with suspension of *T. viride* and *T harzianum* and acacia seed was found as the cheapest substrate for multiplication.

Desai and Kulkarni (2002b) studied the effect of various coloured storage bags for mass multiplication of *Trichoderma viride, T. harzianum* and *Gliocaldium virens.* The storage bags used were black, blue brown, milky white, pink, red po]ythene. Milky white bag recorded highest population of *T. viride* (68 × 106 g^{-1}).

Nakkeeran *et al.* (1997) standardized the packaging and storage conditions for increasing shelf life of talc and vermiculite based formulations of *Trichoderma.* The highest mean population of *Trichoderma viride* was recorded with vermiculite wheat bran acid fermenter. Biomass (VBA FB) formulation could be safely stored as talc based formulations in milky white bags at room temperature (20–30°C) for 75 days.

Prasad and Rageswarana (1984) reported positive result of greenhouse tests of modified granular formulation of *Trichoderma harzianum* and *Trichoderma viride,* on the reduction of chickpea damping off caused by *Rhizoctonia solani.*

Materials and Methods

Production Technology of *Trichoderma* spp.

Selection of Research Site and Participatory Farmers

This experiment was conducted in villages of three districts (Varanasi, Mirzapur and Chandauli) of Uttar Pradesh falling under North Eastern Plane Zone of India. In this region rice-wheat is the main cropping system. At each site, two villages were selected based on their previous performance in the farmer's participatory varietals selection research. A team of scientists constituting plant pathologist, plant breeder and agricultural engineer was associated. Farm household survey and participatory rural appraisal (PRA) were conducted for 65 to 77 farmers in each village to characterize the farmer's agro-ecological and socio-economic status and prevalent farming and crop management practices using a questionnaire. Farmers were consulted for their problems and limitations in *Trichoderma* production technology and its use.

Identification of the Problems and Prioritization

All the participant farmers were imparted training for the production and use of *Trichoderma.* These farmers were advised to produce the formulation and use this technology in their crop production and crop protection. Participatory farmers were called after nine months for the follow up action. Farmers were allowed to discuss various problems relating to the production and adoption of

Trichoderma. Prioritization of their problems was done by consensus, conveyed out through consultative group discussion (COD).

Identification of Substrate

Majority of the participatory farmers were from rice-wheat cropping system area. However, the farmers of other areas (mostly rainfed) grew sorghum, pearl millet, wheat and pulses. Therefore different kinds of easily available substrates *viz.*, rice, wheat straw, rice husk, grains of rice, wheat, sorghum, pearl millet and saw dust were tested for colonization and conidia production of *Trichoderma*. One hundred gram of each these substrates were soaked in water for five to six hours in a tumbler and then transferred to glass bottle, as described by Chaurasia *et al.* (1998). Inoculation was done by placing a bit (5 mm diameter) of the 72 hrs old culture of *Trichoderma harzianum* and incubated at 25°C. When more than 2/3 grains appeared to be colonized, these bottles were shaken to avoid clumping and promot uniform distribution of inoculum. On the basis of early and optimum colonization, substrate was selected for further experiments.

Effect of Sterilization, Pasteurization and Boiling on the Colonization and Contamination by Saprophytes

Sorghum grain was soaked overnight and boiled partially. Excess water of these grains were drained and filled in bottles. Twenty-four bottles, each were sterilized and pasteurized and 24 bottles had boiled seeds only. After cooling, these were inoculated with 4 bits (5 mm diameter) of 72 hrs old culture of *Trichoderma harzianum*. Colonization pattern and per cent contaminants were recorded in each bottle.

Effect of Different Inoculum Concentrations on Colonization and Contamination

Each bottle containing 150 g of sorghum grains was inoculated with different inoculum concentrations *i.e.* 20, 40, 60 and 80 fully colonized sorghum grains (5.35×10^7 cfu/grain^{-1}). Inoculum was mixed thoroughly and bottles were incubated at room temperature (8–24°C). Time taken for the colonization and conidia production in each bottle was recorded.

Effect of Starter Doses of *Trichoderma* on Common Contaminants

Different starter doses of *Trichoderma* spp. and its interaction with common contaminants (*Aspergillus niger*, *A. flavus*, *Penicillium* spp., *Mucor* sp., *Rhizopus stolonifer*, *Fusarium* spp. and *Alternaria* spp.) was tested. Co-inoculation of *Trichoderma* and each contaminant was done following earlier procedure in order to find out the best starter dose. Data related to the colonization and per cent contaminations was recorded.

Drying and Packing of *Trichoderma*

Fully colonized green grains were taken out from the bottles and spread on the polythene sheet under shade for air-drying. When the moisture level in the grains was reduced to 20 per cent, they were then filled in airtight polythene bags and stored at room temperature.

Use of New Formulation of *Trichoderma*

All the participant farmers were supplied *Trichoderma* for seed treatment with an objective to reduce wilt and blight infection in nurseries. *Trichoderma* colonized grains (5g/Kg of seed) were soaked in water (1.10 w/v ratio) and conidial suspension was used for seed treatment as slurry. It was further diluted till it appeared light green. Conidial suspension was spread on the seed in the form of slurry and treated seeds were air dried under the shade. These seeds were sown either in the nursery or directly in the field as per farmer's practices.

Experimental Design

New formulation was tested in paired plot technique where the minimum size of the plot was 100 sq meter and ten plots in a village were considered as replications for the comparison of means. Data related to the effect of *Trichoderma* on different crops were collected by obtaining farmers response on 1–4 scale, where 4 = good, 3 = adequate, 2 = some response, 1 = no response. The parameters considered for recording (as perceived by farmers) were seed germination, plant growth, disease control and yield. Data was also analyzed for test of significance.

Adoption of Technology

The extent of adoption of technology was analyzed by surveying the farmers of participatory and adjoining villages in two subsequent years *i.e.* 2001–2002 and 2002–2003.

Result

Constraints Associated with *Trichoderma* Production and its Use

Figure 14.1, reveals that farmers were not satisfied with quality of carrier based *Trichoderma* supplied by different agencies. Their priority was to get pure product for seed treatment and foliar application. Prioritization of problems revealed that a cheap, pure product without any carrier is important for the adoption of *Trichoderma* by farmers.

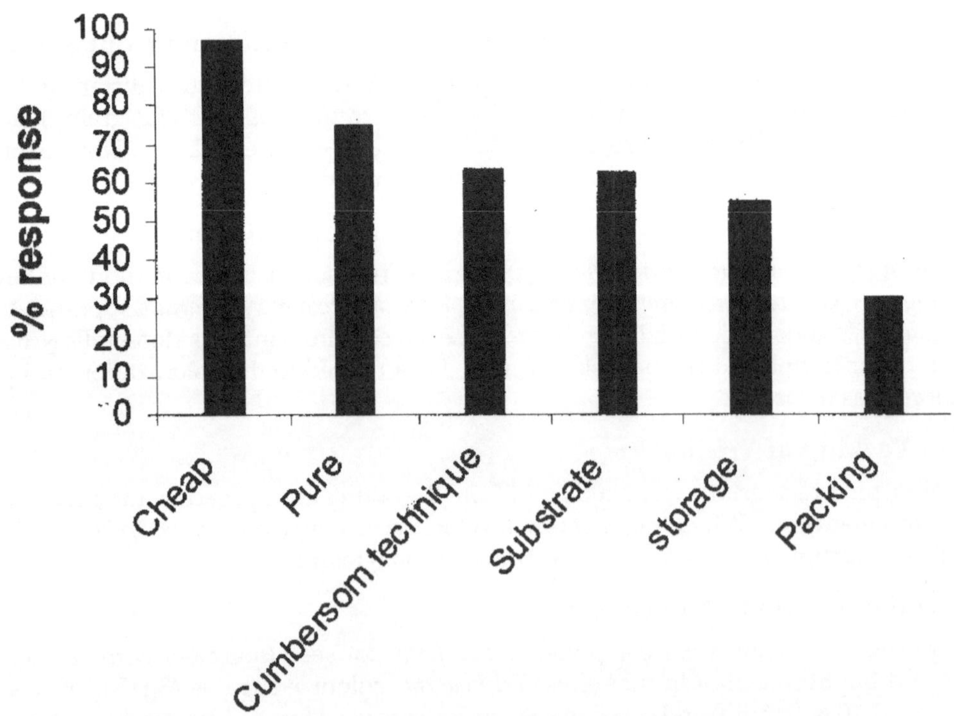

Figure 14.1: Farmers Response Regarding Prioritization of Problems

Identification of Substrate

Out of the eight locally available substrates sorghum and pearl millet grain took only 3–4 days for colonization and were judged best for the colonization of *Trichoderma,* while wheat and barley grains took five days for colonization (Table 14.1). Paddy straw and wheat strawand sawdust were poor substrates and they took more than 35 days to be colonized by *Trichoderma harzianum.* Intensity of mycelium in these substrates was poor as compared to sorghum grain.

Table 14.1: Time Taken for Colonization of *Trichoderma harzianum* on Different Substrates

Substrates	Time Taken (Days)
Sorghum Grain	3
Bajra Grain	4
Wheat Grain	5
Rice Grain	5
Barley Grain	6
Paddy Straw	35.33
Wheat Straw	45.33
Wood Saw Dust	54.66
SEM ±	0.534
L.S.D (0.05)	2.145
L.S.D (0.01)	1.145

Effect of Sterilization, Pasteurization and Boiling of Grain on the Bio-agent Colonization and Contaminations by Saprophytes Contamination Percentage

Table 14.2 shows that there was no saprophytic contamination in the sterilized grains. While, 66 per cent contaminations were recorded in most of the bottles having boiled grain. Contamination percent was 50.75 per cent in case of pasteurized grains.

Table 14.2: Effect of Sterilization, Pasteurization and Boiling of Grain on the Bio-agent Colonization and Percent Contaminations by Saprophytes

Treatment	Percent Contamination (%)
Sterilization	0.00
Pasteurization	50.75
Boiling	66.00
SEM±	1.717
L.S.D (0.01)	3.959
L.S.D (0.05)	2.306

Effect of Inoculum Concentration on the Colonization by *Trichoderma* and Contamination by Saprophytes

Although colonization was noticed on the third day in each of these approaches *i.e.* sterilization, pasteurization boiling, only sterilization was found to be 100 per cent contamination free. Initial higher inoculum concentration of 20 grains (10.7 × 10^8 cfu) was found most appropriate for the rapid colonization of sorghum grains and also to avoid contamination (Table 14.3). However, even at lower inoculum concentrations, contamination was less than 85 per cent (Table 14.4). Among the contaminants bacteria were the most dominant followed by fungi (*Rhizopus* spp., *Aspergillus* spp. and *Penicillium* spp.).

Table 14.3: Effect of Different Inoculum Concentration on Colonization and Contamination

Number of Grains Fully Colonized by Trichoderma spp.	Contaminates Tested	Dominant Fungus Recorded
1 grain	Aspergillus spp.	Aspergillus spp.
1 grain	Penicillum spp.	Penicillum spp.
1 grain	Mucor spp.	Mucor spp.
1 grain	Rhizopus stolonifer	Rhizopus stolonifer
1 grain	Fusarium spp.	Trichoderma spp.
1 grain	Alternaria spp.	Alternaria spp.
10 grain	Aspergillus spp.	Trichoderma spp.
10 grain	Penicillum spp.	Penicillum spp.
10 grain	Mucor spp.	Trichoderma spp.
10 grain	Rhizopus stolonifer	Rhizopus stolenifer
10 grain	Fusarium spp.	Trichoderma spp.
10 grain	Alternaria spp.	Trichoderma spp.
20 grain	Aspergillus spp.	Trichoderma spp.
20 grain	Penicillum spp.	Trichoderma spp.
20 grain	Mucor spp.	Trichoderma spp.
20 grain	Rhizopus stolonifer	Trichoderma spp.
20 grain	Fusarium spp.	Trichoderma spp.
20 grain	Alternaria spp.	Trichoderma spp.
40 grain	Aspergillus spp.	Trichoderma spp.
40 grain	Penicillum spp.	Trichoderma spp.
40 grain	Mucor spp.	Trichoderma spp.
40 grain	Rhizopus stolenifer	Trichoderma spp.
40 grain	Fusarium spp.	Trichoderma spp.
40 grain	Alternaria spp.	Trichoderma spp.

Response of Farmers

All the participatory farmers were satisfied with the performance of new formulation (Table 14.5). The maximum germination score of 3.59 and 3.33 was recorded in case of rice and vegetable growers, respectively. Maximum disease control was also observed in the case of rice (2.79). Farmers growing pulses and wheat under rainfed situation were least benefitted by *Trichoderma* among all the groups. In general, *Trichoderma* exhibited good disease control in this area.

Table 14.4: Effect of Different Inoculum Concentrations on the Bio-agent Colonization and Contamination by Saprophytes Under Boiling Condition

Inoculum Concentration (cfu grain^{-1})	Days to Colonization	Contamination in Each Bottle (per cent)
One grain (5.35 × 10^7)	8	8.50
Five grain (26.75 × 10^7)	6	7.50
Ten grain (53.5 × 10^7)	3	4.75
Twenty grain (107 × 10^7)	3	2.25
Forty grain (214 × 10^7)	3	0.50
Sixty grain (311 × 10^7)	3	0.25
Eighty grain (428 × 10^7)	3	0.25
SEM±		1.27
L.S.D (0.05)		2.66

Table 14.5: Response of Farmers on the Effect of *Trichoderma harzianum* on Crops Raised

Group of Farmers	Mean Score (1–4 Scale) of Traits			
	Seed Germination	Plant Growth	Wilt and Blight Disease Control	Yield
Group 1	3.59	2.00	2.79	2.12
Group 2	2.99	1.73	2.33	1.86
Group 3	3.33	3.66	2.39	2.99
Group 4	2.73	1.59	2.46	2.06
SEM±	0.18	0.13	0.29	0.25
L.S.D. (0.05)	0.39	0.28	0.63	0.55

Group 1: Farmers of Rice-wheat cropping system; Group 2: Farmers of dry land areas; Group 3: Vegetable growers; Group 4: Farmers of upland and irrigated conditions.

Drying and Packing

Viability of this formulation of *Trichoderma* was recorded for 9 months or even more (Table 14.6). Thus these packets could be stored for at least nine months.

Table 14.6: Viability of Packed *Trichoderma* spp. at Different Moisture Levels

Sorghum Seed	Boiled Seed with Fully Colonized Trichoderma spp. (Nos.)	Dry Weight	Moisture (per cent)	Days to Storability
150 gram	305	169.9	16 per cent	9 Months
150 gram	305	186.2	18 per cent	9 Months
150 gram	310	187.7	20 per cent	9 months

Adoption of Technology

The technology proposed was quickly adopted by the farmers and a very high per cent of farmers (81.4 per cent) as indicated in which shows that a very high per cent of farmers used this technology in next two years (Table 14.7). This technology was further encouraged and adopted in the adjoining villages and in the inter location area. Further follow up action revealed that farmers wanted ready made product, since they needed some more time for confidence buildup regarding production of *Trichoderma* by themselves.

Table 14.7: Pattern of Adoption and Production of *Trichoderma* sp.

Group of Farmers	% Farmers Applied	% Farmers Produced Trichoderma	
		Availed Research Station Facilities	Availed their Own Facilities
Group 1	90.90	45.45	15.00
Group 2	83.00	16.66	0.00
Group 3	66.00	44.44	0.00
Group 4	85.70	0.00	0.00

Figure 14.2: Production of Bioagent by Women Bioagent Group

Discussion

Trichoderma is a valuable bio agent used in crop protection and establishment against a large number of plant pathogens (Timothy and Belanger, 2001). It also helps in the decomposition of agricultural wastes (Harper and Lynch, 1985). A number of national and international organizations are promoting use of *Trichoderma* owing to its multifaceted role in crop production and protection. However, *Trichoderma* could not become popular especially in the areas where farmers are resource poor and lack necessary infrastructure. In these area losses due to various diseases in different crops are significant and results due to untimely diagnosis of disease and unavailability of appropriate fungicides. After two years survey and discussion with the farmers it was concluded that the *Trichoderma* supplied by various agencies is of poor quality as proved by their very low cfu g^{-1} content. Due to low population of conidia farmers were not getting desired result and therefore, nearly lost their confidence in the *Trichoderma*. Therefore, an initiative was made to ensure the supply of quality *Trichoderma* to farmers through participatory research. In the first phase (October 2001–February 2002), 25 farmers were given the training of *Trichoderma* production with standard protocol. A team of scientists monitored the process of training and initiated a feedback programme. It was found that initially only three farmers came forward for production. After, discussion with the participatory farmers, a consensus decision emerged that production process is cumbersome and it needed to be simplified. Therefore, the present technique was evolved which is based on the principle of pasteurization and competition for substrate. The initial higher load of inoculum and excellent saprophytic colonizing ability of *Trichoderma*, resulted in rapid colonization of the grains within 3 days and did not allow any other contaminants to grow and multiply. Low contamination may be because of antagonism of *Trichoderma* against other fungal contaminants by its multiple actions. Our results that showed lower contamination at higher level of *Trichoderma* reconfirmed this theory.

The selection of substrate for multiplication of bio-agent was based on its easy availability to the farmers, quick colonization by bio-agent and higher population of *Trichoderma* per unit substrate. In the present study sorghum grain is generally available to the farmers of this area without much difficulty. More over sorghum seed fulfilled all the requirements of efficient *Trichoderma* production. Sorghum grain as a substrate for the production of *Trichoderma* has been also listed by Choube and Singh (2000).

Prasad and Rangeshwaran (1984) also developed a granular formulation of *Trichoderma* based on the wheat bran, kaolin and acacia powder. The present formulation does not require any extra effort in making pellets and was cost effective for the production at cottage level. Drying the grain under shade with the moisture level of 15 per cent and packing in the moisture proof polythene packet is another dimension of research. Since most of the mycelium is present inside the dried grain this ensures safe transportation of the bioagents to long distances and longer storability of the *Trichoderma*. Martinez *et al.* (1997) also studied the storability of *Trichoderma* spp. at room temperature and reported viability up to 24 months while, abundant amount of conidia has capacity to tolerate higher temperature and dryness.

Adoption of *Trichoderma* for the seed treatment was high in rice wheat cropping system. Vegetable growers used *Trichoderma* mostly for raising nursery and treatment of seedlings against various soil borne diseases. There are several reports about the use of bioagent for the control of many soil borne pathogens in the green house and nursery (Timothy *et al.*, 2001; Lewis and Papavizas, 1983; Nelson *et al.*, 1988; Neelamegam and Govindarajalu, 2002; Datnoff *et al.*, 1995). Finding related to increased growth in different crops was similar to as reported by Windham *et al.*, 1986.

Few farmers took self-production of *Trichoderma*, while most of them wanted ready-made material. However, many farmers adopted production on the institution facilities only. This lack of confidence to take up *Trichoderma* production would need an additional effort at a grass root level.

Summary

The major constraint in the use of *Trichoderma* was unavailability of high quality product. This drew farmer's attention towards evolving and standardizing production and use of *Trichoderma*.

Among the various substrates available, sorghum was identified as most appropriate for mass multiplication. Pasteurization of substrate could also be effectively used for the production of *Trichoderma* spp. Contamination in *Trichoderma* production can be minimized by the increasing the amount of starter concentration. *Trichoderma* produced by the present technology can be dried and packed with least effect on quality. This granular product can be used for seed, seedling and nursery treatment and decomposition of rice straw. Use of new formulation reduced the incidence of the disease and increased plant growth and yield. Farmers readily accepted this technology and some of them started production in their villages.

Acknowledgement

Authors are thankful to all the farmers who participated in this research and contributed in development of technology. Special thanks go to the "Rahia Women Bio-agent Group" for the commercialization of the new product.

References

Backman, P.A. and Rodriguez-Kabana, R. (1975). System for the growth and delivery of biological control agents to the soil. *Phytopathology,* 65: 819–821.

Blakeman, J.P. and Fokkema, N.J. (1982). Potential for biological control of plan diseases on the phylloplane. *Ann. Rev. Phytopathology,* 20: 167–192.

Chaube, B.S. and Singh, U.S. (2000). List of growth media used for production of *Trichoderma.* In: *Plant Disease Management: Principles and Practice,* pp. 186

Datnoff, L.E., Nemec, S. and Pemezny, K. (1995). Biological control of Fusarium crown and root of tomato in Florida using *Trichoderma harzianum* and *Glomus intraradices. Biological Control,* 5(3): 427–431.

Desai, S.A. and Kulkarni, S. (2002). Evolution of difference grains and seeds for mass multiplication of bioagents. *Karnataka Journal of Agricultural Sciences,* 15(1): 158–160.

Desai, S.A. and Kulkami, S. (2002). Studies on storage bag colour on mass multiplication of biocontrol agents. *Karnataka Journal of Agricultural Sciences,* 15(1): 172–173.

Hari, K. and Somasekhar, N. (1998). Utilization of sugarcane wastes for the mass multiplication of fungal biocontrol agents. *Cooperative Sugar,* 29(9): 637–638.

Harmby, D. (1983). Suppressive soils. *Ann. Rev. Phytopathology,* 21: 65–85.

Harper, S.H.T. and Lynch, J.M (1985). Colonization and decomposition of straw by fungi. *Trans. British Mycological Society,* 85: 655–661.

Jin, X., Taylor, A.G., and Harman, G.E. (1996). Development of media and automated liquid fermentation methods to produce desiccation tolerant propagules of *Trichoderma harzianum. Biological Control,* 7(3): 267–274.

Kazachkova, L.N. and Dyatlav, V.A. (1992). Production of conithyrim and trichodermin by the submerged surface culture method. *Mikologia Fitopatologia,* 26(2): 122–124.

Lewis, J.A and Papavizas, G.C. (1983). Production of chlamydospores and conidia by *Trichoderma* spp. in liquid and solid growth media. *Soil Biol. Biochem.,* 15: 351–357.

Martinez, B., Fernandez, I., Paula, S.D.E. and Muniz, Y. (1997). Preservation of fungus *Trichoderma* a biological agent of plant disease in zeolite. *Pevistade Protection Vegetal,* 12(2): 123–125.

Nakkeeran, S., Sankar, P. and Jeyaranjan, R. (1997). Standardization of storage conditions to increase the self–life of *Trichoderma* formulation, *Journal of Mycology and Plant Pathology*, 27(1): 60–63.

Neelamegam, R. and Govindarajalu, T. (2000). Integrated application of *Trichoderma viride* and farmyard manure to control damping off of tomato. *J. Biol. Control*, 16(1): 65–69.

Papavizas, G.C. and Lumsden, R.D. (1980). Biological control of soil borne fungal propagules. *Ann. Rev. Phytopathology*, 18: 389–413.

Papavizas, G.C., Dunn, M.T, Lewis, J.A and Ristaino, J.B (1984). Liquid fermentation technology for experimental production of biocontrol fungi. *Phytopathology*, 74(10): 1171–1175.

Prasad, R.D. and Rangeshwama, R. (1984). Granular formulation of *Trichoderma* and *Gliocladium* spp. in biocontrol of *Rhizoctonia solani* of chickpea. *Journal of Mycology and Plant Pathology*, 29(2): 222–226.

Panicker, S. and Jeyaranjan, R. (1993). Mass multiplication of biocontrol agent *Trichoderma* spp. *Indian Journal of Mycology and Plant Pathology*, 23(3): 328–330.

Timothy, C., Paulitz, R. and Belanger, R. (2001). Biological control in greenhouse system. *Ann. Rev. Phytopathology*, 39: 103–133.

Windham, M., Elad, Y and Baker, R. (1986). A mechanism of increased plant growth induced by the *Trichoderma* species. *Phytopathology*, 76: 518–521.

Zhao, L. (1999). Production of *Trichoderma* spp. conidia by a liquid/solid fermentation process. *Chinese Journal of Biological Control*, 15(3): 144–144.

Questionnaire Used for Bench Mark Survey in Villages

१. कृषक का नामः ....

Farmer's Name:

२. पूरा पताः ग्रामः.................. पोस्ट .... जिला .... फोन न॰

Full Address: Village ………….,.Post ……………. Distt ……………. Ph.No.

३. खेती योग्य भूमि (हे॰) ....

Cultivable land (ha):

४ किन–किन फसलों की खेती कितने क्षेत्रफल में करते है धान ( ) गेहूँ ( ) अन्य फसलः ( )

Area under various crops: Paddy ( ) Wheat ( ) Other crop ( )

४. सिंचाई का मुख्य साधन नहर ( ) असिंचित ( ) ट्यबेल ( )

Main source of irrgation: Canal ( ) Unirrigated ( ) tubewell( )

रासायनिक उर्वरकों का प्रयोग करते है हा/नहीं

५. फसलों में लगने वाली मुख्य बिमारियाँ:

(१) उकठा (२) सड़न (३) गलन (४) पत्ती रोग (५)पौधशाला में गलन

क्या आप बिमारियों के नियंत्रण के लिए रसायनों का प्रयोग करते है हा/नहीं

क्या आप रसायनों के प्रयोग से संतुष्ट है: हा/नहीं, दो मुख्य कारण

क्या आप ट्राइकोडरमा का नाम सुना है हा/नहीं

६. इसके बारे में जानकारी का क्या स्रोत है:

७. बाजार में ट्राइकोडरमा कैसे मिलता है: (१) पाउडर (२) घोल (३) दानेदार

८. क्या आपने ट्राइकोडरमा का प्रयोग किया है हा/नहीं फसल का नाम

९. आपको ट्राइकोडरमा के प्रयोग का क्या परिणाम मिला है।

(१) लाभदायक (२) न लाभ न हानि (३) हानि

१०. ट्राइकोडरमा के गुणवत्ता का पता कैसे पता करेंगे

(१) देखकर (२) प्रयोग कर (३) जाँच कराकर (४) पैकेट पर लग स्लिप को पढ़कर

११. क्या आप ट्राइकोडरमा के वर्तमान उत्पाद के गुणवत्ता में सुधार चाहेंगे हा/नहीं (१) (२)

१२. क्या बाजार में उपलब्ध ट्राइकोडरमा का मूल्य उचित है ( ) ज्यादा ( )कम ( ) उचित

१३. क्या आपको कृषि विभाग द्वारा ट्राइकोडरमा उपलब्ध है हा/नहीं

यदि हा तो आपकी राय क्या है (१) .... (२) ....

१९ क्या आप अपने खेत में कम्पोस्ट का प्रयोग करते है हा/नहीं

२० ट्राइकोडरमा का रोग नियंत्रण के अलावा अन्य लाभों की जानकारी है हा/नहीं

२१ क्या आपको ट्राकोडरमा के उपयोग एवम् उत्पादन से सम्बन्धित प्रशिक्षण की आवश्यकता है। हा/नहीं

२२ ट्राइकोडरमा यदि लाभदायक है तो आप इसका प्रयोग क्यों नहीं कर पाते है एक मुख्य कारण (१).

२३ आप इसके प्रयोग को खेतों में बढ़ाने के लिए क्या सुझाव देंगे (१) ....

२४ ट्राइकोडरमा के शोध कर रहे वैज्ञानिकों को आप क्या सुझाव देंगे। (१) ....

©R.C. Ram & Asha Sinha (*Editors*)
***Microbial Diversity: Modern Trends*, 2007**
Daya Publishing House, New Delhi

203–212

Chapter 15

Characterization of Plant Growth Promoting Rhizobacteria (PGPR)

S. Ameer Basha**, B.K. Sarma*, K. Annapurna,** Senthil** and U.P. Singh*

***Department of Mycology and Plant Pathology, Institute of Agricultural Sciences, Banaras Hindu University, Varanasi – 221 005, India**
****Department of Microbiology, Indian Agricultural Research Institute, New Delhi, India**

ABSTRACT

Plant growth promoting rhizobacteria (PGPR) affect growth by direct or indirect mechanisms such as suppression of bacterial, fungal and nematode pathogens by production of siderophores, HCN, chitinase, ACC, antibiotics and volatile metabolites, etc., by induced systemic resistance and/or by competing with the pathogen for nutrients or for colonization space. Characterization of *Pseudomonas aeruginosa* (*Pag*) and *P. fluorescens* (*Pf4* and *CHO*) was performed. The phosphate-solubilization test was done in Pikovskaya's medium. The plates were observed for clearing zone around the colony and the diameter of the clearing zone was measured. All the strains produced siderophores as evident by the formation of yellowish halo zone on CAS plates. The PGPR strains also produced IAA, which was estimated by HPLC. Presence of ACC (1-aminocyclopropane-1-carboxylate) deaminase in the strains was identified with the help of PCR.

Introduction

Plant Growth Promoting Rhizobacteria (PGPR) are soil and rhizosphere bacteria that can benefit plant growth by different mechanisms (Glick, 1995). Their use as natural biofertilizers is advantageous, not only from the economical, but also from the ecological point of view. The beneficial activities of

PGPRs have been established in many crops (Sarma *et al.*, 2000). A large portion of phosphorus in soil is present in insoluble form and therefore, not available for plants (Rossolini *et al.*, 1998). PGPRs can bring the unavailable form to available form for absorption by plants for their better growth thereby increasing the crop yield.

Soil or seed application with PGPR has been used to enhance growth of several crops. Besides promoting plant growth PGPRs can also suppress the growth of plant pathogens. They have been shown to protect plants against fungi, bacteria, viruses and nematodes. Increased plant growth and crop yield have already been established by inoculating seeds with certain specific root colonizing bacteria, particularly, PGPR in several crop plants. The effects of PGPR on plant growth can be mediated by direct or indirect mechanisms (Glick, 1995). The direct effects have been most commonly attributed to the production of plant hormones such as auxins, gibberellins and cytokinins; or by supplying biologically fixed nitrogen. These PGPRs also affect growth by indirect mechanisms such as suppression of bacterial fungal and nematodes pathogens by production of siderophores. Hydrogen cyanide, ammonia, chitinase, 1-amino-cyclopropane-1-carboxylate deaminase (ACC), antibiotics and volatile metabolites, etc., by induced systemic resistance and/or by competing with the pathogen for nutrients or for colonization space (Glick, 1995). Direct interaction of biocontrol agents with the pathogens may involve competition and also antagonistic effects mediated by antimicrobial metabolites and extra-cellular lytic enzymes.

Several new biocontrol mechanisms were described or discussed recently. Some deal with the degradation of pathogenicity factors, plant susceptibility factors and pathogen cell-cell communication systems. Other new mechanisms are concerned with the inhibition of synthesis of pathogenicity factors and the increased susceptibility of the pathogens to stress (Shah *et al.*, 1997). Several workers (Jacobson *et al.*, 1994; Shah *et al.*, 1997; Glick *et al.*, 1998; Shah *et al.*, 1998) reported that most of the PGPRs contain the enzyme 1-amino-cyclopropane-1-carboxylate (ACC) deaminase, which can reduce the amount of ethylene produced by plants, and thereby promote root elongation.

Phosphate solubillizing microorganisms (PSMs) are known to solubilize mineral phosphates chiefly by production of low molecular weight organic acids. Among phosphate solubilizing bacteria (PSB), gluconic acid secretion is the most well-studied mechanism of P-solubilization. Gluconic acid is produced by direct oxidation of glucose via membrane-bound glucose dehydrogenase (Gcd) enzyme. This enzyme is known to act not only on glucose but also on several other sugars such as xylose, arabinose, maltose, etc. and oxidizes them to their corresponding acids.

Keeping this in view, the present experiment was undertaken to isolate and characterize three bacterial strains for their plant growth promotion activities through production of HCN, siderophore, phosphate solubilization, Indole Acetic Acid (IAA) and ACC deaminase and determine their potential for commercial application.

Materials and Methods

Three strains of *Psuedomonas* species, *viz., P. fluorescens* (*Pf4* and *CHO*) and *P. aeruginosa* (*Pag*) (*Pf4* was obtained from Center of Advanced Study in Botany, University of Madras. Chennai; *CHO* strain was obtained from Department of Microbiology, IARI, New Delhi, and *Pag* from School of Biotechnology, Banaras Hindu University, Varanasi) were screened for their plant growth promotion activities by assaying the following attributes: phosphate-solubilization, HCN, Siderophore, IAA and ACC deaminase production. The phosphate-solubilization test was done in Pikovskaya's medium (Pikovskaya. 1948). The ability of the two strains to produce HCN was assessed according to Bikker and Schipper (1987). Siderophore production was tested following the CAS plate assay according to Schwan and Neilands (1987). IAA production test was done according to Sarma *et al.* (2002).

Phosphate Solubilizing Capacity

The phosphate-solubilization test was done on a solid medium described by Pikovskaya (Pikoyskaya. 1948) (Table 15.1). Agar plates were prepared and the bacterial strains were spot inoculated at the center of the plates and incubated for 5–6 days. The plates were observed for clearing zone around the colony and the diameter of the clearing zone was measured

Table 15.1: Pikovskaya's Medium Composition

Ingredients	Quantity
Glucose	10.0 g
Tricalciun phosphate	5.0 g
$(NH_4)_2SO_4$	0.5 g
$MgSO_4.7H_2O$	0.1 g
$MnSO_4$	Trace
$FeSO_4$	Trace
Yeast Extract	0.5 g
Agar	18.0 g
Water	100 mL

HCN Production

The ability of the two (*Pf4* and *Pag*) strains to produce HCN was assessed using the method of Bikker and Schipper (1987). Kings B medium was prepared and amended with 4.4 g/l glycine and sterilize it 25 ml of this medium was poured into each plate. After solidification the isolates were streaked separately in plates. Whatman no. 1 filter paper discs were soaked (9 cm dia) in 0.5 per cent picric acid in 2 per cent sodium carbonate. A soaked disc was placed in the lid of each Petri plate. The Petri plates were sealed with parafilm and incubate at 28 ± 2°C for 4 days. A HCN positive control *i.e. CHO* was kept for comparison of results.

Siderophore Production

The ability of the antagonists to produce siderophores was tested following the CAS plate assay according to the method of Schwan and Neilands (1987). The medium contains an iron-CAS-HDTMA (Hexadecyltrimethylammonium bromide) complex which is blue coloured.

CAS Blue-agar Medium Preparation

Solution A

To prepare 1 liter of CAS blue agar medium, 60.5 mg CAS was dissolved in 50 ml water and mixed with 10 ml iron (III) solution (1 mM $FeCl_3.6H_2O$, 10 mM HCl). This solution was slowly added to 72.9 mg HDTMA dissolved in 40 ml water and the resultant dark liquid was autoclaved.

Solution B

A basal medium containing deferrated 1M sucrose (3 ml)

Deferrated 1M $CaCl_2$ (0.4 ml)

Deferrated 1M $MgSO_4.7H_2O$ (0.8 ml)

Deferrated 2 per cent K_2HPO_4 (10 ml)

Deferrated NaCl (0.2 g)

$NaMoO_4$ (0.005 g)

PIPES (Free acid) 30.24 g and

Agar 15 g in 800 ml water

The pH of the medium was adjusted to 6.8 by the addition of 50 per cent (W /W) NaOH and autoclaved. After cooling to 50°C, 30 ml of deferrated casamino acids 10 per cent was added as sterile solution.

Solution A was finally added to solution B along the glasswall with enough agitation to achieve maximum mixing without formation of foam. 20 ml of medium was poured into each plate. After solidification of CAS medium, spotting of fresh bacterial cultures was done and the plates were incubated at 28+2°C. Presence of yellow halo around the bacterial colony is a positive indication of siderophore production.

IAA Production

Twenty-five ml Kings B broth was poured into each of three 100 ml capacity conical flasks. A loopful of bacterial cultures were scraped from actively growing cultures of *Pf4* and *Pag*, separately and inoculated into the flasks containing Kings B broth. The flasks were incubated at 25°C for 48 h. An equal volume of ethyl acetate was mixed with culture filtrates separately, and after vigorous shaking in a separatory funnel, the ethyl acetate fractions of the culture filtrates were collected separately. The residue was extracted second time and the ethyl acetate fractions were pooled with the previous extract. The fractions were then evaporated under vacuum. Dried samples were resuspended in 1 ml HPLC grade methanol by vortexing and stored at 4°C for further analysis. High performance liquid chromatography (HPLC) of the samples was performance with HPLC system (Shimadzu Corporation, Kyoto, Japan) equipped with two Shimadzu LC-10 AT*VP* reciprocating pumps, a variable UV-VIS detector (Shimadzu SPD-10 AT*VP*) and a Winchrom integrator (Winchron). Reverse phase chromatographic analysis was carried out in isocratic conditions using RP C-18 HPLC column [250 × 4.6 mm id, particle size 5 µm, Luna 5µ C-18 (2), Phenomenex, USA] at room temperature. Running conditions included injection volume: 5 µl, mobile phase: methanol-0.4 per cent acetic acid (80 : 20, v/v), flow rate: 1 ml/min, detection at 290 nm and attenuation response 0.03 AUFS. Samples were filtered through membrane filters (Pore size 0.20 µm, Millipore) prior to injection in sample loop. Indole Acetic Acid (IAA) was used as internal and external standards. Presence of IAA in the samples were identified by comparing retention time (Rt) of the standards and by the co-chromatography. Content of IAA was calculated by comparing peak area of reference compound with those in the samples run under similar elution conditions.

Characterization for ACC Deaminase

Presence of ACC (1-arninocyclopropane-1-carboxylate) deaminase in the strains was identified by growing them in DF salts (Dworkin and Foster 1958) (Table 15.2).

Bacterial cultures were plated onto solid DF salts minimal medium and incubated for 48h at 25°C. These plates were prepared with 1.8 per cent Bacto-Agar which had a very low nitrogen content and were spreaded with ACC (30 µmol/plate) just prior to use. Before streaking with either a loopful of bacteria, the ACC was allowed to dry fully. The inoculated plates were incubated at 35°C.

Table 15.2: DF Salts Medium (per liter) Composition

Ingredients	Quantity
KH_2PO_4	4.0 g
Na_2HPO_4	6.0 g
Mg $SO_4.7H_2O$	0.2 g
Glucose	2.0 g
Gluconic acid	2.0 g
Citric acid with trace elements	2.0 g
$FeSO_4.7H_2O$	1.0 mg
H_3BO_3	10µg, 7.2
$MnSO_4.H_2O$	11.19 µg
$ZnSO_4.7H_2O$	124.6 µg
$CuSO_4.5H_2O$	78.22 µg
MoO_3	10 µg
$(NH_4)_2SO_4$ as a nitrogen source	2.0 g

Characterization for ACC Deaminase with Nested PCR DNA Extraction

A 15 ml culture was grown in LB broth medium: the cells were collected by centrifugation, washed and re-suspended in 0.5 ml of 10 mM Tris-HCl (PH 7.5) and 1 mM EDTA (TE buffer). The bacterial cells were lysed by incubating at 30°C for 20 min with 0.2 mg/ml lysozyme (1 per cent) and Pronnase (0.5 per cent) were added to the suspension and incubated at 31°C for 1 h. The lysate was extracted twice with phenol : chloroform (1 : 1) and twice with chloroform in the presence of 0.3 M sodium acetate. The DNA was precipitated with an equal volume of isopropanol and washed twice with 10 per cent ethanol dried and then dissolved in TE buffer.

Protocol for ACC Deaminase Gene Amplification

Initial denaturation at 95°C:	3 min	
Denaturation at 95°C:	1 min	
Annealing at 50.8°C:	1 min	30 times
Extension at 12°C:	1 min	
Extension at 12°C:	6 min	
Soaking at 4°C:	Till taking out the sample from PCR machine	

First PCR

PCR reaction volume:

Template DNA:	1.0 ul	1 step
Taq buffer (10 X):	1.0 ul	
dNTPs (10mM):	0.25 ul	
Primer 3:	0.25 ul	
Primer 4:	0.25 ul	
Taq polymerase (3U/ul):	0.11 ul	
DH_2O:	7.08 ul	
Total volume:	10.0 ul	

The amplified product that came of this PCR reaction was used as template for the step II PCR using primers P1 and P2. The step II PCR reaction was exacly the same as above; all ingredients were same as in step II except the primers used in step II PCR were P1 and P2. The final amplified product of the ACC deaminase gene was 830 bp. 0.8 per cent Agarose gel was run along with the marker with the PCR product. The gel was run at 60V for 45 min using TBE buffer. The product size was 830 bp. The following primers were used in the present study.

P1 = TA(CT)GC(CG)AA(AG)CG(ACGT)GA(AG)GA(CT)GCAA

P2 = CCAT(CT)TC(AGT)ATCAT(ACGT)CC(GA)TGCAT

P3 = ATGAA(TC)CTG(AC)A(ACGT)CG(ACGT)TT(CT)(GG)
(AC)(ACGT)CG(CGT)TA(CT)CC

P4 = T(TC)AGCC(AG)T(TC)(ACGT)CG(AG)AA(ACGT)
A(GTA)(GA)AA(GA)CT(GA)TA(ACGT)(GC)C

Results and Discussion

Qualitative estimation of secondary metabolites for a specific microbial strain is necessary as it is strain specific. P-solubulization in Pikovskaya's agar plates was canied out for the *Pf4*, *Pag* and *CHO* strains. Only *Pf4* produced clearing zone [Figure 15.1(B)]. This was an indication of its phosphate solubilizing capacity. Similarly, only *Pf4* was positive in relation to the production of IAA and ACC deaminase.

HCN production has been postulated to play an important role in biocontrol of pathogens (Defagao *et al.*, 1990). Both *Pf4* and *Pag* did not produce HCN while the standard HCN positive strain showed positive results (Fig. 15.1A). Other workers have reported HCN production by antagonistic bacteria as well (Saxena *et al.*, 1996; Rangeswaran and Prasad, 1998). HCN is known to inhibit the electron transport, disturbing the energy supply to the cells, ultimately leading to the death of the pathogen (Knowels, 1976).

Specific strains of *Pseudomonas fluorescens* rapidly colonize plant roots of several crops and cause statistically significant yield increase accompanied by reduction in root-zone populations of harmful fungi and bacteria. It is mainly due to their ability to respond to low-iron stress by producing extra cellular, low-molecular-weight (500-1000 Daltons) iron (transport agents, designated siderophores, which selectively complex iron (III) with very high affinity. The function of siderophores is to supply iron to the cell. The microorganisms active in environments with iron at such a low level of availability are those that have at least some ability to produce siderophores. Disease suppressive siderophores produced by certain strains of fluorescent *Pseudomonas* species such as pseudobactin can presumably deliver iron to the plants they benefit; otherwise, these plants would develop iron chlorosis and become susceptible to pathogens (Kong, 1986). *Pf4*, *Pag* and *CHO* invariably form yellow halo zones on CAS plates. *Pf4* strain produced large siderophore induced halo zones on CAS medium but the remaining strains *Pug* and *CHO* produced small halo zones indicating the production of low-affinity siderophore [Figure 15.1(C)]. This implies their low competitiveness with the iron uptake mechanisms and many explain it as weak antibiosis of the strains. *Pf4* strain produce IAA in culture filtrates and showed the gene responsible for ACC deaminase in PCR test while (Figure 15.2), *Pug* gave negative response to IAA and ACC deaminase production.

Antagonistic *Pseudomonas* species produce a wide variety of secondary metabolites that adversely affect the growth or metabolic activity of plant pathogenic fungi. Antibiotics are produced by introduced and indigenous *Pseudomonas* species in plant-associated environments and play an important role in *in situ* interactions with plant. pathogenic fungi. In many crop-pathogen systems, the primary mechanism of biocontrol by fluorescent pseudomonads is production of HCN and antibiotics such as 2,4-diacetylphlorog1ucinol (DAPG), pyoluteorin, pyrrolnitrin, and phenazines such as phenazine-1-carboxylate. Phenazines are pigmented nitrogen-containing heterocyclic compounds that inhibit mycelial growth of several fungal plant pathogens (Turner and Messenger, 1986). Phenazine-1-carboxylic acid (PCA), produced by *P. fluorescens* 2–79 (Ownley *et al.*, 1991; Thomashaw *et al.*, 1990) and *P. aureofaciens* 30–84 (Hamden *et al.*, 1991) plays an important role in the control of take-all disease of wheat caused by *Gaeumannomyces graminis* var. *tritici*.

The performance of *Pseudomonas* species as biocontrol agent may be improved by applying mixtures of strains, which are complementary in their capacity to suppress plant diseases. A combination of *P. fluorescens* CHAO with another well-characterized biocontrol agent was chosen as a model to study how these strains affect each other's expression of a biocontrol trait. In both strains, production of the antimicrobial compound 2,4-DAPG is a crucial factor contributing to the suppression of root diseases.

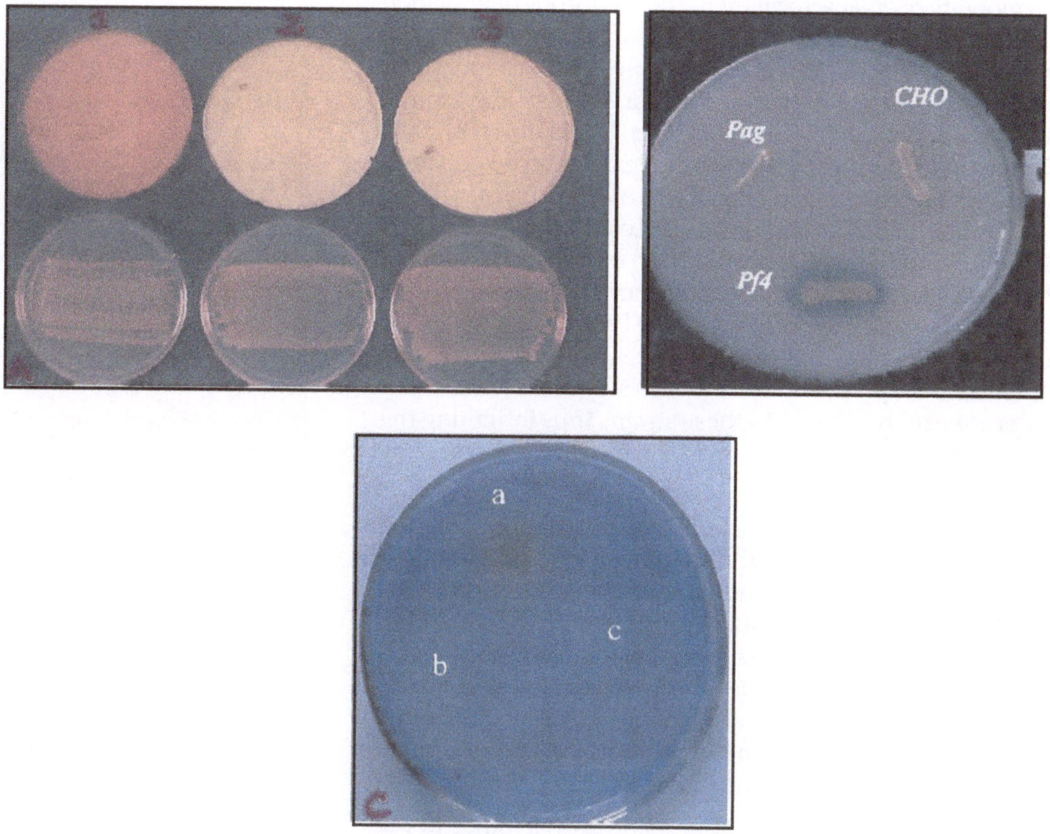

Figure 15.1(A–C): Characterization of PGPR Strains

(A) HCN Production. A.1.CHO strain positive to HCN production, 2 and 3 *Pag* and *pf4* strains negative to HCN production. Ч 3/4th of original size.

(B) Phosphate solubilisation (only *pf4* was positive). Ч 3/4th of original size.

(C) Siderophore production. a (*Pf4*), b (*Pag*) and c (*CHO*) all showed positive reaction to sideropbore production. Ч 3/4th of original size.

2,4-DAPG acts as an intracellular signaling compound inducing the expression of its own biosynthetic genes. It is established that two non-related Pseudomonads may stimulate each other in the expression of an antimicrobial compound important for biocontrol. This interpopulation communication occurs in the rhizosphere, *i.e.*, at the site of pathogen inhibition, and is mediated by the antimicrobial compound itself acting as a signal exchanged between the two pseudomonads (Maurhofer *et al.*, 2003).

Improvement of nutrient availability and other mechanisms (Kapulnik, 1996) by rhizopshere bacteria result in significant increase in plant growth (Suslow *et al.*, 1979). Microorganisms are critical for the transfer of P from the poorly available forms and are important for maintaining P in readily available pools. The present study proved *P. fluorescens* mediated P solubilisation and thereby enhanced uptake by the plants, which resulted in increased plant biomass. The microbe-mediated P mobilisation in plants may be through, (*i*) increase in the surface area of roots, (*ii*) by displacement of sorption equilibria that result in increased net transfer of phosphate ion into the soil solution or an increase in the soil mobility of organic forms of P, and (*iii*) through simulation of metabolic processes that are effective in directly solubilizing and mineralizing P from poorly available forms of inorganic and organic P (Jones, 1998). The microorganisms used in the study were found producing siderophores (Diby Paul *et al.*, 2001) which may in turn release P from complex forms of P through ligand exchange reactions by chelating metal ions associated with the bound P. Processes such as rhizosphere acidification, exudation of organic acids and secretion of phosphatase from plant roots occur in response to P deficiency and are established mechanisms by which the plants require P (Randal *et al.*, 2001). The increased biomass in these PGPR-treated plants may also be attributed to the production of antibiotics, lytic enzymes, HCN and siderophores produced by these strains thus inhibiting the growth of deleterious microorganisms in the rhizosphere (Diby Paul *et al.*, 2001, 2002).

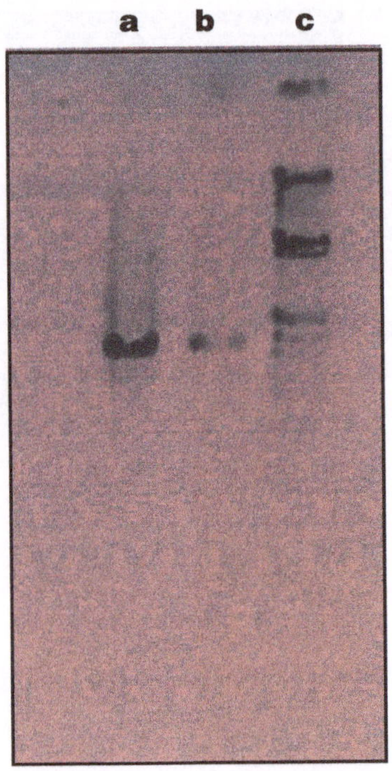

Figure 15.2: ACC (1-amino-cyclopropane-1-carboxylate) Deaminase Activity Using PCR. a. CHO strain, b. *Pf4* and c. marker (830 hp), *Pf4* was positive to ACC deaminase activity when compared with positive strain (*CHO*).

Plant growth promoting bacteria bind to the surface of seed and in response to tryptophan and/or other amino acids exuded from the germinating seeds, synthesize and secrete indole acetic acid (IAA). This IAA may be taken up by the seeds and together with endogenous IAA, stimulates plant cell proliferation, cell elongation or induces the synthesis of 1-aminocyclopropane-1-carboxylic acid (ACC) synthase, the enzyme that catalyzes the conversion of S-adenosyl-L-methionine (SAM) to ACC, the immediate precursor of ethylene (Yang and Hoffinan, 1984).

These bacteria are beneficial to plant growth since in the natural environment plants are often subjected to ethylene-producing stresses. However, it should be emphasized that ACC deaminase-containing PGPR facilitate plant growth to a much greater extent with plants that are ethylene sensitive. Using the procedures described here it is relatively straightforward to isolate and characterize new PGPR strains. These bacteria may be used to facilitate the growth of a variety of plants, especially under stressful conditions. This activity is useful in both agricultural and horticultural settings as well as in environmental cleanup protocols.

Acknowledgement

Shaik Arneer Basha is thankful to the Council of Scientific and Industrial Research (CSIR), New Delhi, India for financial assistance.

References

Bakker, A.W. and Schippers, B. (1987). Microbial cyanide production in the rhizosphere in relation to potato yield reduction and *Pseudomonas* spp. mediated plant growth stimulation. *Soil Biol. Biochem.*, 19: 451–457.

Défago, G., Berling, C.B., Burger, U., Keel, C. and Voisard, O. (1990). Suppression of black root rot of tobacco by a *pseudomonas* strain: Potential applications and mechanisms. In: *Biological Control of Soilborne Plant Pathogen*, (Ed.) Hornby, D. CAB International, pp. 93–108.

Diby Paul, Jisha, P.J., Kumar, A., Anandaraj, M. and Sarma, Y.R. (2002). Induction of mycolytic enzymes in *Pseudomonas fluorescens* by *Phytophthora capsici*. *Indian Phytopathology*, 55(3): 374.

Diby Paul, Kumar, A., Anandaraj, M. and Sarma, Y.R. (2001). Studies on the suppressive action of fluorescent pseudomonads on *Phytophthora capsici*, the foot rot pathogen of Black pepper. *Indian Phytopathology*, 54(4): 515.

Dworkin, M. and Foster, J. (1958). Experiments with some microorganisms which utilize ethane and hydrogen. *J. Bacteriol.*, 75: 592–601.

Glick, B.R. (1995). The enhancement of plant growth by free living bacteria. *Can. J. Microbiol.*, 41: 109–117.

Glick, B.R., Penrose, D.M. and Li, J. (1998). A model for the lowering of plant ethylene concentrations by plant growth-promoting bacteria. *J. Theor. Biol.*, 190: 63–68.

Hamdan, H., Weller, D.M. and Thomashaw, I.S. (1991). Relative importance of fluorescent siderophores and other factors in biological control of *gaeumannomyces garminis* var. *tritici* by *Pseudomonas fluorescens* 2–79 and M4–80R. *Appl. Environ. Microbial.*, 57: 3270–3277.

Jacobson, C.B., Pasternak, J.J. and Glick, B.R. (1994). Partial purification and characterization of 1-aminocyclopropane-1-carboxylate deaminase from the plant growth-promoting rhizobacterium *Pseudomonas putida*. *Can. J. Microbiol.*, 40: 1019–1025.

Jones, D.L. 1998. Organic acids in the rhizosphere: A critical review. *Plant and Soil*, 205: 2544.

Kapulnik, Y. (1996). Plant growth promotion by rhizosphere bacteria. In: *Plant Roots: The Hidden Half*, (Eds.) Waisel, Y., Eshel, A. and U. Kafkafi. Marcel Dekker, Inc., New York, pp. 796.

Knowels, C.J. (1976). Microorganisms and cyanide. *Bacteriol Rev.*, 40: 652–680.

Leong, J. (1986). Siderophores: Their biochemistry and possible role in the biocontrol of plant pathogens. *Ann. Rev. Phytopathol.*, 24: 187–209.

Maurhofer, M., Baehler, E., Notz, R., Vicente, M. and Keel, C. (2003). Cross-talk between 2,4-diacetylphloroglucinol-producing biocontrol pseudomonads on wheat roots. In: *Proceedings of 6th International PGPR Workshop*, Calicut, 5–10 October, pp. 403.

Ownley, B.B., Weller, D.M. and Thomashaw, L.S. (1991). Influence of *in situ* and *in vitro* pH on suppression of *Gaeumannomyces graminis* var. *tritici* by *Pseudomonas fluorescens*, 2–79. *Phytopathology*, 82: 178–184.

Pikovskaya, R.E. (1948). Mobilisation of phosphorus in soil in connection with vital activity of some microbial species. *Microbiolagia*, 17: 362–370.

Randall, P.J., Hayes, J.E., Hocking, P.J. and Richardson, A.E. (2001). Root exudates in phosphorous acquition by plants. In: *Plant Nutrient Acquition: New Perspectives*, (Eds.) Ae, N., Arihara, J., Okada, K. and Srinivasan, A. Springer–Verlag, pp. 71–100.

Rangeswaran, R. and Prasad, R.D. (1998). Screening and selection of rhizobacteria for biological control of *Sclerotium rolfsii* and *Rhizoctonia solani*. *National Symposium on Development of Microbial Pesticides in Insect Pest Management*, November 12–13. HAL. Pune, 41 pp.

Rossolini, G.M., Shipa, S., Riccio, M.L., Berlutti, F., Macaskie, L.E. and Thaller, M.C. (1998). Bacterial non-specific acid phosphatases: Physiology, evolution and use as tools in microbial biotechnology. *Cell Mol. Life. Sc.*, 54: 833–850.

Sarma, B.K., Singh, U.P. and Singh, K.P. (2002). Variability in Indian isolates of *Sclerotium rolfsii*. *Mycologia*, 94(6): 1051–1058.

Sarma, Y.R., Rajan, P.P., Diby, Paul, Beena, N. and Anandaraj, M. (2000). Role of Rhizobacteria on disease suppression in spice crops and future prospects. In: *Seminar on Biological Control with Plant Growth Promoting Rhizobacteria for Sustainable Agriculture*. University of Hyderabad, Hyderabad, April 3–4.

Saxena, A., Sharma, A. and Johri, B.N. (1996). Bacterial characterization of a growth promoting fluorescent pseudomonad from Rajniganda rhizosphere. In: 37th *Ann. Conf. Assoc. Microbiol.*, India, December 4–6, IIT, Chennai, pp. 135.

Schwan, B. and Neilands, J.B. (1987). Universal Chemical Assay for the detection and determination of siderophores. *Anal. Biochem.*, 160: 47–56.

Shah, S., Li, J., Moffatt, B.A. and Glick, B.R. (1998). Isolation and characterization of ACC deaminase genes from two different plant growth-promoting rhizobacteria. *Can J. Micrbiol.*, 44: 833– 843.

Shah, S., Li, J., Moffatt, B.A. and Glick, B.R. (1997). ACC deaminase genes from plant growth-promoting bacteria. In: *Plant Growth-Promoting Rhizobacteria: Present Status and Future Prospects*, (Eds.) Ogoshi, A., Kobayashi, K., Homma, Y., Kodama, F., Kondo, N. and Akino, S. OECD, Paris, pp. 320–324.

Suslow, T.V., Kloepper, J.W., Schroth, N. and Thomas, J.B. (1979). Beneficial bacteria enhance plant growth. *Calif. Agric.*, 33: 15–17.

Thomashaw, L.S., Weller, D.M., Bonsall, R.F. and Pierson, L.S. (1990). Production of the antibiotic phenazine-1-carboxylic acid by fluorescent *Pseudomonas* species in the rhizosphere of wheat. *Appl. Environ. Microbiol.*, 56: 908–912.

Turner, I.N. and Messenger, A.J. (1986). Occurrence, biochemistry and physiology of phenazine pigment production. *Adv. Microb. Physiol.*, 27: 211–275.

Yang, S.F. and Hoffman, N.E. (1984). Ethylene biosynthesis and its regulation in higher plants. *Ann. Rev. Plant Physiol.*, 35: 155–189.

©R.C. Ram & Asha Sinha (*Editors*)
***Microbial Diversity: Modern Trends*, 2007**
Daya Publishing House, New Delhi

213–223

Chapter 16

Screening of Sunflower Genotypes for Resistance or Tolerance Against Alternaria Blight

A.V.V. Reddy**, *C.V.C.M. Reddy**, *U. Rajeswari**, *B. Sinha and *C. Appunu*****

***Division of Genetics and Plant Breeding, Directorate of Oilseeds Research, Hyderabad – 500 030**

****Department of Genetics and Plant Breeding, Institute of Agricultural Sciences, Banaras Hindu University, Varanasi – 221 005**

ABSTRACT

Alternaria blight caused by *Alternaria helianthi* is a disease of leaf, stem and capitulum of cultivated sunflower in majority of areas throughout the world. In this study, genotypes were screened on field and lab environments to evaluate for resistance to *A. helianthi.* Disease severity was recorded by visually examining each leaf after seven days after inoculation as per cent of leaf area occupied by the sporulating fungus using the pictorial key method in each replication and the average computed. The disease intensity for hybrids ranged from 1.29 to 55.00 per cent. DCMS 37B, DRS 9, DCMS 37A × DRS 9, DCMS 37A × DRS 16 and DCMS 15 × DRS 15 were found to be resistant to *Alternaria* blight both under field and lab conditions and therefore has the potential to reduce yield losses because of this disease in the field.

Keywords: *Sunflower, Resistance, Leaf spot, Alternaria.*

Introduction

Sunflower is infected by a large number of disease causing pathogens resulting in severe economic losses in yields. Occurrence of wide spread diseases was identified as one of the major constraint for the low productivity levels in India. Alternaria leaf blight disease caused by *Alternaria helianthi* (Hansf) Tubaki and Nishihara. The pathogenic species of Alternaria cause seed transmitted diseases which induce pre and post emergence mortalities and seed lining blight/rots in the field (Maiti, 1987, Mayee 1988 and 92). By and large the incidence and intensity of these diseases are influenced by season, varieties/hybrids grown, stage of the crop, region and climatic conditions etc., the exact estimates of yield losses due to diseases of sunflower not known. The potential and actual losses estimated for the major diseases revealed that in case of Alternaria blight it was 90 per cent.

Alternaria Blight

Alternaria blight of sunflower was recorded first in Uganda in 1943. In India the disease is widespread and reported from all the sunflower growing areas (Anilkumar *et al.*, 1976, Mukeshwar *et al.*, 1974). The disease has been reported to reduce the seed and oil yields by 27 to 80 per cent and 17 to 35 per cent 14 respectively (Balasubramaniyam and Kolte, 1980). A negative correlation has been established between disease intensity (25–96 per cent) and yield and oil components. The disease also affects the quality of sunflower seeds adversely affecting the seed germination and vigour of seedlings. The loss in seed germination varies from 23 to 32 per cent (Kolte *et al.*, 1979).

Symptoms

Initially the symptoms appear as circular to oval spots of dark brown to black colour, varying between 0.2 to 5 mm in diameter. The spots surrounded by necrotic chlorotic zone with gray-white necrotic centre marked with concentric rings. The spots first appear on the lower leaves and as the plant grows the spots are formed on petioles, stem, ray florets and head. Under humid conditions, the spots enlarge in size and coalesce resulting in blighting of leaves and sometimes rotting of flower heads. Soil borne inoculum of the fungus is reported to cause foliar symptoms as well as seedling blight and was reported that *A. helianthi* of sunflower is seed borne and 22.9 per cent seed transmissible. The fungus located in the endosperm region in about 65 per cent of infected seeds and in the embryo in 25–30 per cent of infected seeds reported associated of *A. alternata* with 78 per cent of sunflower seed samples.

Pathogen

Earlier the pathogen causing leaf blight was recognizing as species of *Helminthosporium* but later confirmed as *Alternaria helianthi,* because of the formation of a few longitudinal septa. Mycelium is septate, rarely branched, brown, 2.5–25 µm width; conidiophores are cylindrical, scattered or gregarious. Pale gray to yellow, upto 5 septa, geniculete, 25–800 µ × 8–11 µ. Conidia cylindrical to ellipsoid rounded at ends without breaks, pale grey yellow, 1–11 transverse septa with constitutions at septa, 40–110 µm × 13–28 µm.

Epidemiology

Disease development is favored by 25–27°C temperature with atleast 12 hours of wet foliage. Extended leaf wetness of three to four days can cause serious losses with enlargement and coalescing of spots (Mayee, 1992). Occurrence of the disease in winter crop also had been reported (Jasbir Kaur *et al.*, 1991). The pathogen survives in winter as mycelium in the infected plant residues (Nagaraju *et al.*, 1994) for about 20 weeks in soil under dry conditions. The fungus produces a toxic metabolite, presumably a phenolic compound in culture as well as in infected leaf tissue. The toxin when sprayed on leaves produces necrotic spots on leaves, but such spots are not characterized by yellow halo. The toxin is found to inhibit seed germination of sunflower as well as root elongation.

Materials and Methods

Field Screening

During crossing programme, the two CMS lines and twenty eight testers along with susceptible check L–101 were screened for their reaction against *Alternaria* blight under natural conditions in field during *Kharif*, 2003.

The intensity of disease in the field was estimated from five randomly selected plants in each genotype which tagged with labels, at flowering stage of the crop. On an average 60 leaves were selected at random from selected plants and disease severity was recorded by visually examining each leaf as per cent leaf area damaged using pictorial key of Allen *et al.* (1983) in each relationship and the average was computed.

Lab Screening

During evaluation of F_1s along with parents, detached leaf technique was employed to identify resistant against *Alternaria* blight under laboratory conditions in *rabi*, 2003–04. These lines artificially inoculated with pathogen extracted from leaf blight affected plants and disease severity was recorded.

The general laboratory techniques followed for the present study was described by Aneja (1993) and Dhingra and Sinclair (1993) for sterilization, isolation and preparation of media and maintenance of fungal cultures.

Cleaning of Glassware

The glassware used in the present study comprised of petriplates, conical flasks, measuring cylinders, pipettes and test tubes of Borosil make.

Glassware were first sterilized with a detergent and then placed in cleaning solution of the following composition.

Potassium dichromite: 60g

Concentrated sulphuric acid: 60ml

Distilled water: 1000ml

The glassware were kept in the cleaning solution for 24 hours, later thoroughly washed with tap water and finally rinsed with distilled water and oven dried.

Sterilization

Glassware *i.e.*, petriplates were wrapped in butter paper and sterilized in hot air oven at 160°C for 1½ hour. Inoculation needles, forceps, cork borers and blades were sterilized by passing through flame after dipping in alcohol.

Culture Medium and its Composition

Sunflower Leaf Extract Medium (SLEM) of the following composition was used as the basal medium for the experiment.

Leaves of sunflower cv. CO-4 = 200g

Sucrose = 20g

Agar = 20g

Distilled water = 1000ml

Preparation of SLEM

Sunflower Leaf Extract Medium was prepared by boiling 200g of chopped leaf bits of sunflower cv. CO-4 in 500ml of water for about 10-15 minutes. In another 500ml of water, 20g of agar was melted. Both these were filtered through the double-layered muslin cloth in to a third container and 20g sucrose was added to the filtrate. The final volume of medium was made upto one litre by adding water and the medium was sterilized in an autoclave at 15 psi (121.6°C) for 20 minutes.

Source of Plant Material

The leaves of sunflower cv. CO-4 were obtained from Research Farm, DOR, Hyderabad. The leaves of fifty six F_1S developed during Kharif, 2003 along with parents were used for screening against *Alternaria* blight during *rabi*, 2003–04.

Isolation of the Pathogen

The pathogen *Alternaria helianthi* was isolated from leaves of infected sunflower plants collected from fields of the DOR, Hyderabad. The leaves were washed thoroughly with sterilized distilled water and small pieces of (3, mm^2) diseased leaf spots adjoined with some healthy tissue were cut with the help of a scalpel and surface sterilized by immersing them in 0.1 per cent mercuric chloride solution for 30 seconds in order to eliminate surface contaminants. The leaf bits were washed thoroughly in sterilized distilled water and excess moisture was removed by blotting using sterilized blotting paper. These leaf bits were transferred aseptically on to sterilized petriplates containing Sunflower leaf Extract Medium and Incubated at 27±1°C. The mycelial growth emerging from diseased leaf bits was transferred directly into culture medium (SLEM) with the help of sterilized needle. The culture of pathogen was purified by single spore isolation.

Single Spore Isolation

A dilute spore suspension (approximately, 100 spores/ml) prepared with sterile water was poured on water agar (0.2 per cent) in culture plates and was allowed to settle down on the agar just enough to form a very thin layer over the surface of agar so as to enable the spores settle quite apart from one another. A single spore was selected and lifted with the help of inoculating needle along with the agar and transferred to culture plated containing SLEM. After obtaining good growth, sub culturing was done to the mass multiplication of the fungus. Pure culture of the organism was maintained to SLEM by periodical transfer into slants.

Identification of Pathogen

The fungus associated with blight disease of sunflower was identified as *Alternaria helianthi* based on the descriptions as given by Ellis (1976).

Microscopic Examination

Temporary aqueous mounts were prepared from the pure culture of fungus and observed for different structures under the compound microscope. The measurements of the spores of the pathogen were taken after calibration of the microscope.

Pathogenicity Tests

Pathogenicity tests were carried out using *Alternaria* blight susceptible sunflower cv. CO-4. The seeds of the cultivar were sown in 18 cm diameter earthen pots containing a mixture of soil and farm yard manure in 4 : 1 ratio and maintained on glasshouse bench. Three plants were raised in each pot.

Preparation of Inoculum

The inoculum (conidial suspension) of *A. helianthi* was prepared by gentle scraping of the 15 day old culture grown on SLEM in petriplates with the help of a camel hair brush by adding 5ml of sterile distilled water to each petriplate. The spore suspension was filtered through three layers of muslin cloth and Tween 20 (polyoxyethylene sorbiton monoleate) was added @ 0.1 per cent to the suspension to enable uniform spread of inoculum on the sprayed leaves. The resultant conidial suspension was diluted to get a spore load of 1 × 10^3 conidia/ml by using haemocytometer.

Screening of Parents and F_1s for Resistance/Tolerance Against *Alternaria* Blight

The parental lines and F_1s were screened for resistance or tolerance against *Alternaria* blight by using detached leaf technique (Sujatha *et al.*, 1997) during *rabi*, 2003–04.

Detached Leaf Technique

Leaves of forty hybrids and their 14 parental lines (respective B lines of four CMS lines and ten testers) grown in DOR Farm, Hyderabad during *rabi*, 2003–04 were detached with petiole from third node of newly emerged branch at similar physiological maturity.

Detached leaves were immersed in spore suspension of 10^3 conidia/mi of *Alternaria helianthi* prepared from 15 day old culture grown on SLEM. The inoculated leaves were transferred to petriplates (15cm diameter) containing two layers of filter paper with one set of respective controls. For each genotype three replications were maintained. The petriplates containing inoculated leaf sample were incubated at 27±1°C in culture room for a week.

Record of Observations

Disease severity was recorded by visually examining each leaf after seven days after inoculation as percentage of leaf area occupied by the sporulating fungus using the pictorial key of Allen *et al.* (1983) in each replication and the average computed. The scale adopted to indicate the degrees of resistance in sunflower genotypes is furnished below:

Sl.No.	Disease Severity (per cent)	Category	Symbol
1.	0	Immune	I
2.	< 1	Highly resistant	HR
3.	1–5	Resistant	R
4.	5–25	Moderately Resistant	MR
5.	25–50	Susceptible	S
6.	> 50	Highly Susceptible	HS

Results and Discussion

Field and Lab Screening

Fourteen parental lines were screened for their reaction against *Alternaria* blight along with susceptible check L-101 under field conditions. All the fourteen genotypes showed significant differences for percent disease severity and it ranged from 4.50 to 29.50 per cent (Table 16.1).

Fourteen parental lines and forty hybrids were screened to search for resistance or tolerance under laboratory conditions using detached leaf technique (Table 16.2). Among the fourteen genotypes tested under field conditions, only one showed resistance, ten showed moderate resistance and the rest of three shown to be susceptible. Under laboratory conditions, two parental lines and three hybrids were resistant, while three parental lines and fifteen hybrids recorded moderate resistant reaction. Only one hybrid and none of the parental lines were highly susceptible and the remaining parents and hybrids were found to be susceptible.

Table 16.1: Reaction of Parental Lines Based on Disease Severity (per cent) Against *Alternaria* Blight Under Field Conditions in Sunflower

Genotype	Disease Severity (%)	Disease Reaction
DCMS 36A	29.50 (32.89)	S
DCMS 37A	22.85 (28.55)	MR
DCMS 41A	13.56 (21.60)	MR
DCMS 42A	26.50 (30.98)	S
DRS 16	15.66 (23.31)	MR
DRS 9	4.50 (12.24)	R
DRS 12	16.97 (24.32)	MR
LTRR 341	23.50 (28.99)	MR
RCR 8297	25.27 (30.17)	S
DRS 11	24.50 (29.66)	MR
DRS15	19.50 (26.20)	MR
NDLR-1	20.50 (26.91)	MR
RHA 272-1	23.50 (28.31)	MR
DRS 3	22.50 (28.31)	MR
L-101 (Check)	55.50 (20.63)	HS
SEm±	0.8306	
CD (0.05)	2.4151	
CY (per cent)	5.4000	

HS: Highly Susceptible; S: Susceptible; MR: Moderately Resistant; R: Resistant

Figures in parenthesis are arcsin transformed values.

Table 16.2: Grouping of Parental Lines Based on Disease Severity (per cent) Against *Alternaria* Blight Under Field Conditions in Sunflower

Category	Genotype
Resistant (1–5 per cent)	DRS 9
Moderate Resistant (5–25 per cent)	DCMS 41A, DRS 16, DRS 12, DRS 15, NDLR-I, DRS 3, DCMS 37A, LTRR 341, RHA 272-1, DRS 11
Susceptible (25–50 per cent)	RCR 8297, DCMS 42A, DCMS 36A

The results are in accordance with Amaresh and Nargund (2000) and Ahmed *et al.* (2001). The differences in disease reaction of genotypes under field and laboratory conditions is due to environmental conditions and availability of inoculum load.

Table 16.3: Reaction of Parental Lines Based on Disease Severity (per cent) Against *Alternaria* Blight Under Lab Conditions in Sunflower

Genotype	Disease Severity (%)	Disease Reaction
Parents		
DCMS 36B	33.50 (35.36)	S
DCMS 37B	3.52 (10.81)	R
DCMS 41B	45.00 (42.12)	S
DCMS 42B	37.50 (37.75)	S
DRS 16	8.66 (17.11)	MR
DRS 9	2.20 (8.52)	R
DRS 3	26.23 (30.80)	S
DRS 12	9.05 (17.50)	MR
LTRR 341	42.60 (40.73)	S
RCR 8297	39.10 (38.69)	S
DRS 11	31.70 (34.25)	S
DRS 15	8.12 (16.54)	MR
NDLR-1	29.30 (32.76)	S
RHA272-1	41.75 (40.24)	S
Hybrids		
DCMS 36A × DRS 16	10.20 (18.62)	MR
DCMS 36A × DRS 9	13.53 (21.57)	MR
DCMS 36A × DRS 3	27.08 (31.34)	S
DCMS 36A × DRS 12	19.00 (25.83)	MR
DCMS 36A × LTRR 341	38.60 (28.40)	S
DCMS 36A × RCR 8297	42.50 (40.68)	S
DCMS 36A × DRS 11	32.10 (34.50)	S
DCMS 36A × DRS 15	22.56 (28.35)	MR
DCMS 36A × NDLR-1	33.70 (35.48)	S
DCMS 36A × RHA 272-1	14.73 (22.56)	MR
DCMS 37A × DRS 16	2.80 (9.6)	R
DCMS 37A × DRS 9	1.92 (7.96)	R
DCMS 37A × DRS 3	34.91 (36.21)	S
DCMS 37A × DRS 12	13.30 (21.38)	MR
DCMS 37A × LTRR 341	34.04 (35.69)	S
DCMS 37 A × RCR 8297	31.70 (34.26)	S
DCMS 37A × DRS 15	4.95 (12.85)	R
DCMS 37A × NDLR-1	45.00 (42.12)	S
DCMS 37A × RHA 272-1	32.31 (34.63)	S
DCMS 41A × DRS 16	13.50 (21.55)	MR
DCMS 41A × DRS 9	12.12 (20.37)	MR
DCMS 41A × DRS 3	44.80 (42.01)	S
DCMS 41A × DRS 12	12.90 (21.04)	MR

Contd...

Table 16.3–Contd....

Genotype	Disease Severity (%)	Disease Reaction
DCMS 41A × LTRR 341	24.20 (29.46)	MR
DCMS 41A × RCR 8297	33.10 (35.11)	S
DCMS 41A × DRS 11	33.70 (35.47)	S
DCMS 41A × DRS 15	14.80 (22.61)	MR
DCMS 41A × NDLR-1	55.00 (47.87)	HS
DCMS 41A × RHA 272-1	31.50 (34.13)	S
DCMS 42A × DRS 16	13.21 (21.30)	MR
DCMS 42A × DRS 9	13.65 (21.67)	MR
DCMS 42A × DRS 3	28.50 (28.50)	S
DCMS 42A × DRS 12	15.92 (23.51)	MR
DCMS 42A × LTRR 341	27.76 (31.79)	S
DCMS 42A × RCR 8297	34.20 (35.78)	S
DCMS 42A × DRS 11	25.50 (25.50)	S
DCMS 42A × DRS 15	20.50 (20.50)	MR
DCMS 42A × NDLR-1	43.50 (43.50)	S
DCMS 42A × RHA 272-1	34.80 (34.80)	S
SEm±	0.5869	
CD (0.05)	1.6268	
CV (per cent)	2.1381	
	3.45	

HS: Highly Susceptible; S: Susceptible; MR: Moderately Resistant; R: Resistant

Figures in parenthesis are arcsin transformed values.

Table 16.4. Grouping of 54 Genotypes Based on Disease Severity (%) Against *Alternaria* Blight in Sunflower

Category	Genotype
Resistant (1–5%)	DRS 9,DCMS 37B, DCMS 37A × DRS 9, DCMS 37A × DRS 16, DCMS 37A × DRS 15
Moderate Resistant (5–25%)	DRS 15, DRS 16, DRS 12, DCMS 37A × DRS 16, DCMS 41A × DRS 9, DCMS 41A × DRS 12, DCMS 42A × DRS 16, DCMS 37A × DRS 12, DCMS 41A × DRS 16, DCMS 36A × DRS 9, DCMS 42A × DRS 9, DCMS 36A × RHA 272–1, DCMS 41A × DRS 15, DCMS 42A × DRS 12, DCMS 36A × DRS 12, DCMS 42A × DRS 15, DCMS 36A × DRS 15, DCMS 41A × LTRR341
Susceptible (25–50%)	DRS 3, NDLR–1, DRS 11, DCMS 36B, DCMS 42B, RCR 8297, RHA 272–1, LTRR 341, DCMS 41B
	DCMS 42A × DRS 11, DCMS 36A × DRS 3, DCMS 42A × LTRR 341, DCMS 42A × DRS 3, DCMS 41A × RHA 272–1, DCMS 37A × RCR 8297, DCMS 36A × DRS 11, DCMS 37A × RHA 272–1, DCMS 41A × RCR 8297, DCMS 41A × DRS 11, DCMS 36A × NDLR–1, ' DCMS 37A × LTRR 341, DCMS 42A × RCR 8297, DCMS 42A × RHA 272–1, DCMS 36A × RCR 8297, DCMS 2A × NDLR–1, DCMS I 37A × DRS 11, DCMS 41A × DRS 3, DCMS 37A × NDLR–1
High susceptible (> 50%)	DCMS 41A × NDLR–1

Isolation and Identification of the Pathogen

Isolation of the pathogen *Alternaria helianthi* (Hansf.) Tubaki and Nishihara was made from the infected leaves of sunflower cultivar CO-4 showing typical leaf spot/blight symptoms on leaf extract medium of cultivated sunflower CO-4 (SLEM), purified by single spore method and maintained on sunflower leaf extract media. Growth of the fungus when incubated at 27±1°C on sunflower leaf extract medium was observed after two days after plating. The rate of growth of culture was very slow on the medium. Maximum colony growth was obtained twenty one days after plating. Thereafter growth of the culture is ceased. The culture was grayish white with brown pigmentation. Sporulation was observed at the centre of colony after twenty one days of incubation. Microscopic examination of the fungal culture revealed branched, septate, smooth and olivaceous brown coloured mycelium. Conidiophores were cylindrical, either solitary or in fascicles, pale grey to brown in colour, straight or curved, simple or branched, septate, upto 120 µm long and 8–11 µm broad. Conidia were solitary, straight or slightly curved, cylindrical long, ellipsoid, rounded at ends, sub hyaline to golden brown, smooth, with 2–12 transverse and occasional one or more longitudinal septa, often constricted at the septa, 45–145 µm long, 10–30 µm thick in the broadest part.

On the basis of cultural and morphological characteristics, the fungus associated with spot/ blight disease in sunflower was identified as *A. helianthi* (Hansf.) Tubaki and Nishihara and the characteristic features tallied with CMI description of *A. helianthi* (Ellis, 1976).

Pathogenicity Tests

Pathogenicity was proved by spray inoculation with conidial suspension (1×10^3 conidia/mi) of the pathogen on susceptible sunflower cv. CO-4. The initial symptoms were observed 48 hours after inoculation on leaves in the form of oval to circular spots surrounded by yellowish halo measuring 1–2 mm in diameter. These spots gradually increased in size, which coalesced forming bigger spots of about 3–5 mm in diameter with angular margins. The pathogen could be re isolated from the infected leaves and the fungus obtained resembled the original isolate.

Screening of Parents and F_1s for Resistance or Tolerance Against *Alternaria* Blight

Fifty six F_1s along with parents (respective B lines of four CMS lines and twenty eight testers) were screened under laboratory conditions using detached leaf technique to search for resistant or tolerant genotypes against *Alternaria* blight. Based on disease severity per cent, disease reaction of parents and F_1s were presented in Table 16.3. The disease intensity for hybrids ranged from 1.29 to 55.00 per cent. It was observed that resistant (DCMS 37B and DRS 9) or moderate resistant (DRS 15, DRS 16 and DRS 12) genotypes were crossed, their F_1 hybrids gave resistant (DCMS 37A × DRS 9, DCMS 37A × DRS 16 and DCMS 37A × DRS 15) and moderate resistant (DCMS 36A × DRS 16, DCMS 41A × DRS 12, DCMS 41A × DRS 9, DCMS 42A × DRS 16, DCMS 37A × DRS 12, DCMS 41A × DRS 16, DCMS 36A × DRS 9, DCMS 42A × DRS 9, DCMS 36A × RHA 272-1, DCMS 41Ax DRS 15, DCMS 42A × DRS 12, DCMS 36A × DRS 12, DCMS 42A × DRS 15, DCMS 36A × DRS 15 and DCMS 41A × LTRR 341) disease reaction.

This shows that genes for resistance to *Alternaria* blight are dispersed differently in genotypes and in hybrid combinations, the favorable alleles get accumulated to give resistant disease reaction. So, it can be concluded that resistance to *Alternaria* blight was polygenically controlled and dominance phenomenon was expressed in hybrids. The present findings are in agreement with the earlier investigations of Ahmed *et al.* (2001) and Nagaraju *et al.* (1992), who suggested that hybrids performed better in response to their resistance to leaf spot and such resistance was derived either from the CMS or restorer lines used in their production.

References

Agrawat, J.M., Chippa, H.P. and Mathur, S.J. (1979). Screening of sunflower germplasm against *Altenaria helianthi*. *Indian Journal of Mycology and Plant Pathology*, 9: 85–86.

Ahmed, S., Sheoran, R.K. and Saini, M.L. (2001). Screening of sunflower CMS lines, restorers, populations and hybrids against *Alternaria helianthi* (Hansf.). *Journal of Oilseeds Research*, 18(1): 120–122.

Allen, S.J., Brown, J.F. and Kochman, J.K. (1983). Production of inoculum and field assessment of *Alternaria helianthi* in sunflower. *Plant Disease*, 67: 665–668.

Amaresh, Y.S. and Nargund, V.B. (2000). Screening of sunflower genotypes against *Alternaria* Leaf Blight and Rust. *Karnataka Journal of Agricultural Sciences*, 13(2): 468–469.

Aneja, K.R. (1993). *Experiments in Microbiology, Plant Pathology and Tissue Culture*. Wishwa Prakasham, New Delhi, pp. 471.

Anilkumar, T.B., Ura, S.D., Seshadri, U.S. and Hegde, R.K. (1974). *Alternaria* Leaf spot of sunflower. *Current Science*, 43: 93–94.

Balasubramanyam, N. and Kolte, S.J. (1980). Effect of *Alternaria* blight on yield components, oil content and seed quality of sunflower. *Indian Journal of Agricultural Sciences*, 50: 701–706.

Chattopadhyay, C. (1999). Identification of sources resistant to *Alternaria* blight of sunflower (*Helianthus annuus* L.). *Journal of Mycology and Plant Pathology*, 29: 402–407.

Dhingra, O.D. and Sinclair, J.B. (1993). *Basic Plant Pathology Methods*. CBS Publications and Distributors, New Delhi, pp. 335.

Ellis, M.B. (1976). *Dematiaceous hyphomycetes*. Commonwealth Mycological Institute, Kew, Surrey, England, pp, 494.

Freidt, W. (1992). Present state and future prospects of biotechnology in sunflower breeding. *Field Crops Research*, 30: 425–442.

Jagadish, G.V., Virupakshappa, K., Venkatramana and Ramaiah. (1994). *Handbook on Diagnostic Characteristics of Parental Lines and Hybrids of Sunflower* (*Helianthus annuus* L.). University of Agricultural Sciences, GKVK, Bangalore, pp. 3–6.

Jamadar, M.M., Nayakar, N.Y. and Desai, S.A. (1999). Field reaction of sunflower cultivars to *Alternaria* leaf spot and rust. *Journal of Maharashtra Agricultural Universities*, 24: 87–88.

Jasbir Kaur, Chahal, S.S. and Aulokh, K.S. (1991). Effect of different crop seasons on the influence of *Alternaria helianthi* in sunflower seeds. *Plant Disease Research*, 6: 83–84.

Kolte, S.J., Balasubramanyam, N., Tewari, A.N. and Awasthi, R.P. (1979). Field performance of fungicides in the control of *Alternaria* blight on sunflower. *Indian Journal of Agricultural Sciences*, 49(7): 555.

Leite Leite, R.M.V.B., Trezzi, M.M., De Oliviera, Arias, C.A.A. and Castiglioni, V.B.R. (1999). Reaction of sunflower genotypes to *Alternaria helianthi* in the state of Parana, Brazil. *Helia*, 22: 151–157.

Lipps, P.E. and Herr, L.J. (1986). Reactions of *Helianthus annuus* and *H. tuberosus* plant introductions to *Alternaria helianthi*. *Plant Disease*, 70: 831–835.

Maiti, S. (1987). *Diseases of Sunflower, Sesamum, Safflower, Niger, Castor and Rapeseed*. Plant Protection Assessment of India, Hyderabad, pp. 275–292.

Mayee, C.D. (1992). New approaches to disease management in kharif oilseed crops. *Journal of Oilseeds Research*, 9: 119–126.

Mirza, M.S. and Hoes, J.A. (1996). Screening for resistance in sunflower against *Alternaria helianthi*. *Helia*, 19: 87–92.

Morris, J.B., Yang, S.M. and Wilson, L. (1983). Reaction of *Helianthus* species to *Alternaria helianthi*. *Plant Disease*, 67: 539–540.

Mukeshwar, P.M., Lambat, A.K., Nath, R., Majumdar, A., Rao, I. and Chandra, K.J. (1974). Blight disease of sunflower caused by *Alternaria helianthi* (Hans), Tubaki and Nishihara in India. *Current Science*, 43: 346.

Nagaraju, A.J., Goud, I.S., Virupakshappa, K., Jayaramaiah, H. and Patil, S.A. (1997). Evaluation of sunflower hybrids against *Alternaria* leaf spot and rust diseases under natural field conditions. *Karnataka Journal of Agricultural Sciences*, 10: 265–266.

Nagaraju, A.J., Jagadish, B.N. and Virupakshappa, K. (1992). Reaction of cytoplasmic male sterile and restorer lines of sunflower to *Alternaria helianthi*. *Indian Phytopathology*, 45: 372–373.

Nagaraju, A.J., Ravikumar, R.L., Gangappa, E. and Jayarame, G. (1993). Sources of field resistance to *Alternaria* leaf spot in sunflower. *Journal of Oilseeds Research*, 10(1): 148–150.

Nagaraju, Jayarame Gowda, Gangappa, E. and Virupakshappa, K. (1994). Integrated management of *Alternaria* leafspot of sunflower. In: Prasad, M.V.R *et al.*, Indian Society of Oilseeds Research, Hyderabad, pp. 125–128.

Panse, V.G. and Sukhatme, P.V. (1985). *Statistical Methods for Agricultural Workers*. Indian Council of Agricultural Research, New Delhi, pp. 361.

Prasad, D.T., Gangappa, E. and Channakrishnaiah, K.M. (1996). *In vitro* technique to screen genotypes for *Alternaria helianthi* tolerance. In: *Disease Tolerance in Sunflower Symposium*, I. (Ed.) Andre Pouzet. 14th International Sunflower Conference, Beijing, China, 13 June, pp. 128–137.

Shobharani, T. and Ravikumar, R.L. (2002). Evaluation of S_1 progenies from populations moderately tolerant to *Alternaria* blight in sunflower. *Crop Research*, 24(1): 77–80.

Shobharani, T. and Ravikumar, R.L. (2003). Reaction of interspecific lines to *Alternaria* leaf and stem blight in sunflower. *Helia*, 26(38): 115–120.

Sujatha, M., Prabhakaran, A.J. and Chattopadhayay, C. (1997). Reaction of wild sunflowers and certain interspecific hybrids to *Alternaria helianthi*. *Helia*, 20: 15–24.

Velazhahan, R. and Narayanasamy, P. (1994). Resistance in sunflower genotypes to rust and leaf spot. *Madras Agricultural Journal*, 81(1): 43–44.

Velazhahan, R. and Narayanasamy, P. and Jeyarajan, R. (1991). Evaluation of sunflower germplasm for field resistance to *Alternaria helianthi*. *Madras Agricultural Journal*, 78: 143–144.

©R.C. Ram & Asha Sinha (*Editors*)
***Microbial Diversity: Modern Trends*, 2007** **224–233**
Daya Publishing House, New Delhi

Chapter 17

Effect of *Glomus fasciculatum* AM Fungus and Nitrogens on Biomass Yield, Chlorophyll, Juice and Sucrose Contents of *Saccharum officinarum* L.

Kamal Prasad** & R.S. Bilgrami**

Department of Botany, Maharshi Dayanand Saraswati University, Ajmer - 305 009, Rajasthan, India

****University Department of Botany, B.R. Ambedkar Bihar University, Muzaffarpur - 842 001, Bihar, India**

ABSTRACT

The response of sugarcane (*Saccharum officinarum* L.) variety B.O. 109 towards *Glomus fasciculatum* AM fungi and different source of nitrogen's has been evaluate individually or dually by performing experiment under green house conditions. Plants in pots were maintained to grow till 12 months and data pertaining to various parameters such as dry weight, plant height, stem thickness, number of nodes/plant, leaves/plant, chlorophyll, juice and sucrose contents. The biomass yield, chlorophyll, juice and sucrose contents were always betters in AMF inoculated plants than in control plants.

Different sources of nitrogen however, exhibited triggering effect on biomass, Chlorophyll, juice and sucrose contents. Among different sources of nitrogen employed, maximum plant dry weight (84.20–302.10 g) height (65.85–200.97 cm), stem thickness (4.47–9.09 cm), number of nodes/plant (7–25), leaves/plant (13–42), chlorophyll 'a' (0.825–1.295 mg/g), chlorophyll 'b'

Corresponding Author: E-mail: Kamalprasadra@yahoo.com.in

(0.794–1.365 mg/g), total chlorophyll (1.619–2.660 mg/g), juice/plant (155.80–325.60 ml) and sucrose (8.85–17.80 per cent) were observed when AMF inoculated plants were grown in soil supplemented with urea. Minimum plant dry weight (71.40–279.10 g), height (59.75–192.32 cm), stem thickness (4.28–8.68 cm), number of nodes/plant (7–23), leaves/plant (12–40), chlorophyll 'a' (0.786–1.225 mg/g), chlorophyll 'b' (0.731–1.346 mg/g), total chlorophyll (1.517–2.571 mg/g), juice/plant (1.49–317.60 ml) and sucrose content (8.65–17.52 per cent) were estimated when AMF inoculated plants were grown in soil supplemented with calcium nitrate. A rapid increase in rate of biomass was recorded upto 6–9 month but it was significantly reduced.

Introduction

Sugarcane (*Saccharum officinarum* L. family Poaceae) is most important cash crop cultivated in many parts of India in different soils. It fulfills the sugar requirement in our country and through out world. AM fungi improve plant growth mainly through increased uptake of P, especially in low P soil (Bowen, 1978). The positive growth response is usually slower in soil where applied P is fixed rapidly (Abbott *et al.*, 1979). Although sugarcane is known to harbour AMF association naturally in its root systems (Chandrakant *et al.*, 1989 a; Prasad, 1993). Significant variation in its intensity in different soils were observed (Chandrakant *et al.*, 1989b) and in different varieties under the same field soil conditions (Chandrakant *et al.*, 1989a,). An effective mycorrhizal association is required for every crop of P uptake and other nutrients especially in soils of low fertility. The AM fungi are widely distributed and associated with more than 90 per cent of land plants with the exception of a few families.(Harley and Smith, 1983). The special relationship established between plants and fungi implies a high degree of structural, physiological and biochemical integration from which both the partners benert (Azcon Anguilar and Bago, 1994).

The AM fungal symbiosis is receiving attention because of its increasing range of application in practical fields like sustainable agriculture (Bethlenfalvay and Schuepp, 1994). In general, agricultural practices, tendencies to use high dose phosphate, nitrogen and other fertilizers decrease the colonization and affectivity of AMF (Menge *et al.*, 1978, Braunberger *et al.*, 1991; Prasad, 1993). In addition, growing non mycorrhizal crops disturbs the ecology of the AMF in soil by blocking the carbon supply to the fungus and the biochemical changes in rhizosphere prevent germination of propagules and decrease their viability (Stahl *et al.*, 1988).

The possibility of inoculation of AMF to cultivated crops with selected strains of AMF which are more effective than those already present in the cultivated soils has generated considerable interest among the workers (Smith and Smith, 1981; Read 1986; Prasad, 1993). Such a possibility could be exploited tremendously for obtaining better biomass and yield of economically important plants.

The plant's growth response is usually dependent on the quantity of AMF formed early in the crop's growth season. Since different AM fungi differ in their ability to form efficient AMF with different crop plants (Reddy and Singh, 1990). The present investigation deals with the effect of native *Glomus fasciculatum* are nitrogen's on biomass yield, chlorophyll, juice and sucrose contents of *Saccharum officinarum* cv. B.O. 109.

Materials and Methods

AMF Spores Isolation Identification, and Inoculum Preparation

Sugarcane cultivated field was surveyed in North Bihar and AMF spore were isolated from the rhizosphere. Initially rhizospheric soil sample were brought to laboratory in polythene bags and stored at 4°C for analysis. Single AMF spores were collected using the modified wet sieving and decanting method of Daniels and Skipper (1982) and Gerdman and Nicolson (1963). AMF spores were surface sterilized using 2 per cent Chloramine T and Streptomycin sulphate (200 g/ml) for 20 minutes and washed for several times by sterile distilled water. Single AMF spores cultures were raised from them by the standard funnel technique using maize on a host plant. Diagnostic slides of spores were prepared using polyvinyl alcohol lactoglycerol (PVLG) as mountant. Identification of AM fungi was done using relevant literature (Mortan and Benny, 1990; Schenck and Perez, 1990). Dominant species of *Glomus fasciculatum* were selected for inoculum preparation. For inoculum preparation of AMF, surface sterilized seeds of *Zea mays* and sterilized substrate (Soil and Sand; 1 : 1 v/v) were used. The substrate containing spores, sporocarps and root pieces served as a stock culture of AMF inoculum.

Estimation of AMF Fungal Infection in Roots

Freshly collected roots were washed in water, cleaned with 10 per cent KOH, acidified with IN HCI and stained in 0.05 per cent trypan blue (Phillips and Hayman, 1970). Quantification of AMF root colonization was carried out using the slide method (Giovannetti and Mosse, 1980).

Pot Experiment

Sandy loam type soil having the following composition: Organic matter 1.85 per cent; Organic carbon 1.08 per cent; Available nitrogen 174.0 kg/ha; Available P–10.6 $\mu g\ g^{-1}$, Available K–3.90 $mg\ g^{-1}$; pH–8.2; Ec–0.45 m mhoh/cm determined by the standard method (Jackson, 1973) was used in the experiment. The soil was crushed to pass through a 4 mm sieve and autoclaved at 121°C for 2 hrs on 3 consecutive days. Experimental pots (60 × 50 × 50 cm diameter) were filled with 50 kg of sieved and autoclaved sandy loam soil.

Plant Inoculation Method

Two thin layers of *G. fasciculatum* (250 spores and sporo carps, bits of hyphae and infected root segments with substrate per plant) inoculum were laid across the pots at the depth of 15cm and 25cm below the soil surface at the time of packing (Amijee *et al.*, 1986).

The treatments included were:

1. Uninoculated Control Soil
2. Soil + AMF (*G. fasciculatum*) alone
3. Soil + AMF + Urea
4. Soil + AMF + Ammonium Sulphate
5. Soil + AMP + Calcium nitrate

Different sources of nitrogen were amended in soil @ 50 kg/ha at different intervals

Procurement of Sample

Sugarcane variety B.O. 109 seeds were obtained from Sugarcane Research Institute, Pusa, Samastipur.

Seed Treatment and Planting

Surface sterilized seeds of B.O. 109 variety of sugarcane (Showing maximum AMF infection (77 per cent) during preliminary survey) were allowed to grow in soil having AMF inoculum to develop infection in roots. The potted plants receiving the above treatments were left in greenhouse benches having temperature 25–35°C, humidity 70–80 per cent, and light 15000 Lux–19000 Lux, in a randomized complete block design to minimize any positional effect with six replicate per treatments.

Biomass Estimation

The biomass of different plants was determined by, weighing over dry process of plant by measuring their height and thickness as well as by counting their nodes and leaves. The increase in total biomass was calculated on the basis of their readings. The data were recorded after 3, 6, 9 and 12 months of plant growth.

Chlorophyll Extraction and Estimation

The chlorophyll extraction from fresh green leaves of sugarcane and its qualitative estimation was done with slight modification by using double spectrophotometric method as described by Anderson and Boardman (1964).

Juice and Sucrose Content Extraction and Estimation

The juice content of fresh sugarcane was extracted by hard operated crusher machine and its volume was measured by measuring cylinder. The sucrose content of juice was analysed by Brix Hydrometer and Poloriscope. The necessary correction of temperature was done and purity was determined as co-efficient of purity.

$$\text{Purity Co-efficient} = \frac{\text{Pol. (\%)}}{\text{Corrected brix reading}} \times 1100$$

The analysis was completed according to the procedure adopted by Spence and Meade (1955).

Results and Discussion

The data presented in Table 17.1 show that maximum density spores/100g soil and maximum percentage of root infection was recorded at the age of 6 month followed by 3 months, 12 months and 9 months of plate growth. The percentage of root infection followed the same trend as AMP infestation in soil.

Table 17.1: Effect of AM Fungus and Nitrogens on AMF Spore Population and Root Infection of Sugarcane

Treatments	AMF Spores/100 g Soil and Root Infection (per cent) After							
	3 Months		6 Months		9 Months		12 Months	
	AMF Infected	AMF Free	AMF Infected	AMF Free	AMF Infected	AMF Free	AMF Infected	AMF Free
Control (without nutrient)	210	63	321	88	125	44	218	65
Urea	215	65	330	90	130	45	227	69
Ammonium sulphate	210	64	326	89	128	44	220	67
Calcium nitrate	205	63	320	87	125	42	219	66

Value of 't' between AMF infected and AMF free observation

(* significance $p > 0.05$)

Control: 4.810*; Urea: 5.251*;

Ammonium sulphate: 4.903*; Calcium nitrate: 4.900*

Vigour Comparison

The soil containing AMF and nitrogen significantly increased the vigour of plant. The Table 17.2 reflect that dry weight, stem height, thickness, node and leaf formation in AMF inoculated plants increased when grown in different sources of nitrogen. Maximum increase was recorded in plants treated with urea followed by ammonium sulphate and calcium nitrate during different phases of plant growth. The effect of AMF was found to be more apparent than the effect of nitrogen on plant biomass. The AMF inoculated plants always showed better response than the uninoculated plants. Similar results were also recorded on various other plants by Mengel and Kirkby, 1982; Sunil Kumar and Lakhanpal, 1989. The present study has also revealed that AM fungi, when inoculated by the layering method at the time of seed sets plantation are more effective and enhancing to the growth rate of sugarcane (Reddy *et al.*, 2004.) Thus, screening experiments are always useful in selecting the most efficient AMF for the crop to achieve maximum benefit. Other workers have also reported the superior performance improved growth (Mosse, 1975; 1977; Manjunath and Bagyaraj, 1982) by showing a certain degree of host preference (Mosse, 1975). This is largely based on the intensity of association and their influence on growth.

Table 17.2: Effect of AM Fungus and Nitrogens on Biomass and Growth Response of Sugarcane

Parameter/Treatments	Biomass and Growth Response of Sugarcane After							
	3 Months		6 Months		9 Months		12 Months	
	AMF Infected	AMF Free	AMF Infected	AMF Free	AMF Infected	AMF Free	AMF Infected	AMF Free
Dry Weight (g)								
Control (without nutrient)	70.30	45.26	138.40	83.20	222.30	138.40	269.30	163.30
Urea	84.20	60.23	158.20	103.15	251.20	174.20	302.10	209.40
Ammonium sulphate	74.40	53.26	144.20	99.45	232.20	158.10	289.20	179.30
Calcium nitrate	71.40	52.10	141.40	95.30	222.10	154.30	279.10	169.30
Stem Height (cm)								
Control (without nutrient)	62.72	38.32	104.79	73.77	173.96	131.23	194.96	150.62
Urea	65.85	60.75	107.97	101.36	177.85	161.75	200.97	187.26
Ammonium sulphate	63.73	53.39	105.96	90.37	174.99	150.65	193.98	176.39
Calcium nitrate	59.75	49.05	101.75	84.23	169.73	142.35	192.32	166.96
Stem Thickness								
Control (without nutrient)	4.42	3.48	6.50	5.30	8.60	7.35	8.75	7.59
Urea	4.47	4.21	6.65	6.26	8.85	8.36	9.09	8.53
Ammonium sulphate	4.29	4.20	6.35	6.20	8.50	8.22	8.71	8.36
Calcium nitrate	4.28	4.18	6.30	6.15	8.47	8.20	8.68	6.32
No. of Nodes/Plant								
Control (without nutrient)	6	4	11	7	19	13	23	16
Urea	7	6	13	10	22	16	25	19

Contd...

Table 17.2–Contd...

Parameter/Treatments	Biomass and Growth Response of Sugarcane After							
	3 Months		6 Months		9 Months		12 Months	
	AMF Infected	AMF Free	AMF Infected	AMF Free	AMF Infected	AMF Free	AMF Infected	AMF Free
Ammonium sulphate	7	5	12	9	20	15	23	18
Calcium nitrate	7	5	12	9	20	15	23	18
No. of Leaves/Plant								
Control (without nutrient)	13	7	19	13	35	21	39	25
Urea	13	9	22	17	37	28	42	32
Ammonium sulphate	13	8	21	16	36	25	41	30
Calcium nitrate	12	8	20	15	34	26	40	30

Value of 't' of between AMF infected and AMF free treatment (significant $p > 0.001$).

Dry weight: 5.234*; Stem height: 5.942*; Stem thickness: 4.204*;

No. of nodes: 8.996; No. of leaves: 9.297*

Chlorophyll Content

Chlorophyll contents were found to be high in leaves of AMF inoculated plants than that in leaves of control (Table 17.3). Maximum chlorophyll contents (Chlorophyll 'a', 'b' and total Chlorophyll mg/g leaves) were recorded in plants treated with urea and AMF followed by ammonium sulphate, calcium" nitrate and control (without nutrients) after different phase of plant growth. The chlorophyll contents was maximum between 3 to 6 months and minimum between 9 to 12 months in plants inoculated with *G. fasciculatum.* Same trend of increasement was observed in control and AMF uninoculated plant's (Table 17.3). The synthesis of chlorophyll is associated with AMF and NPK and other nutrients. The low uptake of these nutrients by uninoculated plants may decrease the chlorophyll concentration in the sugarcane leaves. Chlorophyll contents increased in AMF inoculated sugarcane because AMF play an important role in uptake of different nutrients (Prasad and Bilgrami, 2001). Nitrogen play an important role in the synthesis of chlorophyll as it is present in ring structure of tetrapyrrole (Black, 1968; Shivapulra *et al.*, 2004).

It is observed that when the population of AMF spore in soil and the percentage infection in roots increase the chlorophyll contents. Maximum infestation in soil and infection in roots were recorded between 3 to 6 months when the contents of chlorophyll was also recorded highest (Table 17.4). AMF play a important role in chlorophyll synthesis by absorbing essential elements needed for its synthesis (Prasad, 1993).

Juice and Sucrose Contents

The data registered in Table 17.4 show that there was increase in volume of juice and in the percentage of sucrose content in sugarcane shoots of both AMF inoculated and uninoculated plants. However, the increase was more significant in AMF inoculated plant as compared to uninoculated plants. Among the various sources of nitrogen employed, the amount of juice and the percentage of sucrose was highest in urea followed by ammonium sulphate, calcium nitrate and control (without nutrient) after a month of plant growth. The rate of increase was higher upto 9 months of plant growth but it showed down afterwords. Maximum AMF infection in roots and infestation (Table 17.1) in soil was also recorded in this nutrient. So juice (155.80–325.60 m/plant) and sucrose (8.85 per cent–17.80 per cent) synthesis in plants might have stimulated due to higher uptake of nutrients by AMF. Increase in Juice and sucrose percentage in sugarcane by nitrogen's have also been reported by Kanwar and Kochar, 1960; Mirsa *et al.*, 1964; Naidu *et al.*, 1982.

Table 17.3: Effect of AM Fungus and Nitrogens on Chlorophyll Contents of Sugarcane

Parameter/Treatments	Chlorophyll 'a', 'b' and Total Chlorophyll (mg/g) After							
	3 Months		6 Months		9 Months		12 Months	
	AMF Infected	AMF Free	AMF Infected	AMF Free	AMF Infected	AMF Free	AMF Infected	AMF Free
Control (without nutrient)								
Chlorophyll 'a'	1.240	0.905	1.245	0.950	0.781	0.581	0.635	0.580
Chlorophyll 'b'	1.361	0.878	1.370	0.961	0.726	0.562	0.621	0.561
Total Chlorophyll	2.601	1.783	2.615	1.881	1.507	1.143	1.256	1.141
Urea								
Chlorophyll 'a'	1.295	1.053	1.232	0.988	0.875	0.764	0.825	0.664
Chlorophyll 'b'	1.365	1.191	1.442	0.895	0.807	0.723	0.794	0.631
Total chlorophyll	2.660	2.244	2.674	1.883	1.682.	1.487	1.619	1.295
Ammonium Sulphate								
Chlorophyll 'a'	1.230	0.988	1.250	0.875	0.856	0.768	0.818	0.677
Chlorophyll 'b'	1.360	0.895	1.375	0.807	0.804	0.706	0.767	0.627
Total chlorophyll	2.590	1.883	2.625	1.682	1.660	1.471	1.555	1.304
Calcium Nitrate								
Chlorophyll 'a'	1.225	1.050	1.245	0.850	0.838	0.755	0.786	0.678
Chlorophyll 'b'	1.346	1.185	1.366	0.782	0.795	0.680	0.731	0.643
Total chlorophyll	2.571	2.235	2.611	1.632	1.633	1.435	1.517	1.321

Value of 't' of between AMF infected and AMF free treatment, (* significance $p > 0.05$)

Control: 3.097*; Urea 3.368*;

Ammonium sulphate: 2.877*; Calcium nitrate 2.287*

The present data also reflect that the metabolic activities associated with juice and sucrose synthesis remained highest between 6 to 9 months of plant growth. Increase in juice and sucrose contents in sugarcane by AMF inoculation is an interesting phenomenon which may help in better recovery of sugar from sugarcane. The extra nutrients may trigger the synthesis of juice and sucrose but the effect of AMF was always superior over the effect of nutrients used. It was observed that the joint effect of AMF and fertilizers was always superior than the effect of AMF or fertilizer alone. Indeed the dose of fertilizers should always be within the prescribed limits, which vary for soil to soil.

Table 17.4: Effect of AM Fungus and Nitrogens on Juice and Sucrose Contents of Sugarcane

Parameter/Treatments	Juice Content (ml/plant) and Sucrose Content (pol. per cent) in Cane After							
	3 Months		6 Months		9 Months		12 Months	
	AMF Infected	AMF Free	AMF Infected	AMF Free	AMF Infected	AMF Free	AMF Infected	AMF Free
Control (without nutrient)								
Juice	145.50	66.30	192.90	96.70	271.50	145.40	299.90	175.701
Sucrose	8.40	6.15	11.45	8.45	15.28	12.55	17.50	14.40
Urea								
Juice	155.80	85.50	205.90	108.80	285.70	158.70	325.60	193.80
Sucrose	8.85	6.72	11.55	8.80	15.42	12.44	17.80	14.90
Ammonium Sulphate								
Juice	150.80	78.90	201.50	106.90	280.50	155.50	318.60	187.70
Sucrose	8.77	6.70	11.52	8.77	15.35	12.37	17.55	14.58
Calcium Nitrate								
Juice	149.50	76.30	199.50	105.60	276.90	153.50	317.60	184.70
Sucrose	8.65	6.55	11.43	8.79	15.25	12.39	17.52	14.38

Value of 't' of between AMP infected and AMP free treatment (* significant $p > 0.05$)

Control: 2.636*; Urea: 2.636*;

Ammonium sulphate: 2.645*; Calcium nitrate 2.647*

Acknowledgement

The authors expresses deep sense of gratitude to Head, University Department of Botany, B.R. Ambedkar Bihar University, Muzaffarpur for providing laboratory facilities, to Director, SRI, Pusa, Samastipur, for sparing seeds Sugarcane.

References

Abbott, L.K., Robson, A.D. and Parker, C.A. (1979). Double symbiosis in legumes: the role of mycorrhizas. In: *Soil Microbiology and Plant Nutrition,* (Eds.) Droughton, W.S., Joh, C.K. Rujaroa, J.C. and Lim, B. University of Malaya Press, Kuala Lumpur, pp. 176–181.

Amijee, F., Stribley, D.P. and Tinker, P.B. (1986). The development of endomycorrhizal root system VI. The relationship between development of infection and intensity of infection in young Leck roots. *New Phytologist,* 102: 293–301.

Anderson, J.M. and Boardmann, N.K. (1964). Studies on greening of dark brown bean plants VI. Development of photochemical activity. *Aust. Journal Biological Science,* 17: 93–101.

Azcon-Anguilar, C. and Bago, B. (1994). Physiological characteristics of the host plant promoting an undisturbed functioning of the mycorrhizal symbiosis. In: *Impact of Arbuscular Mycorrhizas on Sustainable Agriculture and Natural Ecosystems,* (Eds.) Gioninazzi, S. and Schhejyo, H. Brikhauser Verlage, Basel, Switzerland, pp. 47–60.

Bethlenfalvay, G.J. and Schuepp, H. (1994). Arbuscular mycorrhizas and agrosystem stability. In: *Impact of Arbuscular Mycorrhizae on Sustainable Agriculture and Natural Ecosystems,* (Eds.) Gianinazzi, S. and Schuepp, H. Birkhauser Verlage, Basel, Switzerland, pp. 117–132.

Braunberger, P.G., Miller, M.H., and Peterson, R.L. (1991). Effect of phosphorus nutrition on morphological characteristics of varcicular arbuscular mycorrhizae colonization of maize. *New Phytologist,* 119: 107–113.

Black, C.A. (1968). *Soil Plant Relationship* (2nd Ed.). John Wiley and Sons Inc., pp. 517.

Bowen, G.D. (1978). Mycorrhizal role in tropical plants and ecosystem. In: *Tropical Mycorrhizal Research,* (Ed.) Micola, P. Oxford University Press, London, pp. 105–190.

Chandrakant, T.K., Moore B.B. and Hapase, D.G. (1989a). A note on incidence of vesicular arbuscular mycorrhizae in sugarcane (*Saccharum officinarum* L.). Department of Science and Technology Assistance, Part 2: 137–139.

Chandrakant, T.K., Zende, N.A. and Moore, B.B. (1989b). Vesicular arbuseular mycorrhizae association of sugarcane in relation to soil properties. Department of Science and Technology Assistance, Part 1: 130–134.

Daniels, A.B. and Skipper, H.D. (1982). Methods for the recovery and quantitative estimation of propagules from soil. In: *Methods and Principals of Mycorrhizae Research,* (Ed.) Schenck, N.C. APS Press, St. Pane, Minnesota, pp. 25–36.

Gerdmann, J.W. and Nicolson, T.H. (1963). Spore of mycorrhizae *Endogone* species extracted from soil by wet sieving and decanting. *Tras. Brit. Mycol. Soc.,* 46: 235–224.

Giovannetti, M. and Mosse, B. (1980). An evaluation of technique for measuring vesicular arbuscular infection in roots. *New Phytologist,* 84: 489–500.

Harley, J.L. and Smith, S.E. (1983). *Mycorrhizae Symbiosis.* Academic Press, London, 483 pp.

Jackson. M.L. (1973). *Soil Chemical Analysis.* Prentice Hall of India, New Delhi.

Kanwar, R.S. and Kochar, A.S. (1960). Effect of NPK fertilizers on the mineral uptake and juice quality of sugarcane. In: *Proc. Conf. Sugarcane Res. Workers,* (India), 4: 74–82.

Manjunath, A. and Bogyaraj, D.J. (1982). Vesicular arbuscular mycorrhizae in three plantion crops and cultivars of field bean. *Current Science,* 51: 707–708.

Mengel, K. and Krikby, E.A. (1982). *Principles of Plant Nutrition.* Internal Potash, Inst., Switzerland.

Menge J.A., Steirle, D., Bagyaraj, D.J., Jhonson, A. and Leonard, R.T. (1978). Phosphorus concentration in plants responsible for inhibition of mycorrhizae infection. *New Phytolosist,* 80: 575–578.

Mirsa, G.M., Kapoor, P.C. and Jouhan, S.C. (1964). Effect of NPK on yield and quality of Gur. *Indian Sugar,* 9: 32–39.

Mosse, B. (1975). Specificity in VA mycorrhizae. In: *Endomycorrhizae,* (Eds.) Sanders, F.E., Mosse, B. and Tinker, P.B. Academic Press, London, pp. 468–484.

Morton, J.B. and Benny, B.L. 1990. Revised classification of arbuseular mycorrhizae fungi (Zygomycates). A new order Olomales, two new suborders, Olomineas, Oigasporineal and two new families, Acaulosporaceae and Gigasporaceae with and emendation of Olomaceae. *Mycotaxon,* 37: 471–491.

Mosse, B. (1977). The role of mycorrhizae in legume nutrition on marginal soils. In: *Exploiting Legume Rhizobium Symbiosis in Tropical Agriculture,* (Eds.) Vincent, J.N., Whitney, A.S. and Bose, J. University of Hawaii Publishers, Hawaii, pp. 275–292.

Naidu, V.C., Murthy, S.S., Janardhan, P.R. and Reddy, P.M. (1982). Effect of NPK on the yield and quality of sugarcane varieties Co. 419. *Maharashtra Sugar,* 7: 51–56.

Phillips, J.M. and Hayman, D.S. (1970). Improved procedures for clearing roots and staining vesicular arbuscular mycorrhizal fungi for rapid assessment of infection. *Trans. Brit. Mycol. Soc.,* 55: 158–161.

Prasad, K. (1993). Ecological factor affecting vesicular arbuscular mycorrhizal infection in sugarcane. *Ph.D. Thesis.* B.R.A. Bihar University, Muzaffarpur.

Prasad, K. and Bilgrami, R.S. (2001). Impact of *Glomus fasciculatum* (VAM) and phosphates on biomass yield and chlorophyll contents of *Saccharum officinarum* L. In: *Bioinoculants for Sustainable Agriculture and Forestry,* (Eds.) Reddy, S.M., Reddy, S.R., Singarachary, M.A. and Girisham, S. Scientific Publishers (India), Jodhpur, India, pp. 53–60.

Read, D.J., Koucheki, H.K. and Hodgson, J. (1986). Vesicular arbuscular mycorrhizae in natural vegetation systems. I. The occurrence of infection. *New Phytologist,* 77: 664–683.

Reddy, N.C. and Singh, R.S. (1990). Prevalence of native VAM fungi and their performance in infectivity of local crops. In: *Current Trends in Mycorrhizae Research,* (Eds.) Jalali, B.L. and Chand, H. Haryana Agricultural University, Hisar, India, pp. 103–104.

Reddy, C.N., Bharati, B.K., Raj Kumar, H.G. and Sunanda, D.N. (2004). Infectivity and efficacy of four native vesicular arbuscular mycorrhizae fungi on Sugarcane (CV. Co-419). *Mycorrhizae News,* 16(2): 9–12.

Schenck, N.C. and Perez, Y. (1990). *Manual for the Identification of VA mycorrhizal-fungi,* 3rd edition. Synergistic Publications, Gainesville, USA.

Shivapulra, S.S., Patil, C.P., Swamy, G.S.K. and Patil, P.B. (2004). Cumulative effect of VAM fungi and vermicompost on nitrogen, phosphorus, potassium and chlorophyll content of Papaya leaves. *Mycorrhizae News,* 6(2): 15–16.

Smith, F.A. and Smith, S.E. (1981). Mycorrhizal infection and growth of *Trifolium subterraneum* use of sterilized soil as a control treatment. *New Phytologist,* 88: 299–309.

Spencer, G.L. and Meade, G.P. (1955). *Cane Sugar Handbook.* John Wiley and Sons, London.

Stahl, P.D., Williams, S.E. and Christian, M. (1988). Efficacy of native vesicular arbuscular mycorrhizal fungi after severe soil disturbance. *New Phytologist,* 114: 1–38.

Sunil Kumar and Lakhanpal, T.N. (1969). Physical and chemical status of mycorrhizae and non-mycorrhizae plants in *Picea smithians.* In: *Mycorrhizae for Green Asia,* (Eds.) Mahadevan, A., Roman, N. and Natarajan, K. Alamu Publication, Chennai, India, pp. 175–178.

©R.C. Ram & Asha Sinha (*Editors*)
***Microbial Diversity: Modern Trends*, 2007** **234–263**
Daya Publishing House, New Delhi

Chapter 18

Morphology and Pathogenicity of Isolates of *Fusarium moniliforme* Causing Pokkah Boeng Disease of Sugarcane in Maharashtra

A.S.Patil*, Hema Singh*, S.R. Sharma* and G.P. Rao**

***Department of Plant Virology, Vasantdada Sugar Institute, P.O. Majari, Pune – 412 307, Maharashtra**

****Division of Plant Pathology, Sugarcane Research Station, Kunraghat, Gorakhpur – 273 008, Uttar Pradesh**

ABSTRACT

Sugarcane, a cash crop, is cultivated in almost all states of India. It is playing a major role in the economy of Maharashtra. Among the different reported diseases on sugarcane in Maharashtra, the Pokkah boeng is very important. Initially the incidence of the disease was considered minor but later it has been found increasing in different parts of Maharashtra states and now-a days, it is cause of major concern. The Pokkah boeng disease of sugarcane was reported first by Patil and Hapase(1987) in Maharashtra. In the present chapter, we have discussed the economic importance, stages of infection and symptomatology of Pokkah boeng disease. The comparison or morphological. physiological and pathogenicity of different prominent isolates of *Fusarium moniliforme* associated with Pokkah boeng disease collected from various places in Maharashtra is discussed.

Corresponding Author: E-mail: gprao_gor@rediffmail.com

Introduction

Pokkah boeng, Javanes term denoting malformed distorted top, was originally described in 1986 by Wakker and Went in Java, however, no causal organism was established. The disease, occurred on the commercial varieties of sugarcane namely Loethers. Striped Preanger and Black Cheribon, as well as on varieties of lesser importance. Bolle (1927) was the first to demonstrate. by repeated isolation and inoculation studies, that the disease is caused by the fungus *Fusarium moniliforme* Sheldon.

Distribution

Pokkah boeng became a major sugarcane disease in Java with the spread of commercial variety POJ 2878 and was extensively investigated in that country by Bolle (1927, 1930, 1936) and by Dillewijn (1948, 1950). The occurrence of Pokkah boeng disease has been recorded in almost all countries where sugarcane is grown commercially (Whittle and Irawan, 2000). This disease was first reported in India during 1983-84, season in Maharashtra (Patil and Hapase, 1987). Now it is occurring throughout India wherever sugarcane is cultivated.

Economic Importance

The greatest losses from Pokkah boeng have occurred in Java where susceptible varieties have been grown in climate in which the hot and dry season is followed by a wet season. Under these conditions leaf infection develops rapidly and even resistant varieties may at times show typical leaf symptoms. Counts on POJ 2878 showed as many as 10.6 to 38 per cent of all stalks had dried because of the disease (Martin *et al.*, 1961). Small outbreaks of this disease have been reported but this has been restricted to seedlings.

Causal Organism

Fusarium moniliforme J.L. Sheldon was identified as causal organism or the Pokkah boeng disease in India (Patil and Hapase, 1987; Patil and Jadhav, 1995).

In the present investigation symptoms, morphology of pathogen, pahtogenicity of isolate, variability in cultural behavior, physiological studies (Nutritional), biochemical studies of Pokkah boeng disease pathogen *Fusarium moniliforme* were studied and the results are discussed below.

Symptoms

The symptoms of Pokkah boeng disease of sugarcane observed in the state of Maharashtra is similar to those described by earlier workers Bolle, 1927; Priode, 1929; North, 1932; Martin *et al.*, 1961; Reimers, 1971. The earliest symptom of Pokkah boeng is a chlorotic condition towards the base of the young leaves and occasionally on other parts of the leaf blades.

In general, features of Pokkah boeng are the development of a chlorotic condition towards the base of young leaves accompanied with a distortion and shortening of the affected leaves, a distortion of the stalk with external and internal lesions and in acute cases, the death of the stalk observed (top-rot).Under field conditions, the disease may develop many variations from the general symptoms, but the final, result is usually a malformed or damaged top and stalk (Martin *et al.*, 1961). The severity of symptoms varies with the susceptibility of the variety *e.g.* Co 7219, Co671, and Co 80 14. The general symptoms of Pokkah boeng disease were mainly of two types; chlorotic phase and acute phase or top rot phase. However, recently, it was also observed that the top rot phase (Acute phase) was always associated with knife cut disease symptoms. Hence, on the basis of symptomology, causal organism and pahtogenicity, the knife cut disease is regarded as an advanced further stage of Acute Phase.

Chlorotic Phase

The earliest symptom of Pokkah boeng was a chlorotic condition towards the base of the young leaves and occasionally on the other parts of the leaf blades. Frequently the malformation or distortion of the young leaves was accompanied by a pronounced wrinkling, twisting and shortening of the leaves (Plates 18.1 & 18.2). The base of affected leaves was seen often narrower than that of normal leaves. In the affected mature leaves irregular reddish stripes and specks were observed within chlorotic parts and these were also seen in green portions of the lower sections of the leaf blade. The reddish area sometimes developed into lens or rhomboid shaped holes, which showed no definite arrangement, or formed ladder like lesions, with dark edges in longitudinal rows. The tissues along the edges and the tips of the leaves developed irregular areas, which turned dark reddish brown in colour producing a burned appearance. In some cases, young leaves at spindle did not unroll normally; a crinkling the leaves was observed. Leaf sheaths were also observed with chlorotic conditions in some cases. Later irregular narcotic areas of a reddish colour similar to leaf blades were observed on the leaf sheath. Ladder like lesions were also found on leaf sheaths and midribs. The severity of symptoms varies with the susceptibility of the variety and with existing environmental conditions governing the development of the causal organism.

Acute Phase

The most advance and serious stage of Pokkah boeng is top-rot; the young spindle is killed and the entire top of the plant dies (Plate 18.3). Leaf infection sometimes continues downward and penetrates the stalk by way of growing point. In advance stage of infection, the entire base of the spindle and even growing point showed malformation of leaves, pronounced wrinkling and twisting and rotting of spindle leaves. Red specks and stripes also appeared and the whole base of the spindle gets rotted and dried up very fast and finally it formed a top-rot of tender tissues of the apical part of the cane. Sprouting of the buds (Plate 18.3) was also observed in severe infection. In such cases, the apical part of the stalks was seriously damaged. Occasionally symptoms of knife cut were observed in the acute phase of disease characterized by one or more narrow transverse cuts in the rind of the stalk in such a uniform manner as if the tissue was removed with a sharp knife (Plate 18.3). This is an exaggerated stage of typical ladder lesion of Pokkah boeng disease.

Knife Cut Phase

Knife cut disease of sugarcane was observed in association with Pokkah boeng disease on cultivars Co 7219, CoC 671 and Co 8014. The general symptoms of Pokkah boeng are the development of chlorotic phase and acute phase (Top rot or PB3 stage). However, under field conditions the disease develops many variations in symptomology. In advance stage, the infection in spindle continues downwards and penetrates the stalk through growing point causing top rot infection. Frequently, the acute phase (top rot) is accompanied by a pronounced wrinkling, twisting, sprouting of buds, shortening of internodes and ladder like lesions. These symptoms, were noticed on cane tops and stalk during field survey (Plate 18.4).

Previously the acute phase was considered to be the most advanced and serious stage of Pokkah boeng disease. However, critical observations on symptomology revealed that the acute phase (top rot) of Pokkah boeng is always associated with knife cut infection. Such canes exhibited staggered appearance. On stripping off the leaves, large horizontal conspicuous angular cuts were noticed which were similar to those of knife cut and ease. In general, the infection occurs uniformly on the cane, which are apparently responsible for the spread of disease.

Plate 18.1: Leaf Symptoms of Pokkah Boeng, Chlorosis with or without Red Markings and with or without Wrinkling and Splitting of Leaves

Plate 18.2: Ladder Shaped Lesions on the Spindle Leaves and Twisting Appearance of Spindle

Plate 18.3: Top Rot Phase of Pokkah Boeng, Rotting of Spindle Leaves, Pronounced Red Strips with White Spore Mass

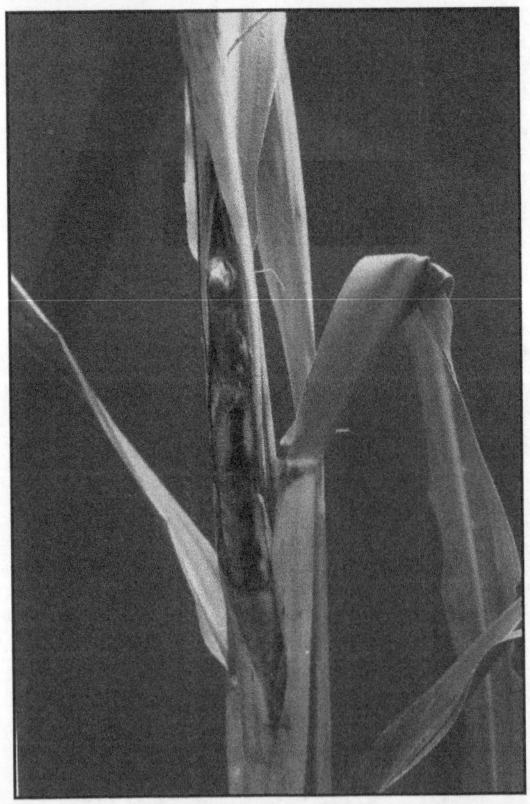

Plate 18.4: Knife Cut Phase of Pokkah Boeng, Conspicuous Angular Cuts Mainly Observed on Stalks and Rarely on Leaf Sheath, Exhibited a Staggered Appearance

The pathogens were isolated on PDA from the Pokkah boeng and Knife cut infected lesions separately and identified as *Fusarium moniliforme* Sheldon. Pahtogenicity and cross inoculation studies proved that the same organism was responsible for causing Pokkah boeng and knife cut diseases under controlled conditions.

On the basis of symptomology, causal organism and the pathogenic behavior, it is surprised that the knife cut disease is not a separate disease, however, it may be regarded as an advanced stage of Pokkah boeng disease with which it is always associated (PB4 state).

On the basis of symptomology of Pokkah boeng disease, four stages of the infection were identified:

PB1–only leaf symptoms, chlorosis with or without red markings (stripes) and with or without wrinkling and splitting of the leaves. Chlorotic Phase I (Plate 18.1).

PB2–Ladder shaped lesions on the spindle leaves pronounced yellowing, wrinkling of spindle twisting or tangling appearance of spindle and marked red stripes Chlorotic Phase II (Plate 18.2).

PB3–Top rot by Pokkah boeng, rotting of spindle leaves, pronounced red strips with white spore mass, necrosis sprouting of buds. Acute Phase (Plate 18.3).

PB4–Infection near the top of the cane. Large, horizontal, conspicuous angular cuts as if made with knife, mainly observed on stalks and rarely on leaf sheath, exhibited a staggered appearance. Knife cut disease stage (Plate 18.4).

The similar finding about the disease symptoms were reported by Dillewijn and Han Lioe Hong (1956). Eira *et al.* (1973) and Highy *et al.* (1977) also described the same symptoms for the *eti'* disease ranging in three stages. Aggressive strains resulted in the development of PB3 phase while non-aggressive produced PB I phase. Martin *et al.* (1961) reported the exaggeration of PB3 " stage in the form of knife cut symptoms. Verma *et al.* (1984) reported knife cut disease on sugarcane and could not notice any association with Pokkah boeng.

Other symptoms included rotting of the stalk and knife cut lesions on the stalk. More recently wilt symptoms are found to be associated with *Fusarium moniliforme* (Waraitch, 1982). Rotting of the stalk and the Pokkah boeng sometimes treated separately in the literature are the part of the same *Fusarium* complex in sugarcane since mycologists and pathologists have been unable to distinguish the causal agents of the two diseases.

The fact that the *Fusarium moniliforme* may also cause wilt suggests that variability of the organism may influence the symptoms produced, as stated by Bourne (1961).

Isolates

During the survey in 1991 and 1992, samples of naturally infected leaves of sugarcane varieties from various locations of nine agroclimatic zones of Maharashtra were collected and preserved under laboratory conditions for further work.

Isolations were made from such diseased samples on potato dextrose agar (PDA) with usual tissue isolation technique. Monosporic pure cultures were maintained on PDA and Czapek's Dox agar media. It was observed that there were wide variations in symptomology, morphology and growth characters of different isolates in various regions of Maharashtra. Hence, it is decided to collect all the diseased samples of different varieties of various agroclimatic zones for further work. It is interesting to note that so for 32 PB isolates have been collected and classified under seven groups, 30 to 32 isolates of *Fusarium* (PB) obtained from various sugarcane varieties *viz.*, Co 7219, CoC 671, Co 8014, Co 85246 and Co 7527 collected from different agroclimatic zones of Maharashtra.

Identifications of these isolates of *Fusariunl* have been done on the basis of morphological and taxonomically characters and the same were classified into seven groups.

Out of 62, only 32 PB isolates of *Fusarium sp.* maintained after their scrutiny and the same were grouped under seven main heads (Group): FMI, FM 2, FM 3, FM4, FM5, FM6 and FM7 (Table 18.1).

Table 18.1: Morphological Grouping of 32-isolates and their Classification

Group I COC 671 FM 1	Group II CO 7219 FM 2	Group III CO 8014 FM 3	Group IV CO 85246 FM 4	Group V CO 7527 FM 5	Group VI CO 8021 FM 6	Group VII CO 1340 FM 7
PB-1	PB-35	PB-33	PB-14	PB-40	PB-32	PB-41
PB-7	PB-38	PB-34	PB-38	PB-50	–	–
PB-12	PB-13	PB-36	PB-39	–	–	–
PB-18	PB-17	PB-37	–	–	–	–
PB-31	PB-19	PB-42	–	–	–	–
PB-23	PB-43	PB-46	–	–	–	–
PB-45	PB-44	PB-48	–	–	–	–
PB-47	PB-29	PB-49	–	–	–	–
–	PB-30	–	–	–	–	–
Total 8	9	8	3	2	1	1

The detailed morphological studies of these seven groups (FM) are presented in Table 18.2. After scrutiny, only three distinct groups (FM 1, FM 2 and FMJ) were selected for further identity and confirmation. Following three isolates were investigated in detail in our study.

Table 18.2: Comparative Study of Morphology and Taxonomy of *Fusarium moniliforme*

Isolate	Microconidia	Macroconidia	Colour of Pigments	Growth
Fusarium monilliforme	6–10 μm (length)	24–40 μm length)	–	–
F. moniliforme	5–12 × 1.5–2.5 μm	25–36 × 2.5–3.5 μm, 3–4 septa	White to greyish violet or greenish magenta (old)	–
F. moniliforme var. intermedium	5–12 × 1.5–2.5 μm	25–27 × 3–4 μm	Tomentress or white or rarely palc white	
Group of PB Isolates				
F1 Group	7.14 × 1.15–3.5 μm (10.40 × 3.01) μm	8.75–17.5 × 1.75–3.5 μm	Rosy buff colour	Fluccose myceliun thick circular growth
F2 Group	7–10.5 × 0.87–3.5 μm (7.94 × 3.01 μm)	8.75–17.5 × 1.75–3 5 μm (11.47 × 1.84 m)	Rosy buff colour	Thick circular growth
F3 Group	7–17.5 × 1.75–4.37 μm	10.5–24.5 × 1.75–4.37 μm	White pale	Scanty submerged growth
F4 Group	7–10.5 × 1.75–3.5 μm	11.37–19.25 × 3.5–4.37 μm	Rosy pink	Profuse circular growth
F5 Group	3.5–13.13 × 1.75–3.5 μm (7.55 × 2.84 μm)	7–14.81 × 2.63–3.5 μm (10.06 × 3.39 μm)	Pale whitish colour	Profused growth
F6 Group	7.14 × 2.63–3.5 μm (8.82 × 5.18 μm)	10.50–18.37 × 3.5–4.37 μm (13.56 × 3.72 μm)	Rossy buff	Profused growth

Isolate FM 1–Isolated from CoC 671 sugarcane variety of early group.

Isolate FM 2–Isolated from Co 7219 of mid-late group.

Isolate FM 3–Isolated from Co 8014 of mid-late group.

For detailed studies like symptomology, taxonomy etc., only 3 varieties *viz.*, COC 671, Co 7219 and Co 8014 were selected since they were relatively more susceptible and distinctive in nature. The variety wise symptoms of the disease are as under.

CoC 671–The infection to PB2 and PB3 stage; in rare cases knife cut phase (PB4) was observed.

CoC 7219–Mostly PB2 and PB3 stages were observed. Wrinkling and shriveling of leaves and yellowing were pronounced in the chlorotic phase with ladder type symptoms.

Co 8014–All four stages PB1, PB2, PB3 and PB4 were observed. This variety was found highly susceptible to this disease with top rot lesions and knife cut symptoms.

Morphology of the Pathogen

Three isolates of Pokkah boeng were selected from different localities and the causal organism *i.e. Fusarium moniliforme* was studied for morphology characteristics. These isolates were selected from Variety CoC 671 (FMI); Co 7219 (FM 2); Co 8014 (FM 3). The fungus was isolated and grown on P.D.A. petriplates. The morphological characteristics of the three isolates *viz.*, FMI, FM 2, FM 3 of *Fusarium* sp. are presented in Tables 18.3 and 18.4.

Table 18.3: Measurement of Mycelium, Conidiophore and Conidia in Micron

	Observations of 8 Microscopic Fields									
	1	2	3	4	5	6	7	8	Mean	Range
FM 1–COC 671										
Mycelium (Width)	3.50	3.50	2.63	1.75	2.63	1.75	1.75	1.75	2.41	1.75–3.50
Conidiophore (Length)	49.00	57.75	38.50	52.50	66.50	45.50	42.00	42.00	49.22	38.5–66.5
Microconidia (Length)	7.00	7.00	7.00	8.75	7.00	7.00	8.75	8.75	7.66	7.00–8.75)
Microconidia (Breadth)	0.88	1.75	1.75	1.75	1.75	1.75	1.75	1.75	1.64	0.88–1.75
Macroconidia (Length)	14.00	10.50	14.00	10.50	14.00	17.50	8.75	14.00	I2.91	8.75–17.50
Macroconidia (Breadth)	1.75	1.75	3.50	1.75	1.75	2.63	1.75	1.75	2.08	1.75–3.50
FM 2–CO 7219										
Mycelium (Width)	3.50	4.38	4.38	3.50	5.25	7.00	6.13	3.50	4.70	3.50–7.00
Conidiophore (Length)	77.00	87.50	80.50	77.00	24.50	42.00	70.00	35.00	61.69	24.5–87.5
Microconidia (Length)	8.75	10.50	7.00	10.50	7.88	10.50	14.00	10.50	9.95	7.00–14.00
Microconidia (Breadth)	3.50	2.63	2.63	3.50	1.75	3.50	2.63	3.50	2.95	1.75–3.50
Macroconidia (Length)	12.25	10.50	17.50	10.50	14.00	to.50	12.25	8.75	I2.03	8.75–17.50
Macroconidia (Breadth)	1.75	1.75	2.63	2.63	1.75	2.10	2.45	1.75	2.10	1.75–2.63
FM 3–CO 8014										
Mycelium (Width)	4.38	3.50	2.63	3.50	4.38	2.10	2.63	3.50	3.33	2.10–3.50
Conidiophore (Length)	70.00	63.00	42.00	35.00	38.50	42.00	21.00	63.00	46.81	21.0–70.0
Microconidia (Length)	8.75	10.50	7.00	11.38	14.00	10.50	17.50	10.50	11.27	7.00–17.50
Microconidia (Breadth)	3.50	1.75	2.63	3.50	3.50	1.75	3.50	4.38	3.06	1.75–4.37
Macroconidia (Length)	17.50	14.00	14.00	14.88	10.50	14.88	21.88	10.50	14.77	10.50–21.88
Macroconidia (Breadth)	3.50	3.50	3.50	4.38	3.50	3.50	3.50	1.75	3.39	1.75–4.38

Table 18.4: Comparative Morphological Observations of Three Isolates of *Fusarium moniliforme* (FM 1, FM 2, FM 3)

Sl.No.	Characters	Isolate No. 1 FM 1	Isolate No. 2 FM 2	Isolate No. 3 FM 3
1.	Morphology of Mycelium	Aerial mycelium, Pinkish-white branched Hyline to greenish septate, sometimes very distinct septa	Aerial mycelium, rosy pink pigmentation, hyphae hyline, septate branched, septa not very clear	Submerged slightly pale, whitish pigmentation, hyline hyphae, branched septate septa not very distinct
2.	Width of mycelium (μm)	1.75–3.50 mean 2.40	3.5–7.00 mean 4.70	2.1–4.37 mean 3.32
3.	Growth on PDA in mm	90	90	85
4.	Morphology of conidiophore	Borne laterally, slightly swollen, cells unbranched, conidia are borne on monophialides or polyphialides, sometimes on false heads	Lateral in position, conidiophore in the form of conidiogenous cells, many conidia on false heads on polyphialides	Conidiophores in the form of swollen conidiogenous cells, conidia on false heads on polyphialides
5.	Size of conidiophore (length in μm)	38.5–66.5 mean 49.21	42.5–87.5 mean 57.31	21.0–70.0 mean 42.43
6.	Morphology of microconidia	Single celled Microconidia in the form of long chains, borne on conidiogmous cells or phialides, hyline curved, very few are straight	Single cell microconidia, unseptate, hyline, slightly curved, borne on conidiogenous cells	Single cell, unseptate, hyline in the form of long chains, slightly curved or straight, more microconidia comparision to macroconidia
7.	Size of microconidia (length × breadth) μm	7.00–8.75 × 0.87–1.75 7.94 × 1.55	7.00–14.0 × 1.75 –3.50 9.95 × 2.73	0.7–17.5 × 1.75–4.37 11.26 × 3.06
8.	Morphology of macroconidia	Borne on conidiophores, septate, 3–5 septa hyaline, falcate to slightly broad at the base, globular cell at the end	Borne on conidiogenous cells, or false heads on polyphialides, hyaline, slightly tapering at one end, globular cell at the other end, 2 to 7 septa	Very long hyaline, curved, tapering at one end, globular cell at other end, 3 to 7 septate borne on false heads or polyphialides
9.	Size of macroconidia (in μm)	8.75–17.5 × 1.75–3.50 11.47 × 1.84	8.75–17.50 × 1.75–2.62 12.03 × 2.09	10.5–21.87 × 1.75–4.37 15.84 × 3.50

External Morphological Characters

The growth on potato dextrose agar was usually completed in seven days. The colour of the colonies showed slight variation. For FM 1 and FM 2 isolate the colonies of pathogen showed typical rosy buff colour of *Fusarium* while FM 3 isolate showed whitish pale colour of colony. FM 1 and FM 2 isolate showed sub aerial to aerial mycelium with very thick floccose growth. While FM 3 isolate the mycelium was aerial for 3–4 days but later it got submerged leaving a pressed surface on P.D.A. The mycelium is scanty and in the form of fine thread like hyphae. The colonies showed circular growth with slightly wavy margin. The growth on P.D.A. was very luxuriant. FM 1 and FM 2 isolate showed max. 90 mm. growths on P.D.A. while FM 1 showed slightly less than FM 1 and FM 2 *i.e.* 85 mm (Table 18.4). In all the three isolates good sporulation was observed. Pigmentation started in the culture after

10 days. FM land FM 2 showed purple pigmentation after maturity, while FM 3 showed pale whitish pigments in the petriplates.

Microscopic Examinations

Microscopic observation revealed that mycelium was hyaline, slender richly branched, septate when young and not varying abruptly in diameter, though swollen segments were found in old cultures. Sometimes the individual filaments united to form corenial strands, usually only two or three hyphae take part in formation. The width of mycelium ranged in. between 1.75–7.00 µm for all the three isolates.

In the present study morphology of the imperfect stage of fungus *Fusarium moniliforme* was observed with the help of seven days old culture and found that the fungus produced two types of spores *i.e.* microconidia and macroconidia.

In this fungus great variability in spore size and septation was seen depending on the medium used, age of culture, host from which obtained.

Microconidia

The microconidia are produced on simple lateral conidiogenous cells for conidiophores. They are simple or branched. Unbranched or loosely branched conidiophores terminate in to conidiogenous cells. Which may be monophialides or polyphialides. Production of polyphialides is a reflection of a tendency toward sympodial proliferation of the conidiogenous cells. Microconidia are non-septate clavate with a flattened base measured 7.94–11.26 × 1.55–3.06 µm and found in chains (Table 18.4).

Macroconidia

Pionnate sporodochia purplish red or brown on the surface were seen with an reddish brown or purplish red on the reverse of the petriplates. Macroconidia variable in size were found. Shape and septation also showed lot of variation but mostly they were relatively straight and slender with thin walls. Septa sometimes visible but usually very indistinct. All the isolates produced straight to falcate macroconidia, with pedicillate base and somewhat curved constricted tip. There were three to nine but mostly three to five septate 11.47–15.84 × 1.84–3.50 µm. Comparative morphological characters for three isolates are presented in Table 18.4.

Prasad and Padwick (1939) examined the morphology of the fungus and observed spores with one, two and three septa and size of spores ranged 6–14 × 3.5 µm, 14–21 × 2–4 µm and 19–28 × 2–5 µm respectively.

Snyder and Hansen (1940) described the morphology of the fungus *Fusariun oxysporum* which showed aerial mycelium, white purple or violet in colour. However, they have not studied various aspects of mycelium conidial and chlamydospores measurement. Chattopadhyay and Gupta (1967) observed spore size on PDA and reported 9.3–29.7 µm in length and 2.7–6.0 µm in breadth.

In the present study the mycellium of *Fusariun, moniliforme* was generally dense and delicately floccose to felted with a powdery appearance due to formation of Microconidia. Similar description is reported by Booth (1971). Microconidia were formed in chains and measure 5–12 × 1.5–2.5 µm fusiform to clavate and occasionally become one septate. Such types of observations were recorded in Isolate FM 1 and FM 2.

Macroconidia rarely formed if present, borne on lateral conidiophores and measure 20–24 × 3.5–4 µm. Chlamydospores were absent in both mycelium and conidia.

Neish and Leggett (1981) reported Microconidia size 5–12 × 2–3.5 µm and formed in chains and Macroconidia (3–9 septate) 25–75 × 3–4 µm capable to produce polyphialides (Stable character to distinguish *Fusarium moniliforme* var. intermedium. The present study also revealed the same observation about the morphology of Isolate FM 3 (Co 8014) confirming the fungus as *Fusarium moniliforme* var. *intrermedium* while two isolates *i.e.* FM 1 (CoC 671) and FM 2 (Co 7219) were also confirmed as *Fusarium moniliforme* sheldon. Nelson *et al.* (1983, 1990) and Burgess *et al.* (1988) Tiedt and Joost (1988a) and Liddell (1991) also have described morphological characters (ultra structure) of the genus *Fusarium.*

On the basis of morphology and taxonomy of 3 isolates it is clearly revealed that FM 1 (CoC 671) and FM 2 (CoC 7219) isolates were similar to each other while FM 3 isolate (Co 8014) was distinctly separated from the above isolates in their behaviour.

Hence, it is assumed that there are two types of pathogens, *viz.*, one group comprises FM 1 and FM 2 and the other group, FM 3 which are responsible for causing Pokkah boeng disease in Sugarcane in India (Maharashtra). Therefore, the detailed studies of these 3 isolates have been undertaken for further confirmation in the present investigation.

Mode of Entry of Pathogen

Electron microscopy of infected leaves revealed that after lodging of conidia and incubation period of minimum one month at the time of germination, the walled bulliform cells of the epidermis are attacked and soon collapse. Then the older cells of the epidermis are attacked. From the epidermal cells, the hyphae enter the underlying tissues. Change in the structure of stomata was also observed in infected leaves samples. There was no evidence found for the entry of pathogen through stomata.

Pathogenicity of Isolates

The pahtogenicity of *Fusarium* species isolated from diseased leaves of sugarcane varieties *viz.*, Coc 671, Co 7219 and Co 8014 successfully tested on healthy leaves of vigorously growing plants in field by following methods:

1. Foliage inoculation by spraying the spore suspension.
2. Inoculation by hypodermal syringe on standing cane.
3. Inoculation by nail injury.
4. Inoculation by crock- borer on standing cane.

Comparative Virulence of Isolates Among Themselves

In these experiment, twelve sugarcane varieties were inoculated artificially by FM 1, FM 2 and FM 3 isolates under controlled condition at the four-month age crops. The development of infection of PB stage of Pokkah boeng on each varieties were recorded after 60 to 90 days of inoculation. The results are presented in Table 18.5. The FM 1 and FM 2 isolates had a wide range of infection of PB stage on each varieties except Co740 and CoM 7125 where the disease is totally failed to infect FM 1 and FM 2 isolates are seems to more virulent than FM 3 isolate which is more host specific in nature. It is clearly showed that FM 3 isolate is different from FM 1 and FM 2 isolate in their varietal reaction in nature. PB I stage of infection was only developed on Co 8021 and Corn 7714 whereas varieties Co 7201 and Co 85246 were found to be more susceptible to all the three isolates of Pokkah boeng disease (Table 18.6).

Table 18.5: Comparative Virulence of Three Isolates of Pokkah Boeng Disease

Isolates	Varieties		
	Co 7219	CoC 671	Co 8014
FM 1	+++	+++	+
FM 2	+++	+++	+
FM 3	+	+	+++

Note: FM 1: Isolated from COC 671; FM 2: Isolated from CO 7219; FM 3: Isolated from CO 8014; +: PB 1 stage of infection of Pokka boeng infection; ++: PB2 stage of infection; +++: PB3 stage of infection.

Table 18.6: Cross Inoculation Studies of Three Isolates of Pokkah Boeng Disease of Sugarcane

Sl.No.	Varieties	FM 1 Isolate (COC 671)	FM 2 Isolate (CO 7219)	FM 3 Isolate (CO 8014)
1.	COC 671	+++	+++	+
2.	CO 7219	+++	+++	+
3.	CO 8014	+	+	+++
4.	CO 7527	+++	+++	++
5.	CO 8021	0	+	+
6.	CO 740	0	0	0
7.	COM 7125	0	0	0
8.	CO 7704	++	++	+
9.	CO 86032	+++	+	+
10.	CO 7201	+++	+++	++
11.	COM 7714	+	+	+
12	CO 85246	+++	+++	+++
13.	COM 7605	+++	+++	0
14	COM 7611	+	+	++
15.	CO 7634	++	+++	0

Note: +: PB 1 stage of infection chlorotic phase I; ++: PB 2 stage of infection chlorotic phase II; +++: PB 3 stage of infection acute or top rot phase.

Varietal Reaction

Screening of sugarcane varieties and promising genotypes against Pokkah boeng disease caused by *Fusarium moniliforme* were undertaken at VSI research farm under artificial and natural field conditions. Three budded ten sets of ten-month-old crop were taken for planting of each variety. Twenty-one varieties were planted at VSI research farm in 1991–92 seasons and inoculated after four months of planting by spraying as well as by hypodermal syringe inoculation technique (Han, 1956) of 8 days old culture of *Fusarium moniliforme* isolated from Co 7219. The observations were noted after 45 to 60 days after inoculation. The same experiment was also conducted in 1992–93 planting season, 24 cultivars were tested artificially against this disease. Results of these are presented in Table 18.7 and 18.8.

Table 18.7: Effect of pH on Growth and Sporulation of *Fusarium moniliform* (FM 1 Isolates)

Sl.No.	Initial pH	Average Weight of mycellium (mg)	Final Average pH	Sporulation
1.	3.00	380.00	6.43	+++
2.	4.00	400.00	5.30	+++
3.	4.50	410.00	5.76	+++
4.	5.00	420.00	6.08	+++
5.	5.40	433.00	7.10	+++
6.	6.00	440.00	5.60	+++
7.	6.50	480.00	6.92	+++
8.	7.00	460.00	6.38	+++
9.	8.00	440.00	7.22	+++
10.	9.00	390.00	7.59	+++
	SE	5.98		
	SED	8.46		
	CD 5 per cent	17.79		

++++: Excellent; +++: Good; ++: Moderate; +: Poor

Out of 21 cultivars typical leaf symptoms of Pokkah boeng were produced on 15 cultivars *viz.*, Co 8021, Co 8153, Co 8019, Co 7634, Co 7704, Co 7506, Co 62175, Co 775, CoM 7609, Co 419, CoM 7611, Co 7201, Co 671, Co 1527 and Co 85246. Whereas, six cultivars *viz.*, Co 740, CoM 7125, TS-1, TS-8, Co 7712 and CoM 7114 did not produce any reaction, Co 7219 was kept as standard. Thirty new genotypes under final varietal trial were planted in the field and were also inoculated artificially by culture (FM 2 Isolate from Co 7219) after four months in February- March 1992. However, the symptoms of Pokkah boeng were not reproduced probably due to high temperature and low humidity dry weather) at field level in the month of February–March. The results indicate that the pathogen could not develop in high temperature (above 35°C) and low humidity (below 70 per cent).

In 1992–93 seasons, twenty-four cultivars were screened artificially. Out of 24, seventeen varieties *viz.*, Co 7219 (standard), CoC 671, Co 7527, Co 8014, Co 7201, Co 8021, Co 8338, 80R41, Co 85246, Co 8341, Co 7620, Co 7708, CoM 7714, Co 8011, Co 87024, Co 658 and Co 419 showed symptoms of Pokkah boeng disease within 60 days after inoculation. However, Co 7219, CoC 671, Co 8014, and Co 85246 showed PB3 (Top rot) and PB4 (Knife cut disease) stage of infection. Cultivars *Viz.*, Co 8021, 80R41, Co 7620, CoM 7714, Co 87024, and Co 8011 produced only PBI stage (chlorotic phase 1) whereas seven cultivars, *viz.*, Co 740, CoM 7125, Co 85004, Co 85001, Co 8102, Co 85018, and Co 87044 did not produce any symptoms of Pokkah boeng disease of sugarcane.

Cultural Behaviour

Various synthetic and non-synthetic (vegetable) media were employed to study the cultural behavior of three isolates of *Fusarium* spp isolated from sugarcane. In all fourteen media were included in this study in which six were synthetic and rest were non-synthetic type. These media were prepared in distilled water using standard formulae for their various ingredients as described in material and methods. Three isolates of *Fusarium* species *i.e.*, FM 1 (CoC 671), FM 2 (Co 7219) and FM 3 (Co 8014) were used for this study. The fungus was grown in triplicate plates and cultures were inoculated at room temperature (25–28°C.) with diurnal light.

Table 18.8: Comparative Carbon Nutritional Requirement of *Fusarium moniliforme* (FM 1, FM 2, FM 3) in Liquid Medium

Sl.No.	Test Source	My. Dr wt (mg)	Spor.	pH	My. Dr wt (mg)	Spor.	pH	My. Dr wt (mg)	Spor.	pH	Av. My. Dr wt (mg)
(A)	**Monosaccharides**										
1.	Xylose	780	+++	7.67	890	++++	8.63	780	++++	6.60	817
2.	Fructose	870	++++	7.95	860	++++	8.28	800	++++	6.86	843
3.	Glucose	850	++++	8.16	810	++++	8.57	850	++++	6.86	837
4.	Galactose	800	++++	6.83	630	+++	6.79	620	++++	7.22	683
5	Mannose	350	++	5.95	420	+++	5.93	270	++++	5.61	347
6.	Arabinose	810	++++	7.79	710	++++	7.52	630	++++	7.65	712
7.	Dulcitol	530	+++	7.20	400	+++	7.76	420	++++	8.88	450
8.	Mesoinositol	850	++++	7.97	790	++++	7.81	640	++++	6.86	760
9.	Rhamnose	710	++++	7.92	550	++++	7.56	610	++++	7.27	623
10.	Ribose	460	+++	7.16	430	+++	6.56	500	++++	6.22	463
(B)	**Dissacharides**										
11.	Cellobiose	830	++++	8.41	780	++++	8.11	530	+++	5.63	713
12.	Maltose	760	++++	7.14	680	+++	7.10	620	+++	6.96	687
13.	Sucrose	860	++++	8.28	780	++++	8.77	730	++++	6.77	790
14.	Lactose	630	++++	7.83	720	+++	7.96	700	++++	6.59	683
15.	Trehalose	660	+++	8.31	590	++++	40	560	+++	6.45	603
16.	Melibiose	790	++++	7.62	870	++++	6.69	760	++++	7.11	807
(C)	**Polysaccharides**										
17.	Dextrine	730	+++	6.77	660	+++	6.56	780	+++	7.29	723
18.	Starch	880	++++	7.56	910	++++	7.81	730	++++	6.79	840
19.	Salicin	250	+++	6.36	200	+++	6.35	210	++	6.24	220
20.	Inulin	650	+++	7.15	620	+++	7.19	510	+++	6.72	593
(D)	**Oligosaccharides**										
21	Raffinose	340	++	7.88	400	++	8.61	360	++	6.36	367
(E)	**Sugar Alcohols**										
22.	Mannitol	720	++++	7.19	680	++++	8.27	520	++++	7.16	640
23.	Sorbitol	580	++++	8.32	440	++++	8.18	420	++++	7.16	480
	Control	63	+	5.89	53	–	5.70	50	–	5.76	55
	SE	7.28			6.98			8.48			
	SED	10.30			9.87			11.99			
	CD at 5%	20.72			19.8			24.10			

Note: Sporulation: ++++: Excellent; +++: Good; ++: Moderate; +: Poor.

Synthetic Media

All the isolates of *Fusarium* species shield did respond very well with all synthetic media. Excellent growth was seen in Czapeck Dox, Richard's solution and Asthana and Hawker's agar media. Profuse sporulation was seen in Czapeck dox and Richard's solution. FM 1 and FM 2 isolates showed less sporulation in nutrient agar inspite of well spread growth in terms of colony diameter than FM 3 isolates. Sabouraud's agar also showed less sporulation in comparison to Czapeck dox and Richard's solution agar and Asthana and Hawker's medium. Isolates FM 1 and FM 2 are similar to each other in their growth character while FM 3 showed distinct character in growth. There were also little morphological differences in each isolate (Figure 18.1).

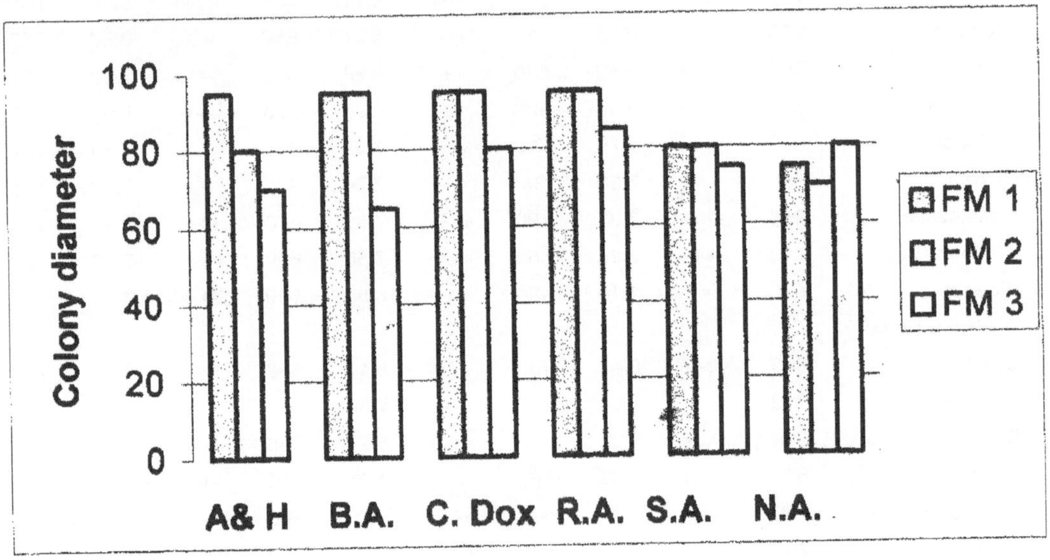

Figure 18.1: Cultural Behaviour on Synthetic Media

Non-synthetic Media

All the non-synthetic media showed good growth except Onion agar where sporulation was little less and slow. Cane juice agar supported the maximum growth. Profuse luxuriant growth with Arial cottony mycelium was observed in cane juice agar. Potato dextrose and Host decoction agar also shows good growth. Bean's agar and carrot agar also supported good growth but it was thin in compression to potato dextrose and host decoction agar. Aerial whitish pink thin hyphae were seen in bean's agar and carrot's media. There was no difference in colony characters in all the three Isolates (Figure 18.2).

The same results were observed by Shaikh (1974), who studied the different Media for the growth of *Fusarium*, reported that different synthetic and non-synthetic media had profused influence on the cultural characteristics. In the present study, FM 1 and FM 2 isolated have similarities to each other in cultural behavior as of *F. moniliforme* while FM 1 isolate showed the separate distinct character.

Yogeshwari (1948), Parsad and Mathur (1964), studied the growth of *Fusarium moniliforme* on different media and reported that fungus growth was best on Richard's medium, Jaurihar and Mehta (1976) also selected Richard's medium as base medium for *Fusarium moniliforme.* Agarwal and Singh (1974) reported modified Czapeck dox agar as highly selective medium for *Fusarium moniliforme.*

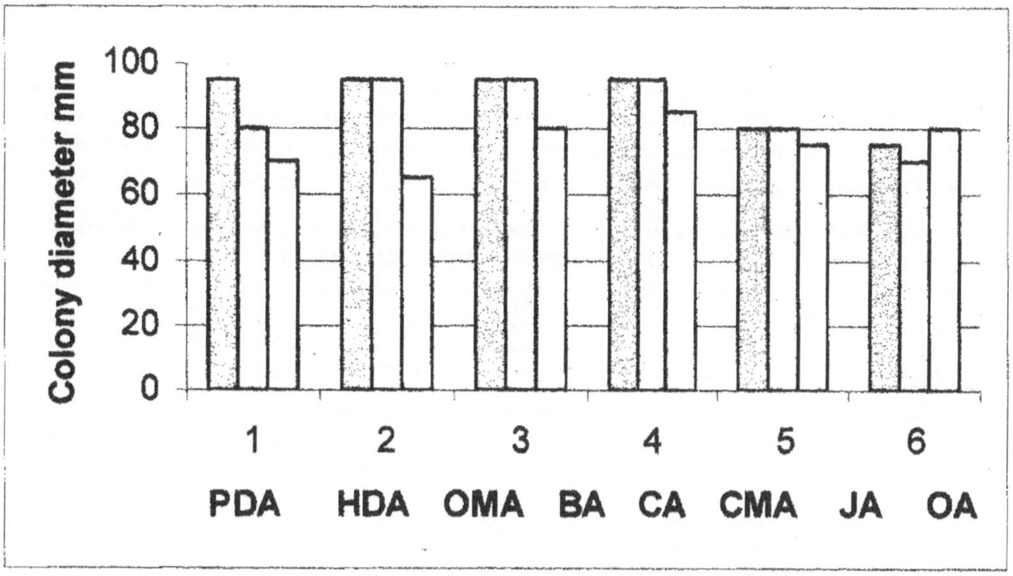

Figure 18.2: Cultural Behaviour on Non-synthetic Media

Abro and Khan (1982) reported light and gravity treatment for zonation in culture of *Fusarium moniliforme.* Mradela and Silva (1986) reported pigmentation in *Fusarium moniliforme* due to heterokaryosis. Patil and Hapase (1988) also suggested Richard's medium as best for the growth of *Fusarium moniliforme.* Sahab *et al.* (1987) reported that a virulent isolate of *Fusarium* species grew faster on PDA than Czapeck dox or Richard's agar medium with liquid or solid medium with greater spore production. Arshed *et al.* (1992) reported malt extract as best medium for the growth of *Fusarium moniliforme.*

Physiological Studies

The *in vitro* studies of the pathogen were carried out to the growth and sporulation of *Fusarium moniliforme,* the causal organism of Pokkah boeng disease. In the present study physiological aspects of the fungus in respect to environment and nutritional factors were studied. Among the environmental factor temperature and hydrogen ion concentration were studied. The growth factors with the respect to nutritional factors such as carbon, nitrogen, amino acids, and stimulatory factors like vitamins and trace elements were analysed to have better understanding on the physiology and disease development.

Detailed physiological investigations were carried out with three isolates (FM 1, FM 2, FM 3) of *Fusarium* spp. causing Pokkah boeng disease of sugar cane.

Environmental Factors

Effect of Temperature

Temperature is an important factor determining the geographical distribution of fungi. Temperature influences germination, growth, reproduction and all other vital activities of an organism. Among the environmental factors, temperature is the most important physical factor affecting the metabolic activity of fungi. There is a certain range of temperature within which a fungus can grow and this range of temperature varies with different fungi. The fungi also have cardinal temperature as in the case of their organisms. The temperature favorable for growth may not be for sporulation and vice-versa. Both these factors may have separate cardinal temperature. Bisby (1943) recognized temperature as an important natural factor governing the distribution of fungi.

Considering the importance of the effect of temperature on fungi the present experiment was carried out with three isolates of *Fusarium* spp. Petriplates containing equal amount of P.D.A. (15 ml) were inoculated in duplicate sets with the fungal culture and incubated at various temperatures. The radial measurements were recorded after seven days. The data is presented in Table 18.9.

Table 18.9: Effect of Temperature on Growth and Sporulation of Three Isolates of *Fusarium moniliforme* (FM 1, FM 2, FM 3)

Temp. Index	Isolate FM 1			Isolate FM 2			Isolate FM 3			Avg. Col. Diam. in mm
	Col. Dia. mm	Sporu-lation	Growth Characters	Col. Dia. mm	Sporu-lation	Growth Characters	Col. Dia. mm	Sporu-lation	Growth Character	
5°C	11	+	Aerial Pinkish White Mycelium, Thin growth	14	+	Thin growth, rose buff colour mycelium	9	+	Very faint Growth of mycelium	11
10°C	22	++	Aerial Whitish Pink Mycelium, Wavy margin, circular growth	18	++	Aerial Thin Growth, Rose buff coloured mycelium	14	++	Sub aerial White Mycelium, Wavy margin	18
15°C	43	++	Good growth aerial mycelium rose buff in colour	36	++	Good aerial Growth, rose buff coloured mycelium	24	++	Good growth Aerial White mycelium	34
20°C	53	+++	Good growth Circular, rose buff in colour, wavy outline	54	+++	Good growth subaerial rose coloured mycelium	55	+++	Thin aerial growth, white mycelium	54
28°C	72	++++	Excellent growth, submerged to subaerial rose buff mycelium good sporulation	83	++++	Excellent growth, aerial, cottony, rose buff coloured mycelium, good sporulation	68	++++	Good growth but submerged mycelium whitish pale in colour, good sporulation	74
32°C	40	++	Good growth circular, whitish pink mycelium, wavy margin	44	++	Good growth but submerged whitish mycelium, wavy margin, circular growth	39	++	Good growth, sub merged whitish mycelium wavy margin	41
37°C	10	+	Poor growth Very thin, less sporulation white	12	+	Poor growth, thin less sporulation, mycelium	11	+	Poor growth, very thin white mycelium	11
40°C	–	–	–	–	–	–	–	–	–	–

Sporulation: Less (+) < 10 spores/microscopic field; Moderate (++) 10–19 spores/microscopic field

Good (+++) 20–29 spores/ microscopic fields; Excellent (++++) > 30 spores/microscopic field.

The results of Table 18.9 revealed that three isolates FM 1, FM 2 and FM 3 Pokkah boeng disease of sugarcane were able to grow in a wide range of temperature 10–37°C. All the three isolates showed good growth and excellent sporulation between 20–28°C. The three isolates was found to be at 28°C (± 1°C). Poor growth and sporulation was observed at 5°C and 37°C whereas growth and sporulation was totally ceased below 5°C and at 40°C. On this basis all physiological experiments were carried out at room temperature *i.e.*, 25–28°C.

Clayton (1919) reported that several species of *Fusarium* grew and sporulate in temperature range of 25–30°C. Voorhees (1933) reported that minimum, optimum and maximum temperature for mycelia growth of fungus *Gibberella moniliforme* were 10–14°C, 30°, 35–39°C. Snyder and Hansen (1940) reported that *Fusarium* fungus had an optimum temperature for growth was 5°C and growth ceased at 40°C. Good growth and sporulation occurred in between 20–32°C and the optimum temperature was found 28°C where maximum growth and good sporulation was observed. Sebeck (1952) observed optimum temperature being 27°C and *Fusarium* could grow in the range of 13 to 32°C. Agarwal and Sarabhoy (1978) observed 28°C optimum temperature for growth of *Fusarium oxysporum* was 30–35°C while Bolkan *et al.* (1982) reported optimum temperature of *Fusarium moniliforme* subglutians was 25°C. Joff *et al.* (1973) reported that growth of *Fusarium moniliforme* Sheldon did occur at 40°C, which is been observed in the present study also.

Influence of pH

Growth and reproduction of fungi are profoundly influenced by the acidity and alkalinity of the medium and this is due to the differential activity if the ions at different pH levels. Generally fungi tolerate a wide range of pH but range for maximum growth is narrower and it varies with organism and strains. Most of the previous workers have reported the optimum pH to be less than 7 for fungi studied by them.

In order to study the various pH levels on growth and sporulation of *Fusarium* spp. The experiment was carried out with liquid medium of Richard's solution. Various pH values were determined by using a Beckman pH meter. Various pH values of the media were adjusted in the range of I between 3 and 9 using 0.1 N, NaOH or HCI. The results obtained are presented in Table 18.7. The fungus could grow very well over a wider range of pH ranging from pH 3 to pH 9. In the present study, maximum growth (dry mycelium weight) of all three isolates of *Fusarium* spp. was observed at pH 6. 5. The sporulation was maximum at pH 6.5 followed by pH 7.0, where spores were maximum. The fungus showed good mycelial growth and sporulation at other levels but maximum growth was observed at pH 6.5. Similar results were obtained in all the three isolates of *Fusarium* spp. at different pH levels.

The fungus grows maximum over a certain range of initial pH values of the medium and will fail to grow at high and low extremes under given conditions. In the present study different level of pH were adjusted in the basal medium (Richard's agar medium). Growth was recorded in the range of 3–9 pH but the best growth was observed at pH 6.5. The growth range of 3–9 pH levels has also supported by the Huang and Zhao (1988). Arshad *et al.* (1992) also reported rapi mycelial growths at 6.5 pH. Jaurihar and Mehta (1972) reported pH 4.0 optimum for Richard's medium for the growth or *Fusarium moniliforme* (*Gibberella fujikuroi*) over a wide range of hydrogen ion concentration *viz.*, Neutral, alkaline and also at 3 pH.

Chung and Su (1988) reported autolysis of medium at pH > 8 in PDA liquid medium Kozlowska (1962) reported growth of *Fusarium moniliforme* in between pH 2.0 to 9.0 (optimum being 4–7.07) with which our results are similar to this findings.

Nutritional Factors

Studies of nutrition of fungi form an important branch in the physiology. It helps to determine the exact nutritional requirements of the macro organisms. This further helps in preparation of different media required for their cultivation under artificial conditions, which has special significance in the industrial and applied mycological studies.

In this study, some nutritional requirements of three isolates of *Fusarium* spp. have been undertaken and are reported briefly. The studies include influence of carbon sources, inorganic and organic nitrogen sources, amino acids, vitamins and trace elements.

Carbon Sources

Three isolates of the *Fusarium* spp. were grown in twenty-two carbon sources (along with control) to study their relative effect, the results of which are discussed below. The results obtained with various sources revealed that almost all sugars were utilized equally well by the fungus with some exception. (Table 18.8).

FM 1 Isolate

Among all monosaccharides, fructose, glucose, meso-inositol, arabinose, galactose, xylose, rhamanose were utilized very well by this isolate followed by dulcitol, ribose and mannose. In disaccharides, sucrose and cellobiose, melibiose, maltose, proved better sources than trehalose and lactose. Among polysaccharides, starch was best source followed by dextrin, inulin and salicin. Among oligosaccharides and sugar alcohols, mannitol was better than sorbitol and raffinose. Control (without sugar) showed very less growth and sporulation.

FM 2 Isolate

Among monosaccharides, xylose was the best utilized carbon source, showed maximum mycelial growth and sporulation followed by fructose, glucose, meso-inositol, arabinose, galactose, rhamanose, ribose, mannose and dulcitol. In disaccharides, melibiose was the best source followed by sucrose and cellobiose, lactose, maltose and trehalose. In polysaccharides, starch showed maximum mycelial growth followed by dextrin, inulin. Salicin, which showed less effect on the growth of this, isolate. In oligosaccharides and sugar alcohols, mannitol showed better results than sorbitol and raffinose. Control (without carbon source) showed very little mycelial growth and sporulation was nil in the medium.

FM 3 Isolate

Among monosaccharides, glucose was the best source for the growth and the sporulation followed by fructose, xylose, meso-inositol, arabinose, galactose, rhamnose, ribose, dulcitol and mannose. In disaccharides, melibiose was best source followed by sucrose, lactose, maltose, terhalose, and cellobiose. In polysaccharides, dextrin was best utilised carbon source followed by starch, inulin. Salicin was the poorest carbon source showing least mycelial growth. Among oligosaccharides and sugar alcohols, mannitol was better than sorbitol and raffinose. Control shows very less mycelial growth and sporulation did not occur at all.

In general, starch, fructose, glucose, xylose, melibiose were best utilised carbon sources followed by meso inositol, sucrose, dextrin arabinose, cellobiose, galactose, lactose, maltose, rhamnose, mannitol, terhalose. All other showed moderate growth. Growth was extremely poor in raffinose, mannose and salicin. Control showed very little growth due to the presence of nitrogen source. In absence of carbon source, all the isolates could not sporulate at all. Different forms of carhohydrates are utilised by different fungi according to their choice. Glucose and sucrose proved to be excellent sources of carbon for the different fungi studied by number of workers.

In the present study almost all sugars were utilised equally well by the fungus with some exceptions. Fructose, glucose, xylose, sucrose, arabinose, starch dextrin galactose, Mesoinositol were better utilised test sources than others. The carbon sources utilised by FM 1 and FM 2 isolates are similar to each other while FM 1 isolates has separate growth behavior on carbon nutrition. The results of present study are similar to that of Moore and Chapp (1952) who noticed, that all the isolates of *Fusarium* utilised number of carbon compounds from different sources supported moderate to good growth of *Fusarium* spp. While poor growth was noticed in maltose.

Yogeshwari (1948) observed highest dry weight of fungus *Fusarium moniliforme* with high *CIN* ratio in Richard's medium (sucrose and potassium nitrate).

EI-Bahrami (1977) studied the effect of carbon and nitrogen sources on fifty isolates of *Fusarium moniliforme.* He observed sucrose as the best carbon source followed by fructose, glucose, sorbose, maltose, mannose, xylose and arabinose while lactose as a poor carbon source.

Bolkan *et al.* (1982) studied the effect of carbon source of *Fusarium moniliforme* var *subgtutinans.* They observed best growth on soluble starch and maltose and microconidia were abundant on sucrose, starch, and fructose. Pimentel *et al.* (1985) studied the effect of carbon-source and temperature on the radial growth colony diameter of some isolate of *Fusarium.* Patil and Hapsae (1988) reported starch; xylose, fructose, melibiose, sucrose, glucose and maltose were the best carbon sources for growth"of the fungus. Dextrin was found poor source, which is different from the present study. Arshad *et al.* (1992) reported galactose as the best source of carbon for *Fusarium* spp. However the results of FM 1 land FM 2 were agreed while FM 3 isolate showed variation in growth character.

Nitrogen Sources

Nitrogen is indispensable for the growth of almost all fungi. It has both functional and nutritional importance. Nitrogen occupies an important role position only next to carbohydrates. A particular source of nitrogen is however not equally suitable for every fungus because of various organisms are very specific in choice of nitrogen for their growth and reproduction.

To determine the suitable source of nitrogen for growth and sporulation of *Fusariu,* spp. Sheld, a detailed experiment was carried out with inorganic and organic nitrogen sources and amino acids. All the studies were carried out with the basal medium of Richard's solution. In order study the effect of various nitrogen sources, sources, they were substituted singly for KNO3 of basal medium to supply equivalent amount of nitrogen in each case as described in material and methods.

FM 1 Isolate

The isolate showed maximum growth and sporulation when KNO3 (basal medium) was replaced with ammonium nitrate. The growth was supported with very good sporulation followed by ammonium di-hydrogen phosphate, sodium nitrate, potassium nitrate, ammonium chloride, and ammonium sulphate while urea supported the least growth for the above isolate. The control was kept without nitrogen source. Very little mycelial growth was found in control with absolutely nil sporulation (Table 18.10).

Table 18.10: Comparative Inorganic and Organic Nitrogen Nutritional Requirements of 3 Isolates of *Fusarium moniliforme* (FM 1, FM 2 and FM 3) in Liquid Medium

Sl.No.	Test Source	Isolate FM 1			Isolate FM 2			Isolate FM 3			Average
		My. Dr wt (mg)	Spor.	pH	My. Dr wt (mg)	Spor.	pH	My. Dr wt (mg)	Spor.	pH	
1.	Potassium Nitrate	730	++++	8.43	820	++++	8.02	540	++++	6.19	697
2	Sod. Nitrate	780	++++	6.06	630	++++	7.33	400	+++	6.74	603
3.	Amm. Nitrate	960	++++	6.64	920	++++	6.41	480	+++	6.09	787
4.	Amm. Dihy. Phosphate	790	+++	2.55	940	++++	2.56	490	++	4.34	740
5.	Amm. Sulphate	560	++++	2.80	650	++++	2.47	260	+	2.71	490
6.	Amm. Chloride	710	++++	2.68	580	+++	2.64	360	++	2.84	550
7.	Urea	400	+++	8.07	440	++	8.21	390	++	6.99	410
8.	Control	220	+	6.26	200	–	5.74	120	+	5.98	203
	SE		10.07			9.78			534		
	SED		14.25			13.84			17.55		
	CD at 5%		30.57			29.69			16.20		

Note: Sporulation: ++++: Excellent; +++: Good; ++: Moderate; +: Poor; My.Dr. Wt: Mycelial Dry Weight; Spor.: Sporulation.

FM 2 Isolate

The isolate showed maximum growth with ammonium di-hydrogen followed by ammonium nitrate, potassium nitrate, ammonium sulphate, and ammonium chloride. In this isolate also, urea supported the least growth and sporulation. The control showed scenty growth without sporulation due to carbon source present in the medium without sporulation. The control showed scanty growth without sporulation due to carbon source present in the medium (Table 18.10).

FM 3 Isolate

This isolate showed less growth of dry mycelial mat in comparison to other two isolates. The isolate showed maximum mycelial growth with potassium nitrate and ammonium di-hydrogen phosphate followed by ammonium nitrate, sodium nitrate. Ammonium sulphate, urea and isolates. Sporulation was very in this isolate. The maximum sporulation was found with potassium nitrate (Table 18.10).

In general, ammonium nitrate was the best inorganic nitrogen source followed by ammonium di-hydrogen phosphate, potassium nitrate, sodium nitrate, ammonium chloride, ammonium sulphate and urea for the growth of *Fusarium moniliforme* Sheld. El-Bahrami (1977) also found ammonium nitrate as the best nitrogen source. Ganacharya *et al.* (1976) reported calcium nitrate as best source of nitrogen followed by magnesium nitrate, potassium nitrate and ammonium sulphate and ammonium chloride. In the present study also ammonium sulphate and ammonium chloride were found poor source of nitrogen for the fungus. Isolate FM 1 and FM 2 showed equal utilisation while FM 3 isolate showed variations in utilisation of nitrogen.

Agarwal and Sarabhoy (1978) reported magnesium nitrate followed by sodium nitrates as best source for *Fusarium moniliforme.* Patil and Hapase (1988) reported urea as best source for nitrogen followed by potassium nitrate, sodium nitrate, ammonium nitrate and ammonium sulphate, which are more or less similar to the results of present study. Mehta *et al.* (1991) reported potassium nitrate was best Nitrogen source for pathogen. Arshad *et al.* (1992) reported potassium nitrate as the best Nitrogen source for *Fusarium* species.

While studying the effect of nitrogen sources, different inorganism sources were replaced in basal medium and the results revealed that ammonium nitrate was the best nitrogen source followed by amrrionium dihydrogen phosphate, potassium nitrate, sodium nitrate, ammonium chloride, and ammonium sulphate and urea. All treatments showed good growth in comparison to control (without nitrogen source) (Table 18.10). Isolate FM 3 was different from FM 1 and FM 2 isolate in nitrogen utilization for growth.

Amino Acids

Amino acids are building material for proteins and are the most easily assimilable form of inorganic nitrogen for fungi. Amino acids are nothing but pure form of organic nitrogen, which are easily available to the organism. Different amino acids as reported by many workers.

In the present study, the ability of three isolates of *Fusarium* spp. to assimilate nitrogen from different amino acids was, there fore, studied and also to find out the best source for growth and sporulation 17 different amino acids were selected and incorporated in the basal medium as described in material and methods, triplicate flasks were inoculated and incubated at laboratory temperature (25–28°C) for 15 days. Observations are presented in Table 18.11.

FM 1 Isolate

The data presented in Table 18.11 revealed that the best amino-acid source for the growth and sporulation was found asparagine where maximum dry mycelial weight (1110 mg) was recorded followed by histidine (1090mg) L (+) glutamic acid (1020 mg), L-aspartic acid (1020 mg), alanine (1010 mg), glutamine (980 mg), L-serine (950 mg), Valine (910 mg). The moderate growth was observed in L-phenylalanine, proline, tryptophan, methionine, tyrosine. Amino acids like cystine showed poor response for the growth. Cysteine hydrochloride showed least growth with poor sporulation. Control did not show any sporulation.

FM 2 Isolate

It is revealed from Table 18.11 that this isolate showed maximum growth and sporulation with histidine (1680 mg) followed by glycine (1340 mg) and L-(+) glutamic acid (1260 mg) whereas alanine, aspargine, L-aspartic acid, valine also supported least growth for this isolate while control (without nitrogen-source) did not show any growth and sporulation in the medium.

FM 1 Isolate

It is revealed from the results that FM 3 showed that FM 3 showed less growth in terms of dry mycelial fungal weight in comparision to FM I and FM 2 isolates. The maximum growth was recorded with aspargine (1020 mg) and it supported sporulation too, whereas histidine, glutamine, glycine, L-aspartic acid also showed good growth with very good sporulation. Lucine, valine, L- phenylalanine, cystine, tryptophan, cysteine hydrochloride, proline, L (+) glutamic acid, methionine supported less growth of mycelial mats, tyrosine showed least growth with this isolate with very poor sporulation.

In the present study for the utilization of amino acids for three isolates showed that FM 1 and FM 2 utilized aspargine and histidine very well where growth and sporulation were maximum, while FM 2 isolate showed histidine best source for the enhancement of growth and sporulation. In the same isolate, glycine showed better results than aspargine as another two isolates. Most of the amino acids sources showed better results than aspargine, as another two isolates. Most of the amino acid sources' showed marked effect on the growth of fungal mats, but in all the cases it was not supported by sporulation. It was found that without nitrogen sources the growth was poor and scanty with almost no sporulation. For good growth and sporulation the nitrogen source is also equally important as carbon source, which is basic nutritional requirement of fungi. In general, histidine, aspargine, L (+) glutamic acid, L (–) aspartic acid, glycine, alanine, glutamine were found better sources of amino acids than other used in the experiment. In present study all the amino acids showed marked effect on the growth. Aspargine and histidine were the best-utilized amino acids followed by L-aspartic acid glycine etc. In fact the addition of amino acids in purest form enhanced the growth of fungus to quite an extent.

Table 18.11: Comparative Nutritional Requirements of Amino Acids for 3 Isolates of *Fusarium moniliforme* (FM 1, FM 2 and FM 3) in Liquid Medium

Sl.No.	Test Source	Isolate FM 1			Isolate FM 2			Isolate FM 3			Average
		My. Dr wt (mg)	Spor.	pH	My. Dr wt (mg)	Spor.	pH	My. Dr wt (mg)	Spor.	pH	
1.	Aspergine	1110	++++	7.97	1140	++++	8.00	810	++++	7.46	1020
2.	L (–) serine	950	++++	6.40	900	++++	5.55	320	+++	5.86	723
3.	Leucine	880	++++	5.93	1010	++++	5.28	220	+++	4.95	783
4.	L-aspartic acid	1020	++++	8.68	1140	++++	8.43	430	+++	8.06	863
5	Valine	910	++++	5.49	1110	++++	5.42	280	++	4.80	767
6.	L. phenylanine	630	+++	5.77	750	+++	5.58	340	++	5.26	573
7	Cysteine	330	+++	5.34	460	++	4.63	310	++	4.49	367
8.	Tryptophan	550	+++	6.08	470	++	5.41	220	++	5.04	413
9.	Alanine	1010	++++	5.31	1240	++++	5.78	320	++	5.00	857
10.	Glutamine	980	++++	7.65	970	++++	7.36	470	++	5.42	807
11.	Histidine	1090	++++	8.42	1680	++++	8.35	800	++++	6.37	1190
12.	Proline	580	++++	5.32	670	++++	5.62	330	+++	5.51	527
13.	Glycine	880	++++	7,69	1340	++++	8.26	420	+++	6.15	880
14.	L(+) Glutamic acid	1.020	++++	8.29	1260	++++	8.30	340	+++	6.50	837
15.	Methionine	490	+	3.45	560	++	8.0	270	++	5.10	440
16.	Cystenine hyrochloriode	200	+	5.09	320	+	7.81	280	+	5.07	267
17.	Triosine	450	+++	5.58	530	+++	6.56	210	++	4.39	397
18.	Control	220	+	6.26	200	–	5.74	190	+	5.98	203
	SE		10.43			14.06			6.06		
	SED		14.75			19.88			8.57		
	CD at 5%		30.12			40.60			17.50		

Note: Sporulation: ++++: Excellent; +++: Good; ++: Moderate; +: Poor; My.Dr. Wt: Mycelial Dry Weight; Spor.: Sporulation.

Patil and Hapase (1988) reported serine, aspargine, leucine and aspartic acid were the good sources and methionine, cystine and tyrosine were the poor source for growth sporulation of *Fusarium moniliforme*. There was very poor growth in absence of amino acids.

Vitamins

Vitamins play an important role in promoting the growth and reproduction of fungi. It is an established fact that fungi require minute amount of Vitamins or growth factors for their normal growth, reproduction and other vital processes. Many workers have worked on the role of vitamins on fungal metabolism.

The study was undertaken of *Fusarium* spp. The experiment was carried out as described in material and methods. The results of the present study for ten vitamin sources are presented, in Table 18.12.

Table 18.12: Comparative Vitamin Nutritional Requirement of Three Isolates of *Fusarium moniliforme* (FM 1, FM 2, FM 3) in Liquid Medium

Sl.No.	Test Source	Isolate FM 1			Isolate FM 2			Isolate FM 3			Average
		My. Dr wt (mg)	Spor.	pH	My. Dr wt (mg)	Spor.	pH	My. Dr wt (mg)	Spor.	pH	Mycelial Dry Weight (mg)
1.	Ascorbic acid	760	++++	8.39	630	+++	8.10	550	++++	7.35	647
2.	B Complex	930	++++	8.47	750	++++	8.20	500	++++	7.11	727
3.	Biotin	680	+++	7.52	630	+++	7.48	600	+++	7.34	637
4.	Ca. penthanate	690	+++	7.72	710	+++	7.50	650	+++	7.73	683
5.	Folic acid	580	++++	7.59	550	++++	7.49	570	++++	7.36	567
6.	Pyridoxin	680	++++	7.59	600	++++	7.37	570	++++	7.45	617
7.	Thiamine	710	+++	8.23	620	++++	7.37	430	+++	0.3	587
8.	Riboflavin	1210	++++	8.20	630	++++	7.60	480	+++	7.32	773
9.	Niacin	1000	++++	8.27	470	++++	7.88	500	++++	7.40	657
10.	Vitamin B12	680	++++	7.66	460	++++	7.91	530	++++	7.57	557
11.	Control	980	++++	6.49	990	++++	7.88	530	++++	6.09	833
	SE	11.61		9.11		7.72					
	SED	16.43		12.88		10.92					
	CD at 5%	34.27		26.88		22.79					

Note: Sporulation: ++++: Excellent; +++: Good; ++: Moderate; +: Poor; My.Dr. Wt: Mycelial Dry Weight; Spor.: Sporulation.

FM 1 Isolate

This isolate showed marked enhancement in the growth of mycelial mats when riboflavin was added in the medium followed by niacin. The control was kept without the addition of vitamin source. The growth was very good with control supported by good sporulation. In rest all the vitamin sources; the growth was less in comparison with control.

FM 2 Isolate

This isolate showed marked decrease in the weight of mycelial mats when different, vitamin sources were added along with basal medium. The best growth and profuse sporulation was reported In control. Rest all sources showed less weight of mycelial mats although supported by good sporulation.

FM 3 Isolate

For this isolate calcium penthanate, biotin, pyridoxine showed better response in comparison to control. Rest all sources less growth on mycelial mats in comparison with control. Vitamins did not support good growth in all the cases in terms of mycelial weights.

Results tabulated in Table 18.12, revealed that all the three isolates showed marked variations in the growth of mycelial mats as compared to control except isolate FM I which showed good growth when riboflavin and niacin were added in basal medium than control. Rest two isolate did not respond much to vitamins. On the contrary, the growth was reduced with different test sources of vitamins. The average nutritional requirements are shown.

Lilly and Bamett (1951) reported that although there are many fungi, which are self sufficient in vitamins, there are other partially or completely deficient in one or more vitamins. Vitamins in minute quantities are effective in many of the biochemical reactions of the cells, which obviously reflects their role as catalysts. Many of the vitamins perform their catalytic role as co-enzymes or constituent part of the enzymes (Bilgrami and Verma, 1978).

The study of vitamins as a source of nutrition/stimulators revealed that there was no significant effect of them on growth and sporulation in comparison to control (without vitamin). Relatively better growth was obtained in riboflavin. All other sources *viz.*, Biotin, inositol, folic acid, B complex nicotinic acid could not show much difference.

Bilai and Shoherbina (1977) reported that mycelium or *Fusarium moniliforme, Fusarium gibbosum* and *Fusarium sambucinum* contain maximum nicotinic acid and pyridoxine followed by riboflavin pantothenic acid and biotin where as in the present study also isolates were found to be self sufficient in their vitamin requirements.

Trace Elements

Trace elements are always used in micro quantities to enchance the growth of fungi. The importance of trace elements in growth and reproduction in fungi was recognized by the classical researches of Raulin (1869) Wolfand Emmerie (1930) and Steinberg (1939, 1950). Many related problems of this study such as growth and sporulation of fungi, wilt syndrome, saprophytic activity and colonization 'of moulds on plant debris in soils have been investigated by many workers including Mundkar (1928) Yogeshwari (1948), Sadasivan (1950), Apparao *et al.* (1955), Saraswati devi (1956), Grewal (1955), Tandon (1961) and others. The present study was carried out to observe the effect of trace elements on growth and sporulation of three isolates of *Fusarium* spp. The results are presented in the Table 18.13.

FM 1 Isolate

The control showed the best growth supported by sporulation. Rest all sources like boron, calcium, cobalt, iron, molybdenum, lead, and zinc showed retard ness in the growth of this isolate although the growth was supported by good sporulation. Cobalt showed least growth for this isolate.

FM 2 Isolate

This isolate showed maximum growth with iron as trace element in comparison to control. In other cases like boron, calcium, cobalt, copper, molybdenum, lead and zinc showed less growth in comparison to control.

Table 18.13: Comparative Nutritional Requirement of Trace Elements for Three Isolates of *Fusarium moniliforme* (FM 1, FM 2, FM 3) in Liquid Medium

Sl.No.	Test Source	Isolate FM 1			Isolate FM 2			Isolate FM 3			Average
		My. Dr wt (mg)	Spor.	pH	My. Dr wt (mg)	Spor.	pH	My. Dr wt (mg)	Spor.	pH	
1.	Boron	640	+++	7.52	600	+++	7.44	550	+++	7.47	560
2.	Calcium	850	++++	8.63	760	++++	7.80	500	+++	7.40	713
3.	Cobalt	470	++	6.55	630	+++	6.67	600	+++	6.91	527
4.	Copper	700	+++	6.70	920	++++	6.73	650	+++	7.09	687
5.	Iron	790	+++	7.34	1050,	++++	7.57	570	++++	7.32	793
6.	Molybdenum	590	++++	8.01	790	++++	7.48	570	++++	7.39	630
7.	Lead	510	++++	6.78	980	++++	7.57	430	++++	7.12	673
8.	Zinc	780	++++	9.12	1020	++++	9.11	480	++++	9.00	830
9.	Control	980	++++	8.49	990	++++	6.88	530	++++	6.09	833
	SE	10.68		12.33		6.17					
	SED	15.11		17.45		8.73					
	CD at 5%	32.04		36.99		18.52					

Note: Sporulation: ++++: Excellent; +++: Good; ++: Moderate; +: Poor; My.Dr. Wt: Mycelial Dry Weight; Spor.: Sporulation.

FM 3 Isolate

Zinc showed maximum growth of mycelial mats for this isolate in comparison to control. Copper supported least growth but sporulation was moderate. In all other cases the growth was hampered evidently in comparison to control.

In present study, the trace elements did not show marked changes in the growth and sporulation ofall the three isolates except FM 2, which grew well in presence of iron. In almost all the cases the growth was hampered as compared to control. Average nutritional requirements are shown in Figure. However, the effect of trace elements was not significant in all the three isolates of *Fusarium* spp. (Table 18.13). The present study was carried out to observe the effect of trace elements on the growth and sporulation of three isolates of *Fusarium* spp. Most of trace element could not show marked enhancement on the growth and sporulation of *Fusarium* in comparison to control except iron, which showed better growth of the fungus. Ali and Sonar (1984) also reported the influence of iron on the growth and sporulation of *Fusarium moniliforme.* Enhancement in the growth may be attributed to shift in the qualitative production of amino acids by *Fusarium moniliforme,* significant influence of iron appeared on an increased production of total amino acid to inhibitory role upon phosphate operative in the alkyl amino acid biosynthesis in the medium in comparison to medium without iron. In the medium containing iron total amount of amino acids was however five times to that in control medium.

Effect of amino acids of growth is well recognized due to accumulation of more amino acids in the medium containing iron the growth of fungus is increased considerably and that is why iron could show enhancement in the growth and sporulation of *Fusarium moniliforme* and on other element could support better growth than control.

Micronutrient analysis of healthy and infected leaves reveal that due to impaired photosynthesis and pathogenesis the micronutrients like N, P, K Ca, Mg, Fe, Zn, Cu get accumulated in the cells and they get bound. In this state they are not available for plant metabolism. Accumulation of iron in leaves results in enhancement of the growth of fungus.

References

Agarwal, V.K. and Singh, C.V. (1974). Routine testing of crop seeds for *Fusarium moniliforme* with a selective medium. *Seed Res.*, 21: 19–22.

Abro, H. and Khan, M. (1982). Gravity inducing the endogenous circadian rhythmicity in *Fusarium monilforme. Pakistan Journal of Botany*, 14(2): 163–164.

Arshad, M., Dogar, M.A., Khan, M.A. and Sahi, S. T. (1992). Physiological studies on *Fusarium moniliforme* Sheld a boll rotting fungus of cotton. *Pakistan Journal of Phytopathology*, 2(1–2): 1–7.

Agarwal, D.K. and Sarbhoy, A.K. (1978). Physiological studies on species of *Fusarium* pathogenic to soyabean. *Indian Phytopath.*, 31: 24–31.

Appa Rao, A L., J. Saraswati Devi and Suryanarayanan *et al.* (1955). Growth Requirements of *Piricularia oryzae.* Ber et Cav. *Jour. India Bot. Soc.*, 34: 27–42.

Ali, S.S. and Sonar, M.L. (1984). Influence of iron on the extracellular production of amino acids by *Fusarium moniliforme* Sheld. *Indian Bot. Soc.*, 63: 148–150.

Agnihotri, V.P. (1983). *Diseases of Sugarcane.* Oxford IBH Publishing Co., 363 p.

Agrawal, G.P. (1958). Nutritional sutides on *Curvularia penniseti* I. Influence of nutrient media, pH, temperture and carbon nitrogen ratio. *Phyton.*, 101: 77–87.

Bilgrami, K.S. and Verma, R.N. (1978). *Physiology of Fungi.* Vikas Publ. House, New Delhi, pp. 274–297.

Bilai, V.I. and Shoherernina, S.M. (1977). Dynamics of the vitamins content of different strains of *Fusaria. Mikrobiologichii Zhumal*, 39(5): 597–600. *Review Pl. Patho.*, 57(12): 5312 (1978).

Bolle, P.C. (1927). Een onderzock de oorzak van pokahboeng en toprot. *Arch. Suikerind Ned-Indie*, 3(35): 589–609.

Bolle, P. C. (1930). Different forms of toprot. *Proc. Int. Soc. Sugarcane Tech. 3rd Cong.*, pp. 146–150.

Bolle, P.C. (1936). Idem. Bijdrage II. Over het onstaan van de pokahboeng-infectie. Verh. Leden Proefstation Java-Suikerind, 18: 857–886.

Booth, C. (1971). *The Genus Fusarium.* Commonweath Mycological Institute, Kew, Surrey, England.

Bisby, G.R. (1943). Geographical distribution of fungi. *Bot. Rev.*, 9: 466–482.

Bolkan, H.A. Dianese, J.C., Silva, C.B.D.A. and Arauji, J.C.A.D.E. (1982). Influence of Carbon source light, water potential and temperature on growth and sporulation of *Fusarium moniliforme* Var. Subglutinans. *Revista de microbiologia*, Dep. Pl. Biol., Univ. Brasilia, Brazil, 1 (3): 264–271.

Cochrane, V.W. (1958). *Physiology of Fungi.* John Wiley and Sons Inc., New York, Chapman and Hall Ltd., London.

Chattapadhyay, S.B. and Bhattacharya, S.K. (1966). Physiological studies on incitants of guava wilt: *Fusarium solani* and *Macrophomina phaseoli. Indian Jour. Mycol. Res.*, 4: 22–31.

Chattopadhyay, B.B. and Gupta, P.K. (1967). Studies on wilt diseases of pulses. I. Variation and taxonomy of *Fusarium* spp. Associated with diseases of pulses. *Indian J. Mycol. Res.*, 5: 45–47.

Chuang, T.Y. and Su, H.J. (1988). Physiological study of *Fusarium oxysporum* f. sp. Cubense. *Memoirs of the College of Agric.*, National Taiwan Univ., 28(2): 19–26.

Dillewijn, C. van. (1948). Stalk population, yield composition and *Fusarium* top rot in sugarcane. *Sugar*, 43(1): 28–30.

Dillewijn, C. van. (1950). *Fusarium* Pokkah boeng. *Proc. Int. Sugarcane Techn. 7th Cong.*, 473–498.

El-Bahrami, S. (1977). Survey of some *Fusarium moniliforme* strains from different host plants for compounds possessing gibberellin like activities. Zentralblatt. Fur Bakeriologii, Parasitenkunde. Infectioskran keilin and Hygeine. Tauta Univ. Egypt, 132(2): 178–183.

Ganacharya, N.M. and Wanker, B. N. (1976). Efficacy of manganese sulphate in inducing growth and sporulation of *Fusarium oxysporum* Schl. Causing potato wilt. *Indian Phytopath.*, 29(4): 464–469.

Grewal, J.S. (1955). Effect of nitrogen on growth and sporulation of *Alternaria tenuis strain* B. Causing rot of apple. *Lloydia*, 18: 74–81.

Huang, Z.X. and Zhao, J.H. (1988). Observation on spore gennination and growth of cucumber wilt disease fungus and ultra structural study on fungal colonization in host tissue. *Acta. Agri.*, Shanghai, 4(1): 29–36.

Hughes, C.G. (1967). Pokkahboeng and top rot. *Rower's Quart. Bull.*, 30: 81.

Jaurihar, S.S. and Mehta, P.P. (1976). Growth and sporulation of *Fusarium moniliforme* Sheld on various sources of carbon and nitrogen. Presented at symposium held at Bhagalpur University, 26–28 September (Communication).

Kozlowsk, Czeslawa (1962). Investigation on the biology of *fusarium oxysporum* schl. and attempt to control it. *Prace. Int. Badow Lesu.*, 246: 3–91.

Liddle, C.M. (1991). Introduction of recent advances in Fusarium systematics. *Phytopathology*, 81: 1044–1045.

Lilly, V. G. and Barnett, H. L. (1951). Physiology of fungi, McGrew Hill Book Co. Inc. U. S. A.

Martin, J.P., Han Lioe Hong and Wismer, C.A. (1961). Pokkah boeng. In J.P.Martin, E.V. Abbott and C.G. Hughes (Eds) Sugarcane diseases of the world. Voll. Elsevier, Amsterdam, pp 246–261.

Mradela Filho, 0., Cilva, W. J. Da (1986). Variability in *Fusarium moniliforme* Sheld. Summa Phytopathologica 12(1/2): 125–131.

Moore, H. and Chapp, C. (1952). Physiological study of *Fusaria* causing tomato, cabbage and muskmelen wilts. Mycologia, 44: 525–532.

Mehta, A., Mehta, P. and Chopra, S. (1991). Effect of various nitrogen sources on the production of pectolytic and celluloytic enzymes by *Fusarium oxysporum* and *F, moniliforme.* Zentralblatt fur Mikrobiologie, 146 (6): 393–398.

Mundkar, B.B. (1928). Importance of iron for the growth of fungi (*Fusarium vasinfectum Atkinson*). In: *Proc. 15th Ins. Sci. Cong.*, pp. 230.

North, D.S. (1932). Pokkab boeng. *Proc. Int. Soc. Sugarcane Techn. 4th Cong. Bull.*, p. 100.

Nelson, P.E., Tonsson, T.A. and Marasas, W.F.O. (1983). *Fusarium Species: An Illustrated Manual for Identification*. University Park. P. A., Pennsylvania State University Press.

Nelson, P.E., Burgess, L.W. and Summerbell, B.A. (1990). Some morphological and physiological characters of *Fusarium* species in section Liseola and Elegans and similar species. *Mycologia,* 82(1): 99–106.

Neish, G.A. and Leggett, M. (1981). *Fusarium moniliform* var. *intermedium* a new variety in the Liseola section. *Can. Jour. Bot.,* 59: 288–291.

Patil, A.S. and Hapase, D.G. (1987). Studies on Pokkah boeng disease of sugarcane in Maharashtra. *Indian Phytopath,* 40: 290.

Patil, A.S. and Hapase, D.G. (1988). Nutritional studies on *Fusarium moniliforme* causing Pokkah boeng disease of sugarcane. *Indian J. of Plant Path.,* 6(2): 171–175.

Patil, A.S. and Jadhav, S.B. (1995). Studies on Pokkah boeng and pineapple disease of sugarcane in Maharashtra with their economic losses in yield and quality in sugarcane. *Final Progress Report ICAR,* New Delhi, pp. 442.

Priode, C.N. (1929). Pokkah boeng and twisted top disease of sugarcane in Cuba. *Phytopathology,* 19(4): 343–366.

Prasad, N. and Padwick, W.C. (1939). The genus *Fusarium.* A species of *Fusarium* as a cause of wilt ofgram, *Indian J. Agric. Sci.,* 9: 371–380,

Prasad, N. and Mathur, B.L. (1964). Studies on *Fusarium* wilt of Cumin. *Indian J. Agric. Sci.,* 34(4): 273–277.

Pimentel, C.P.V., Feitosa, M.I. and Chiba, S. (1985). Studies on the cultural variability of some isolates of *Fusarium spp. Arquivos do Instituto Biologico,* 52(1–4): 7–14.

Reimers, J.F. (1971). Pokkah beong a common disease of little consequence. *Cane Grower Quat. Bull.,* 34 (3): 101–102.

Raulin, J. (1869). Estudes Chimique Sur la vegetation. *Ann. Sci. Nat. Ser. V,* 10: 93–229.

Sahab, A.F., Fahim, M.M., Osman, A.R. and Abd El-kadar, M.M. (1987). *In vitro* studies on *Fusarium oxysporum* f.sp. Lupini. *Minutia Journ of Agricul Res.,* 7: 41–52.

Snyder, W.C. and Hansen, H.N. (1945). The spectes concept in *Fusarium* with reference to discolor and other sections. *Amer. Jour. Bot.,* 32(10): 657–666.

Sebeck, O. (1952). Physiological properties of *Fusarium lycopersici* and *Fusarium-vasinfeclum. Phytopathology,* 42: 119–192.

Steinberg, R.A. (1939). Growth of fungi in synthetic nutrient solution. *Bot. Rev.,* 5: 327–350.

Steinberg, R.A. (1950). Growth of fungi in synthetic nutrient solution of some fungi. *Am. J. Bot.,* 37: 711–714.

Sadasivan, T.S. (1950). Some aspects of soil born fungal diseases of plants. *Palaeobotonist.,* 1: 400–407.

Saraswati Devi, L. (1956). *Proc. Indian Acad. Sci.,* 42B: 145–150.

Thind, K.S. and Chand, J.N. (1962). Studies on physiology of *Helminthosporium australiense* Bug. The causal agent of a leaf spot disease of Bajra. *Indian Phytopathology,* 21: 190–194.

Thind, K.S. and Madan, M. (1976). The effect of various carbon sources on the growth and sporulation of *Cephalothecium roseum* Lorda causing pink rot of apple (*Malius sylvestris* Mill). *Proc. Nat. Acad. Sci., India Sect.,* 37B: 373–380.

Tandon, R.N. (1961). Physiological studies on some pathogenic fungi. *U.P. Sci. Res. Comm.*, Allahabad, U.P. India.

Tiedt, L.R. and Jooste, W.J. (1988). Aberrant conidiogenesis in *Fusarium* subglutinans isolate. *Trans. Brit. Mycol. Soc.*, 90: 527–530.

Verma, K.P., Singh, M.P., Upadhyay, U.C. and Singh, H.N. (1984). Knife cut: The report of the incidence of this disease in India. *Sugarcane*, 4: 7–8.

Voorhees, R.K. (1933). *Gibberella moniliforme* on Corn. *Phytopath.*, 23(4): 368–378.

Whittle, P.J.L. and Irawan (2000). Pokkah Boeng. In: *A Guide to Sugarcane Diseases,* (Eds.) P. Rott, R.A. Bailey, J.C. Comstock, B.J. Croft and A.S. Saumtally. CIRAD and ISSCT, Montpellier, France, pp 136–140.

Wolfe, L.K. and Emmerie, A. (1930). *Biochem Z.*, 228: 443–450.

Went, F.A.F.C. (1986). Notes on sugarcane diseases. *Ann. Bot.*, 10: 583–600.

Yogeswari, L. (1948). Trace elements nutrition in Fungi I. The effect of boron, Zinc and manganese on *Fusarium* spp. *Proc. Indian Acad. Sci. Set.* B. 28(6): 177–201.

©R.C. Ram & Asha Sinha (*Editors*)
***Microbial Diversity: Modern Trends*, 2007** **264–268**
Daya Publishing House, New Delhi

Chapter 19

Management of Root-knot Nematode (*Meloidogyne incognita*) on Lentil (*Lens esculenta* Monch.)

Amar Bahadur**, *D.P. Singh** and *Pramila Gupta**

***Department of Mycology and Plant Pathology, Institute of Agricultural Sciences, B.H.U., Varanasi - 221 005**

****Department of Plant Pathology and Nematology, Allahabad Agricultural Institute, Naini, Allahabad - 211 007**

ABSTRACT

Pulses are the most important constituents in the diet of vegetarian besides being rich in protein and other nutrients. The root-knot nematode invades the living tissue of roots of the plants produces gall, which cause damage to crop. The Phytonematodes infecting lentil root-knot nematode is one of important pathogens responsible to causing significant loss to this crop. Due to the adverse effect of root-knot nematode on lentil this experiment was carried out to evaluate the effect of two nematicide (Cabofuran and Phoret) in management of root-knot nematode (*Meloidogyne incognita*) and there by to observe the effect of these nematicide of the growth of crop number of nodulation on root and root galls with a view of manage the losses to lentil (*Lens esculenta*) recorded in control nematode (RICN).

Introduction

Pulses are important of the Indian diets and supply a major parts of the protein requirement some of the essential amino acids. Pulse crops through symbiotic nitrogen fixation in the soil. India the food grain production 1994–95 was 185 million tons was contributed by pulse. Lentil (*Lens esculeta*) is one of the important rabi pulses. It is one of the oldest pulse crops and the most nutrition of the pulse. It has the potential to cover the rise of the dry land agriculture. It is also used as covers check the soil erosion in problematic area. India rank first in the world in respect to production on the lentil crop, which belongs to family Leguminaceae. Lentil is used for human consumption as well as for feeding animals. Nutritionally it is very rich and contains about 25 per cent protein and 60 per cent carbohydrate. It is also rich about in Ca, iron and niacin. Leguminous its significant and fundamental role in agriculture production could be attributed to their ability to atmospheric nitrogen fixation association with root nodules bacteria of the genus *Rhizobium*. Pulse crops are attack by the various pest and disease caused by fungi, bacteria and viruses.

Meloidogyne sp. belongs to Phylum–Nematoda, Class–Sccernentia, Order–Tylenchida, Family–Heterodaridae. Nematode of the genus Meloidogyne is important liming factor in production of this pulse crop. Damage to crop by nematode in the pest was often ignored to the caused such as lack of nutrients different soil moisture partly due to lack of the different information of their occurrence of the pathogenesity. Plant parasitic nematodes are important crop yield reduce and caused by them. They cause Parenchyma and vascular disorder extensive necrosis of the root system suppress. *Rhozobium* nodulation and increase severity of the many soil borne disease infected plant are stunted yellowish suffer from water stress and yield very poorly.

Intensity of the symptoms vary with nematodes population density at sowing time and with environment condition. Among the parasitic nematode of the genus *Meloidogyne* is the most important with an extensive host range and wide distribution in temperature as well as tropical areas of the world. Lentil (*Lens esculenta*) which is one of the most important pulse is attacked by root nematode (*Meloidogyne* sp.). Severe stunting and yellowing of the crop was observed in some field in Delhi, Ghaziabad and Meerut district of the Uttar Pradesh.

In survey in Delhi state, *Tylenchorhychus* sp., *Helicotylechs* sp., *Pratylenchus* sp., *Heplolamus indicus*, *Meloidogyne incognita*, *M. javanica* and *Heterodera cajani* were recorded in lentil field (Mishra and Gure, 1980).

In view of the above present study were conducted in which two mematicides *viz.*, Phoret and Carbofuran were applied in nematode infected field and the effect of the nematode population on lentil was studied.

Materials and Methods

Field Experiment

Experiment was conducted in experimental garden of the Department of Plant Pathology and Nematology, Allahabad Agricultural Institute, Naini, Allahabad in the rabi season 1995–96. The field was ploughed three time. The fertilizer (20 kg Nitrogen, 50 kg Phosophorus, 50 kg Potash per hectare) was applied after the first ploughing and then the field was leveled to maintain the optimum moisture for seed germination. Local variety of lentil seeds were sown 10–15 cm apart at 2–3 cm depths during the second week of October. Row to row distance as 30 cm and each plot was of 2 × 3 m in size with three replications. Design consisting of soil treatment with nematicides was laid out in randomized block design chemicals were applied in their field at the rate of carbofuran 3 per cent @ 2 kg/ha and Phorate 10 g @ 2 kg/ha.

Plant Growth Character

On the effect of the plants growth characters samples were taken shoot length in cm, shoot weight (fresh) in gram, root length in cm, root weight (fresh) in gm. Root nodules and number of gall/root system was observed at the time interval 45, 60 and 90 DAS (Days after sowing).

Results

Meloidogyne goeldi with marked sexual dimorphism female white saccate to elongated pyriform body with elongated neck cuticle marked by fine transverse strain was identified in 1892. The diagrammatic perennial pattern in lip region being sex minute lips on the papillae ampid have not been observed spear more slender than that of larvae or male being. Strong basal knobs median esophageal bulb strongly develop fallowed by a short isthmus basal bulb expending into large flattened lobes reaching back over interior and of intestine and containing large esophageal gland nuclei, ovaries the spacious body cavity consisting of hundred of developing oocytes arranged it random. The male cylindroids verm like with neck tapering to the strangely sclertized head region. The lateral fields marked by four incises extending from near middle of the neck to the plant round terminals which is devoid of bursa derides not observed phasmid offer difficult to see. Usually located slightly interior to latitude of anus lip region being prominent literal lip or checks, which contain the amphidal pouches. Aphids apertures slits like close to oral opening visible only from a face view spear with strongly develop basal knob esophageous with well develop median bulb fallowed by slender isthmus which connect the three elongated gland forming the basal bulb. Expending interior and of intestine the executary pore and hemizoid of intestine. The excretory pore and hemizoid near the latitude of nerve ring twisted one or two body generally twisted (Chitwood, 1949).

Perennial pattern some what oval with a high irregular arch composed closely spaced wary lines in interior portion of arch above anus masked by numerous zig-zag and broken straiae with some time form a whorl there is a little evidence of literal lines but along there area both the dorsal and ventral straiae have tendency to become forked often there are short transverse straiae extending from the inner streak toward vulva males produce from the egg of single female head either one or two testis.

Effect of Chemicals of Plant Growth of Lentil

Shoot length at 45, 60 and 90 DAS was not significant difference (Table 19.1). Shoot weight of lentil was significant difference more in Phorate (T3) compare to carbofuran (T2) and control (T1) which intern were not significantly different. Shoot weight at 90 DAS in the treatments T2, T3 was significantly higher than T1. Treatment T1 and T2 were non significant from each other (Table 19.2).

Table 19.1: Effect of Nematicides on Shoot Length

Treatments	Days of Sampling (DAS)		
	45	60	90
		Centimeter	
Control (T1)	14.0	31.0	42.16
Carbofuran (T2)	14.0	35.2	46.16
Phorate (T3)	13.8	33.0	45.16

Effect on root lent than two chemical treats did not significantly different at 45 and 60 DAS in comparison to control (Table 19.3). Root wt indicate Table 19.4 that none of the two chemical significantly increase or decrease. On the number of root nodulation nine of the two chemical did not significantly effect at 45 and 60 DAS. At 90 DAS number of nodules were significant less in the treatment (T2, T3) as compare to control (Table 19.5). Effect on gall is indicated in Table 19.6. Root gall was significantly reduced with both carbofuran (T2) and phorate (T3) treatments compared to control (T1) at 45, 60 and 90 DAS but there was no significant difference between the two chemicals.

Table 19.2: Effect of Nematicides on Shoot Weight (gm)

Treatments	Days of Sampling (DAS)		
	45	60	90
	Weight in gm		
Control (T1)	0.79	3.17	1.73
Carbofuran (T2)	1.17	3.16	4.02
Phorate (T3)	1.14	5.75	6.88

Table 19.3: Effect of Nematicides on Root Length

Treatments	Days of Sampling (DAS)		
	45	60	90
	Centimeter		
Control (T1)	7.08	10.83	12.66
Carbofuran (T2)	10.08	11.83	11.00
Phorate (T3)	9.66	11.50	12.33

Table 19.4: Effect of Nematicides on Root Weight (gm)

Treatments	Days of Sampling (DAS)		
	45	60	90
	Weight in gm		
Control (T1)	0.18	0.58	0.24
Carbofuran (T2)	0.23	0.43	0.23
Phorate (T3)	0.21	0.43	0.23

Table 19.5: Effect of Nematicides on Nodulation (Number)

Treatments	Days of Sampling (DAS)		
	45	60	90
	Number of Nodules		
Control (T1)	5.66	6.6	4.16
Carbofuran (T2)	11.16	10.8	2.0
Phorate (T3)	8.0	8.5	1.16

Table 19.6: Effect of Nematicides on Galls

Treatments	Days of Sampling (DAS)		
	45	60	90
	Number of Galls		
Control (T1)	5.0	12.17	14.2
Carbofuran (T2)	0.83	3.0	1.83
Phorate (T3)	11.17	13.50	13.0

Discussion

The lentil (*Lens esculenta*) is a very important cultivated crop so to save the plants for root knot (*Meloidogyne incognita*) and to increased the yield management of this nematode is very essential. This experiment was carried out in view of controlling the nematode among by chemical application and effect of the treatments of growth characters of lentil which were taken into consideration were shoot lent. shoot weight root lent, root weight, and number of nodules and gall of various stages of plant growth *viz.*, 45, 60 and 90 DAS shoot length recorded at 45, 60 and 90 DAs showed no effect of nematicides (Carbofuran and Phorat) but showed that carbofuran higher average shoot length fallowed by phorat (Fazad *et al.*, 1991). While no effect in shoot weight of lentil at 45 DAS was observed with application. Nematicide Phorat At was generally more effect at 60 and 90 DAS followed by carbofuran increased the shoot weight. Root lent as recorded at 45 and 60 DAS showed no effect of carbofuran and phoret showed that carbofuran produce higher average root length followed by Phoret application increased the root length were as 90 DAS. Phorat maximum increased in root length followed by carbofuran while no difference in root weight of lentil at 45, 60 and 90 DAS was observed with application of any nematicides while at 45 and 60 DAS none of the nematicides effected root nodulcation at 90 DAS. Nodules were more in carbofuran fallowed by Phorate at the nematicide *viz.*, Carbofuran and Phorat were effective in reducing nematode infection on lentil. Observed by reduction in nematicide number of gall at 45, 60 and 90 DAS these observation confirm similar results in other crop have been reported by Anon (1978).

References

Anonymous (1977). Evaluation of nematicide for the control of meloidogyne incognita in brinjal nursery. Annual Progress Report. All India Coordination Project on Nematicide, Pest on Crop their Control at Tamil Nadu Agriculture University, Coimbatore, pp. 5–6.

Chitwood, B.G. (1949). Root not nematodes Part I: A vision of the genus *Meloidogyne goeldi*, 1887. Proc. Helm. Soc. Wash., 16: 90–104.

Mishra, S.D. and H.S. Gaur (1980). Relation between namatization levels of root-not nematode (*M. incognita*) and the growth of lentile (*Lens esculenla* L.) *Indian J. Entomology*, 42(2): 262–263.

Siddiquie, Fazal M. (1991). Pathogenic effect of *M. incognita* (Kofaid and White 1919) Chitwood 1949 on lentil. *Current Nematology*, 2(1): 51–52.

©R.C. Ram & Asha Sinha (*Editors*)
***Microbial Diversity: Modern Trends*, 2007**
Daya Publishing House, New Delhi

269–273

Chapter 20

Studies on Some Dominant Species of *Aspergilli* Isolated from Soil of Rice-Wheat Cropping System

Asha Sinha, P.S. Kalhapure and Rakesh K. Singh

Department of Mycology and Plant Pathology, Institute of Agricultural Sciences, Banaras Hindu University, Varanasi – 221 005

ABSTRACT

The soil fungi of the rice wheat cropping system was isolated by standard technique in Varanasi region. Five species of *Aspergilli viz., A. niger, A. flavus, A. sulphureus, A. sydowi* and *A. terreus* was described and illustrated.

***Keywords**: Soil fungi, Rice-wheat cropping system, Aspergilli.*

Introduction

In rice-wheat cropping system, wheat is grown just after harvest of rice and disposal of large amount of rice residue like rice straw, stubbles etc. is a serious problem. A large amount of rice residue left after harvesting in soil is metabolised by soil microorganism *i.e.*, bacteria, actinomycetes, fungi, protozoa, micro and macro arthopods etc. Fungi play important role in residue decomposition. Among all organisms fungi play an important role in the recycling of mineral nutrient from plant tissue bulk to plant tissue via herbivore, carnivore and saprophagous food chain (Witkanp 1971). The ecological studies of soil fungal flora was made by Mishra 1995; Sinha *et al.*, 1998; Sinha *et al.*, 1999; Griffith *et al.*, 2000; Park Kyuji *et al.*, 2001; Reichardt *et al.*, 2001; Zou *et al.*, 2001; Ahmeli *et al.*, 2002. This article include studies on some species of *Aspergilli*, which is dominant in the soil of rice-wheat cropping system.

Materials and Methods

The soil of experimental field was typical Gangetic alluvium (Inceptisol). Composite samples were collected three times, *i.e.*, just after cutting of rice, 15 day after cutting of rice and just before sowing of wheat. The soil fungi was isolated by dilution plate technique (Warcup, 1960). The soil sample was crushed and 10 gms of it was suspended in 100 ml of sterilized distilled water. Further dilution series 1 : 10^3, 1 : 10^4, 1 : 10^5 were prepared from it. One ml of each dilution was poured individually in sterilized Petri dishes. Then 15–20 ml of sterilized cool nutrient medium was added. Five replications of each dilution was prepared and all plates were incubated at 25 ± 2°C for a week. From which pure culture of five species of *Aspergilli viz., A. niger, A. flavus, A. sulphureus, A. sydowi* and *A. terreus* were isolated. The fungi was identified by Thorn and Raper (1945); Raper and Thorn (1949); Clement and Shear (1954); Gilman (1975); Subramanian (1971); Ellis (1971) Barnett and Hunter (1972).

Results and Discussion

From the soil five species of *Aspergilli viz., A. niger, A. flavus,* A. *sulphureus, A. sydowi* and *A. terreus* was dominant in the soil was studied in detail (Plates 20.1–20.5).

Aspergillus niger Van Tieghem

Colonies on Martin's medium rapidly growing with abundant submerged mycelium, aerial hyphae usually scantily produced. Conidiophores mostly arise directly from the substranum, varying greatly in length and diameter 210–350 × 7–9 µ. Conidial heads fuscous, blackish brown, varying from small, almost columns masses of few conidial chains. Vesicles globose commonly 25–45 µ in diameter, phialides typically in two series, thickly covering the vesicle, primary varying greatly in length. 7–10 × 2–3 µ Conidia globose 2.5–3.5 µ in diameter.

Aspergillus flavus. Link

Colonies on Martin's medium widely spreading, with floccosity limited to scanty growth of few aerial hyphae in older culture. Conidiophores arise separately from the substranum 500–650 µ long × 5–13 µ in diameter, appear rough or spiny with low magnification. Vesicle 10–30 µ in diameter, primary phialides 8–10 × 3–3.5 µ and secondary 7–10 × 2.5–3.5 µ. Conidia pyriform to almost globose.

Aspegillus terreus Thom

Colonies on Martin's medium forms tints of pinkish–cinnamon rough. Cinnamon to deeper, brown shades in age, spreading conidiophore 150–200 µ long × 5–8 µ diameter. More or less flexuous, with apex enlarged to form vesicle commonly 13–18 µ in diameter bearing phialides in two series upon its dome like upper surface. Conidia elliptical to globose 2.1–2.5 µ in diameter.

Aspergillus sydowi (Bainier and Sartory) Thom and Church

Colonies on Martin's medium blue green, velvety with some aerial trailing hyphae. Conidiophores arise mostly from the submerged hyphae upto 450 × 4–7 µ colourless. Vesicles 12–18 µ in diameter. Phialide radiate in two series. Conidia globose 2.5–3.5 µ in diameter.

Aspergillus sulphureus (Fresenius) Thom and Church

Colonies on Martin medium powdery, sulphur yellow in colour Conidiophores arises from aerial hypahae, up to 190µ in length. Head loose columns of conidial chains, phialides in two series. Conidia globose, smooth 3–3.5 µ in diameter.

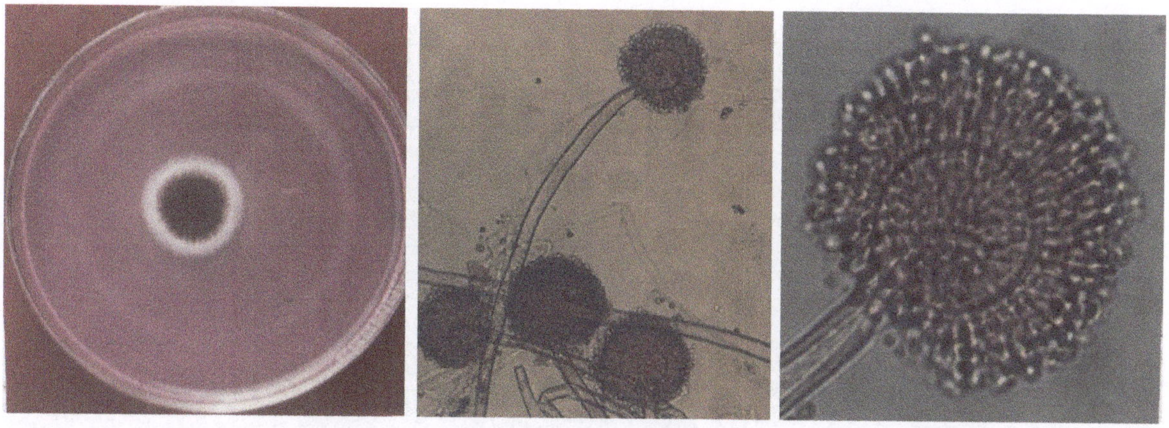

Plate 20.1: ***Aspergillus niger***

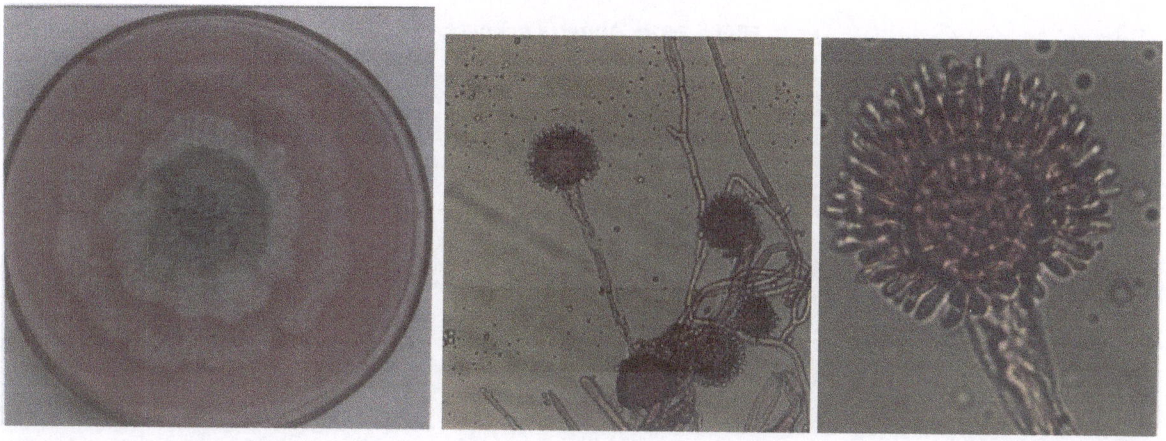

Plate 20.2: ***Aspergillus flavus***

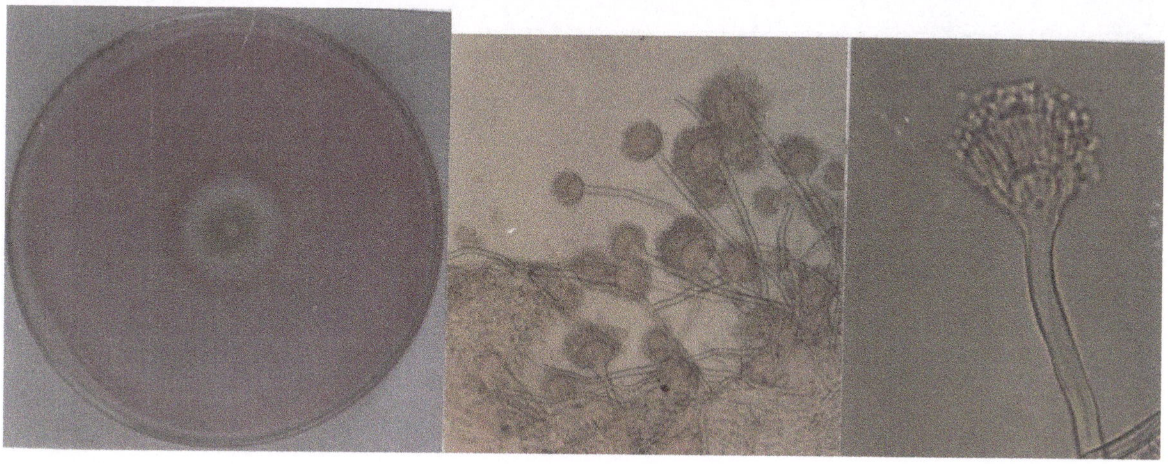

Plate 20.3: ***Aspegillus terreus***

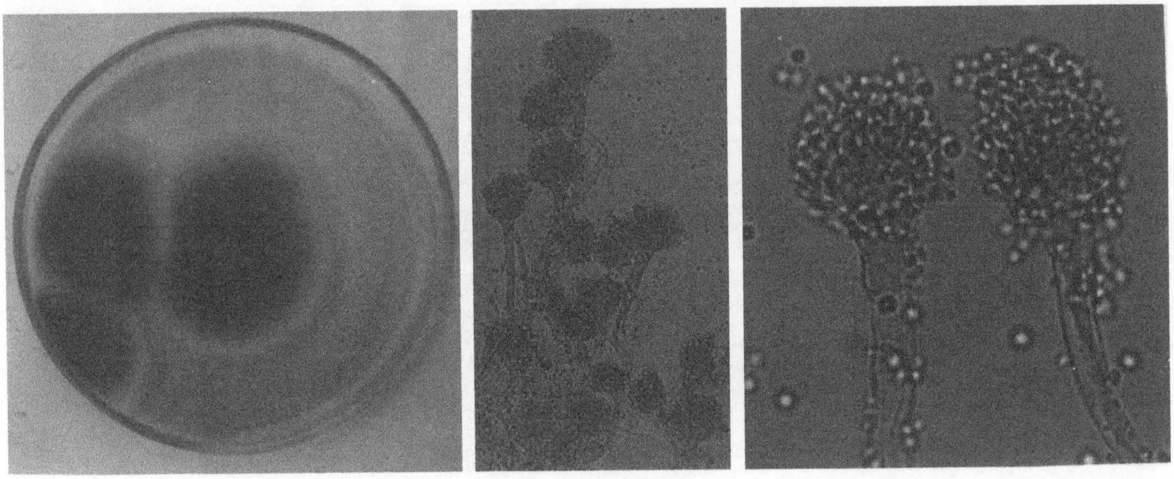

Plate 20.4: ***Aspergillus sydowi***

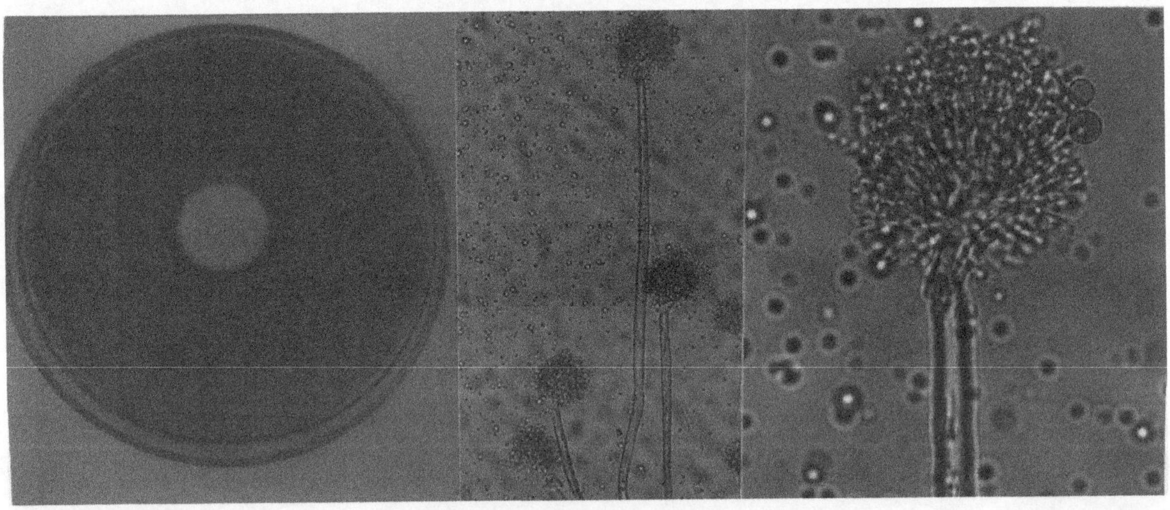

Plate 20.5: ***Aspergillus sulphureus***

Acknowledgements

The authors are grateful to the Head, Department of Mycology and Plant Pathology, Institute of Agricultural Sciences for providing facilities to carry out this work.

References

Ahmed, M. and Das N.K. (2002). Certain effect of shifting cultivation in selected areas of East Garo Hills District Meghalaya and effect on fungal microflora. *Environment and Ecology*, 20(4): 839–843.

Barnett, H.L. and B.B. Hunter. (1972). *Illustrated Genera of Imperfect Fungi*. Burgess Publishing Co., Minnesota.

Clements, E.F. and C.L. Shear (1954). *The Genera of Fungi.* Hofner Publishing Co., New York (U.S.A.).

Ellis, M.B. (1971). *Hypomycetes.* C.N.U., Kew, England.

Griffth, G.S. and Bardgett, R.D. (2000). Influence of resource unit distribution and quality on the activity of sail fungi in particulate medium. *New Phytologist,* 148(1): 143–151.

Mishra, H.M. (1995). Studies on litter, decomposing mycoflora of rice straw. *M.Sc. Thesis,* B.H.U., Varanasi.

Park kyujin Byen Jeungsu, Lee, L.H.O. and Park Hyconsues. (2002). Influence of soil flooding with organic matter amendment at various temperature on changes of microbial population in ginseng replanting field soil. *Research in Plant Disease,* 7(1): 25–30.

Raper, K.B. and C. Thorn. (1949). *A Manual of Pencillia.* Williams and Williams Co., Baltimore, U.S.A.

Reichardt, W. Briones, A. Jesus R. De and B. Padse. (2001). Microbial population shift in experimental rice system. *Applied Soil Ecology,* 17(2): 151–163.

Sinha, A. Singh, A.K. and A.S. Yadav (1999). Fungal decomposition of parthenium leaf in soil. *India J. Weed Sci,.* 30: 76–78.

Sinha, A., Pathak, S.K. and A.S. Yadav. (1998). Studies on mycoflora of decomposing root litter of safflower (*Carthanus tinctorium* L.) in relation to different climatic factors. *Crop Res.,* 16: 265–270.

Subramanian, C.V. (1971). *Hypomycetes.* I.C.A.R., New Delhi.

Thorn, C. and K.B. Rapor. (1945). *Manual of Aspergilli.* William and Wilkins Co., Baltemore (U.S.A.).

Warcup, J.H. (1960). Methods of isolation and estimation of activity of fungi in soil. In: *Ecology of Soil Fungi,* (Eds.) D. Parinson and J.S. Waid. The University Press, Liverpool, pp. 3–21.

Witkamp, M. (1971). Soils as components of ecosystems. *Ann. Rev. Ecol. and Syst.,* 2: 85–110.

Zou, Li., Chem Yongliang, Yan Tingzhen. (2000). Ecological distribution and seasonal changes of soil microorganisms in pure and mixed plantation. *J. Forestry Res.,* 11(2): 106–108.

©R.C. Ram & Asha Sinha (*Editors*)
***Microbial Diversity: Modern Trends*, 2007** **274–279**
Daya Publishing House, New Delhi

Chapter 21

Evaluation of Casing Media for Cultivation of *Agaricus bisporus* (Lange) Imbach

T.S. Shandilya and R.S. Jarial

Mushroom Science Unit, Dr. Y.S. Parmar University of Horticulture and Forestry, Nauni, Solan, H.P.

ABSTRACT

The production of mushroom largely depends upon the casing layer on the top of mushroom beds. Physical properties *viz.*, water holding capacity, pore space, pH and electrical conductivity were determined in case of ten casing substrates namely, vermicompost-leached, vermicompost-cropped, coir pith, filter mud, saw dust and municipal waste (garbage). Vermicompost-normal had the maximum water holding capacity (90 per cent), while, loam soil exhibited minimum (30 per cent) water holding capacity. Per cent pore space was found to be maximum (68 per cent) in case of coir pith while, it was minimum in case of loam soil (40 per cent). The pH of all the casing materials was found to vary between 6.2 to 7.4, while, the conductivity was recorded to be maximum in case of municipal waste (> 10,000 u mhos/cm) and minimum in case of loam soil (120 u mhos/cm). The yield of *Agaricus bisporus* was found to be minimum (12.0 kg/100 Kg compost) when the mushroom beds were cased with saw dust. Different treatment combinations were also used for casing mushroom beds and out of these combinations a casing combination of vermicompost-leached + spent compost (1 : 1 v/v).

Keywords: *Agaricus bisporus, Casing, Physical and chemical properties.*

Introduction

The production of mushroom largely depends upon the casing layer on the top of mushroom beds. Once the compost has been fully colonized by mushroom mycelium, it induces fruit body formation of *Agaricus bisporus* (Atkin, 1973; Edwards, 1974; Hayes and Nayer, 1975; Stoller, 1979; Yeo and Hayes, 1979; Shandilya, 1989). Lambert and Humfeld (1939) were the first to put the casing on scientific line and suggested 5.5 to 8.0 pH of casing medium for better cutination of *A. bisporus*. For the past 30 years, a lot of development has been made on casing materials. Unsterilized top soil was originally used during cultivation but now pateurization of casing materials either by steam or aerated steam has been found essential part of the mushroom growing system. A variety of casing materials are in vouge in different countries depending upon their availability. In USA, top loam soil is being used, whereas pulberized soft marl dug out of queries is used for casing in France. In England, peat and chalk powder are mixed together and marl are used for casing in Holland. However, casing materials available in India vary markedly from those in developed countries.

Mantel (1973) recommended the use of one year old spent compost + sand in the ratio of 4 : 1 (v/v) under Indian conditions with mixing of lime after treatment with Nemagon and Formalin. Hayes *et al.* (1976) and Shandilya (1978) however, reported ineffectiveness of one year old spent compost resulting in higll salinity, leading to crop flushes in patches. The use of farmyard manure (FYM) + loam soil (1 : 1; v/v) for casing mushroom beds has also been advocated (Hayes and Shandilya, 1977). Later Shandilya and Agarwal (1983) suggested an increase in yield with ageing of spent compost. Since peat is not easily available in this part of the country for casing mushroom beds thus research is needed to use agricultural waste materials so that cost of production and continuity of supply of these materials are of prime importance in successful application and acceptance by the growers. Present studies on casing media evaluation are thus a part of series of experiments designed to develop more productive substrate for cultivation of *A. bisporus* which are best suited to local conditions. The article describes various Indian resources used as casing materials and their comparative performance during cultivation of *A. bisporus*.

Materials and Methods

Casing Materials

Various casing materials *viz.*, farmyard manure, (one and a half year old cowdung), three years old spent compost, loam soil, vermicompost, coir pith, filter mud, saw dust and municipal waste were collected and their physical and chemical parameters were determined as per the methods suggested by Flegg (1951) and Flegg (1958), respectively. However, pH of the materials was determined according to Shandilya (1982). These materials were stored in bags at room temperature for further use. It included following parts:

Farmyard Manure (FYM)

This material consisted only one and a half year old cowdung and was free of any bedding material.

Three Years Old Spent Compost

Spent compost aged for three years was collected from the pits specially dug for the storage and weathering of this material near the, mushroom farm.

Loam Soil

It was collected from the local vegetable farm.

Vermicompost

It is the byproduct of earthworms excreta formed out of the cowdung. The conductivity of the normal vermicompost was quite high so it was leached as well as crop was grown on this material so as to lower down the conductivity limit to acceptable level. The three treatments of vermicompost *viz.*, Vermicompost-normal, vermicompost-leached and vermicompost-cropped were used.

Coir Pith

This material was collected from South India and subjected to weathering for one year under normal conditions.

Filter Mud

This was the byproduct of sugar factories (UP) containing minute particles of crushed sugarcane plant tissue and mud.

Saw Dust

It was taken from hard wood of *Eucalyptus* spp. and weathered naturally for one year.

Municipal Waste

This product was collected locally from Municipal Waste Recycling Plant, Salogra, Solan (HP) and used as fresh material for casing.

Pasteurization of Casing Materials

All the collected materials were pasteurized with steam at 65°C for four hours and were used for casing after they cooled down. pH of all casing materials individually as well as in combination was adjusted between 7.6 to 7.8 by adding calcium carbonate after pasteurization.

Mushroom Production

Compost was obtained commercially and filled in growing bags and spawned with U3 strain of *Agaricus bisporus* @ 0.2 per cent. Bags were filled with 13 kg of pasteurized compost. The bags were kept for mycelial colonization of compost at 24 ± 1°C with relative humidity of 70–80 per cent. It was covered with different casing materials individually as well as in combination in different experiments. The casing combinations are given below.

1. FYM + Three years old spent compost (1 : 1, v/v)
2. FYM + loam soil (1 : 1, v/v)
3. FYM + vermicompost-leached (1 : 1, v/v)
4. Vermicompost-leached + spent compost (1 : 1, v/v)
5. Vermicompost-leached +loam soil (1 : 1, v/v)
6. FYM + vermicompost-leached + spent compost (1 : 1 : 1, v/v/v)
7. FYM + vermicompost-leached + loam soil (1 : 1 : 1, v/v/v)

In both the experiments, the number of replications was kept five and each bag was treated as one replication. The room temperature after casing was maintained at 24 ± 1°C till the mycelium grew to the top the casing layer. The temperature was later gradually lowered down to 16-18°C and relative humidity raised to 80-90 per cent. The experiment was carried out for 60 days cropping period. Mushrooms were harvested at the right stage, cut at soil level and weighed separately for each replication. The treatment and yield were expressed as kg/100 kg of the compost.

Results

Water holding capacity and pore space percentage among physical while pH and conductivity among chemical properties are considered as important characteristics of the casing media. Keeping this in view these characteristics were analyzed for ten collected materials and have been recorded in Table 21.1.

Table 21.1: Physical and Chemical Properties of Casing Materials

Casing Media	Water Holding Capacity (%)	Pore-space %age	pH	Conductivity (µ mhos/cm)
Farmyard manure	82	65	7.2	350
Spent compost (3 years old)	77	62	7.2	360
Loam soil	30	40	6.8	120
Vermicompost-normal	90	65	7.4	820
Vermicompost-leached	95	62	7.2	210
Vermicompost-cropped	88	60	7.2	230
Coir pith	62	68	6.2	160
Filter mud	70	63	6.5	180
Saw dust	60	58	6.5	480
Municipal waste (Garbage)	50	52	6.8	>10.000

A perusal of Table 21.1 reveals that farmyard manure (FYM), three years old spent compost had 82 per cent and 77 per cent water holding capacity, respectively as compared to vermicompost-leached which had got 95 per cent water holding capacity being the highest amongst all test materials. Loam soil casing medium though having the minimum salt concentration had the poorest water holding capacity. Pore space percentage of farmyard manure and vermicompost-normal was equal (65 per cent), whereas, spent compost (three years old) had slightly lower value (62 per cent). In coir pith pore space percentage was highest (68 per cent), whereas the loam soil had the minimum (40 per cent) pore space percentage. The pH values in all the test materials varied between 6.2 to 7.4. Vermicompost-normal had the highest pH (7.4) followed by farmyard manure (FYM) three years old spent compost, vermicompost-leached and vermicompost-cropped (7.2). Municipal garbage showed highest salt concentration (> 10,000 µ mho/cm), whereas in loam soil it was 120 µ mho/cm. In vermicompost-normal the conductivity was 820 µ mho/cm. The salinity in vermicompost-leached and vermicompost-cropped was 210 µ mho/cm and 230 µ mho/cm, respectively.

Results given in Table 21.2 indicate that farmyard manure increased the mushroom production (20.0 kg/100 kg of compost) significantly followed by vermicompost-leached (17.0 kg/100 kg of compost) and vermicompost-cropped 915.0 kg/100 kg of compost). Yield obtained in filter mud, coir pith, loam soil, three years old spent compost and municipal waste (garbage) was statistically similar at 5 per cent level of significance. Vermicompost-normal was found to give the lowest production (10 kg/100 kg compost) in the present studies.

The results (Table 21.3) regarding the combinations of farmyard manure + spent compost, farmyard manure + loam soil, farmyard manure + vermicompost-leached, vermicompost-leached + spent compost, vermicompost-leached + loam soil (1 : 1, v/v), farmyard manure + vermicompost-leached + spent compost and farmyard manure + vermicompost-leached + loam soil (1 : 1 : 1, v/v/v) for casing the mushroom beds revealed that maximum yield of button mushrooms was obtained when farmyard manure + vermicompost-leached + spent compost was used in equal volume for casing already spawned run compost. This treatment was significantly better and superior to others in mushroom production (25 kg/100 kg of compost) followed by farmyard manure + vermicompost-leached combination treatment. The farmyard manure + vermicompost-leached when used in combination performed better than their individual use in increasing productivity of *A. bisporus*.

Table 21.2: Comparative Performance of Different Casing Material on Mushroom Yield

Casing Material	Average Yield* (kg/100 kg ready compost)
Farmyard manure	20.8
Spent compost (3 years old)	13.7
Loam soil	14.0
Vermicompost-normal	10.0
Vermicompost-leached	17.0
Vermicompost-cropped	15.0
Coir pith	14.0
Filter mud	14.5
Saw dust	12.0
Municipal waste (Garbage)	13.0
$CD_{0.05}$	1.86

* Average yield of two crops.

Table 21.3: Effect of Different Casing Material Combinations on the Yield of *Agaricus bisporus*

Casing Combination	Ratio (v/v; v/v/v)	Average Yield* (kg/100 kg ready compost)
Farmyard manure + Spent compost	1 : 1	22.00
Farmyard manure + Loam soil	1 : 1	21.80
Vermicompost-normal	1 : 1	24.50
Farmyard manure + Vermicompost-leached	1 : 1	19.80
Vermicompost-leached + Loam soil	1 : 1	18.00
FYM + Vermicompost-leached + spent compost	1 : 1 : 1	25.00
FYM + Vermicompost-leached + loam soil	1: 1 : 1	22.80
$CD_{0.05}$		1.97

The performance of farmyard manure as a casing material improved with the addition of vermicompost-leached and showed a perfect combination with the higher water holding capacity. It allowed to retain relatively the higher moisture content in the mixture along with varied types of bacterial flora (Shandilya, 2000).

Discussion

Pore space percentage and water holding capacity have been reported to have direct relation with the production of *A. bisporus* (Flegg, 1965; Atkin *et al.*, 1937 and Kalberer, 1990). In the present investigations vermicompost treatment had the highest water holding capacity followed by farmyard manure and three years old spent compost. Pore space percentage was highest in farmyard manure and vermicompost-normal followed by three years old spent compost and vermicompost-leached. Water holding capacity and porocity of these materials were observed very near to peat casing material used in Europe (Shandilya, 1982). pH in these materials was at acceptable level 7.0 to 8.0. Some of test materials showed less than 7.0 pH. Lambert and Humfeld (1939) mentioned 7.6 as optimum pH while, De Kleermaeker (1953) postulated that pH is not a limiting factor in the suitability of a soil mixture for mushroom production.

Various materials have been used by different workers for casing mushroom beds in various parts of the world (Mantel, 1973; Hayes and Shandilya, 1977; Garcha and Khanna, 1993, Noble and Gaze, 1995, Lelley *et al.*, 1994 and Wuest and Bayer, 1996). Farmyard manure has been used since long in India as the main substrate for casing. Low yield in vermicompost-normal and municipal waste (garbage) can be related to the salinity of these materials as these were 820 µ mho/cm and 10,000 µ mho/cm, respectively. The present studies are in agreement with the findings of Flegg *et al.* (1961), Shandilya and Agarwala (1983) and Kurtzmans (1995) who have also If stated that the salt concentration was detrimental to fruit body formation and decreases the yield.

References

Atkins, F.C. (1973). *Guide to Mushroom Growing*. Feber and Feber, London.

Allison, W.H. and L.R. Kneebone. (1963). Influence of compost pH and casing pH on mushroom production. *Mush. Sci.*, 5: 81–89.

De Kleermaeker, E. (1953). Some experiments with various casing soils. *Mush. Sci.*, 2: 193–202.

Edward, R.L. (1974). Historical review and physics of casing soil. In: *The Casing Layer,* (Ed.) W.A. Hayes. Maney and Sons Ltd., Leeds, England.

Flegg, P.S. (1951). Casing soil. *Annual Report or the Mushroom Research Station*, Yaxely, 1950: 26–30.

Flegg, P.S. (1958). Some aspects of the compost and casing layer in relation to fruiting of cultivated mushroom. *Report G.C/R.I.*, 1957: 108–115.

Flegg, P.B. (1965). Water the mushroom casing layer by capillary. *Jour. Hort. Sci.*, 40: 150–155.

Garcha, H.S. and P.K. Khanna. (1993). Use of biogas effluent slurry for mushroom cultivation. In: *Biogas Slurry Utilization*. Publ. Consortium of Rural technology, Paragon Enterprises, New Delhi, p. 133–134.

Hayes, W.A., Shandilya, T.T., Munjal, R.L. and Seth, P.K. (1976). Improvements to the soils as casing media in H.P. *Indian J. Mushrooms*, 2: 29–32.

Kalberer, P.P. (1990). Influence of water potential of casing soil on crop yield and on dry matter content, osmotic potential and manni content of the fruit bodies of *Agaricus bisporus*. *Journal. Hort. Sci.*, 65: 573–581.

Kurtzman, R.H. Jr. (1995). Cobalt chloride and ethylene effect on fruiting of *Agaricus bisporus*. *Mycologia*, 87: 366–369.

Lambert, E.B. and Humfield, H. (1939). *Mushroom Casing Soil in Relation to Yield.* U.S. Deptt. Agri. Circ., 509 pp.

©R.C. Ram & Asha Sinha (*Editors*)
***Microbial Diversity: Modern Trends*, 2007** **280–285**
Daya Publishing House, New Delhi

Chapter 22

Strain Improvement in Oyster Mushroom (*Pleurotus* spp.)

Mithilesh Kumar**, Harsh Kumar* and Dayaram**

***Department of Genetics, **Department of Microbiology**
Rajendra Agricultural University, Pusa (Samastipur) – 848 125, Bihar

ABSTRACT

In the present investigation on strain improvement in different *Pleurotus* species, both inter and intra-specific hybridization was attempted among eight species of *Pleurotus*. These species included *Pleurotus florida, P. djamor* (R_{22}), *P ostreatus* (WC 595, WC 739) *P. sajor caju P. pulmanorius, P. sajor caju, P. flabellatus* and *P. columbinus*. Out of total 409 intra-specific made, only 64 compatible crosses were obtained having higher yield than their parents. However, only 12 successful cross combinations were obtained from altogether 187 inter-specific cross made among different *Pleurotus* species. These inter-specific hybrids were found to be better performer than their parents under pilot trials. Thus, ample scope of strain improvement in oyster mushroom through hybridization exists.

Keywords: Pleurotus, Oyster, Inter and intra-specific hybridization mating, Desirable character.

Introduction

Pleurotus species is widely grown on a number of substrates and in a wide range of temperature regime. But inspite of easy method of cultivation its production is very less in Bihar than other parts of country. Under Bihar condition white button mushroom (*A. bisporus*) is grown from November to mid February and summer Button mushroom (*A. bitorquis*) is grown from October to mid December and

February to March, whereas paddy straw mushroom (*Volvariella volvacea*) and milky white (*Calocybe indica*) are grown from April to September in hut. A survey revealed that liking of people are button and oyster mushroom. Most preferred species of oyster is *Pleurotus sajor caju* which is grown from September to March. But during winter (December to January) its yield is very poor and light brownish colour discourages growers. Therefore, there is need to find out a suitable strains which can be grown under natural conditions at higher/lower temperature and may have other desirable characters like sporeless, white colour, better texure, test and high yielding capacity. These qualities are not found in existing species of *Pleurotus*. Recently pink oyster mushroom, *P. djamor* has gained popularity among farmers as well as consumers.

Though earlier attempts have been made for producing quality strains of *Pleurotus* (Engenio and Inderson, 1968 *Eger et al.*, 1976, Bahukhandi and Munjal, 1990, Ghosh and Chakarvorty, 1991, Thakur and Bhandal, 1993). The hybrids developed by these workers did not show much improved characters and hence lacked popularity and adoption among growers. Therefore, the present study was carried out on inter-and intra-specific hybridization among the selected spp. of *Pleurotus i.e., P. florida, P. ostreatus, P. dajamor, P. pulmonarius, P. sajor caju, P. flabellatus,* and *P. columbinus* to develop suitable strains with desired characters.

Materials and Methods

The cultures of *Pleurotus* spp. were obtained from NRCM, Solan, H.P. and subcultured at regular intervals. The spawn was prepared on wheat grains and sporophores were obtained by standard cultivation method. The spore prints were taken from healthy mature sporophore, which served as spore stock for further study (Watling, 1970).

The spore suspension was made by serial dilution from spore print and 10^{-4} dilution (4–5 spores/10 x field) was selected for study. One full loop of this solution was streaked on 2 per cent plain agar in petriplates and incubated at 26 ± 1°C. The germinating spores were isolated and purified on PDA slants. Altogether 22 single spores isolates of *P. florida,* 20 of *P. ostreatus,* 30 of *P. djamor,* 28 of *P. pulmonarius,* 27 of *P. columibinus,* 9 of *P. sajor caju* and 8 of *P. flabellatus* were obtained.

The monokaryons were grown in petriplates and the hybridization between the 7 species (both intra- and inter-specific) was done randomly.

Three mm bits were cut out from 7 days old monocaryon cultures were placed at two opposite ends in petriplate containing PDA.

In this way a total of 267 number of pairing from intra specific and 201 pairs from inter specific crosses were obtained. These isolated pairs were incubated at 26 ± 1°C or for 7–10 days for maturing. Three mm disc-obtained from anastomosing point of the paired monokaryon, if only. The spawn was prepared by standard method and cultivation was done by the method described by Chaudhary (2003).

Data for sporophore colour, size, yield and temperature requirement for cultivation were recorded in different hybrids and compared with parents. Besides these, time taken for substrate colonization, cropping period and total yield in 3/4 flusphes were also recorded.

Results and Discussion

It was observed that the single spore isolates obtained in the 7 spp. of *Pleurotus* have peculiar features by which they were differentiated from dikaryotic mycelium (*i*) The single spore isolates were having slower growth and in some cases the growth was limited to few mm and absence of clamp connection was also observed. They did not produce primordia/fruiting bodies during cultivation. Although some abortive primordia were observed in some monokaryon with further growth (Plate 22.1).

During the mating process between monokoryons the hyphae developing from both the oppositely placed inocula were anastomosed and showed clear difference in appearance between individual parents and mating hyphae, when tested they were fertile.

The non mated pairs belonging to three categories have same special morphological characters. In first categories the hyphae of the monosporous pairs were intermingled giving an apperance of rating, but after testing failed to produce fruiting bodies, were sterile. Most of the non-mating pairs belonged to this category. In some cases the hyphae of both paired monospores grew but did not mate and remain separated, considered as neutral. In other cases a clear inhibition zone was developed between paired isolates and their growth occurred in different directions opposite to each other. This type of reaction is termed as inhibitory.

Whereas in fertile isolates clamp connection, dikaryotization and sporophere formation were observed. In this study the sterile: fertile ratio were recorded as 187 : 12 in interspecific crosses (Table 22.1) and 409 : 64 intra specific crosses in (Table 22.2). The hybrids obtained showed characters corresponding to their parents. There was hardly any difference between hybrids and their parents in their vegetative growth characters. The sporopheres in all the combinations were categorized in two types. Some of the hybrids were similar to their either of the parents and some of them had blending of characters from both the parents. In the first set of mating type (PF × Pd) (hybrid-1) were phenotypically possessing blending of characters between the two parents. The similarities were observed in appearance, colour, size and texture of fruiting bodies and temperature requirements for cultivation of hybrids in inter specific hybridization.

Table 22.1: Intraspecific Crosses in *Pleurotus* spp.

Sl.No.	Pleurotus Species	No. of Monospores	No. of Crosses	No. of Success Crosses
1.	P. florida	31	74	10
2.	P. ostreatus (WC 739)	28	68	18
3.	P. ostreatus (WC-595)	08	27	04
4.	P. djamor (R_{22})	39	89	11
5.	P. pulmonarius (WC 537)	42	80	08
6.	P. columbinus	27	34	09
7.	P. sajor caju	09	15	02
8.	P. flabellatus	08	22	02
	Total	**162**	**409**	**64**

In intra-specific crosses a total of 64 successful crosses were obtained and they were phenotypically similar to their parents. However, the yield potential were higher than parents.

Pleurotus is heterothallic and tetrapolar (Anderson, *et al.*, 1973, Bahukhandi and Muunjal, 1990, Thakur and Bandal, 1993). A single basidiocarp produces 4 types of basiodiospores with two different mating factors termed as A and B. Dikaryon produced by the basidiospores having different incompatiblity factors which can be detected on the basis of presence of clamp connection and conjugate

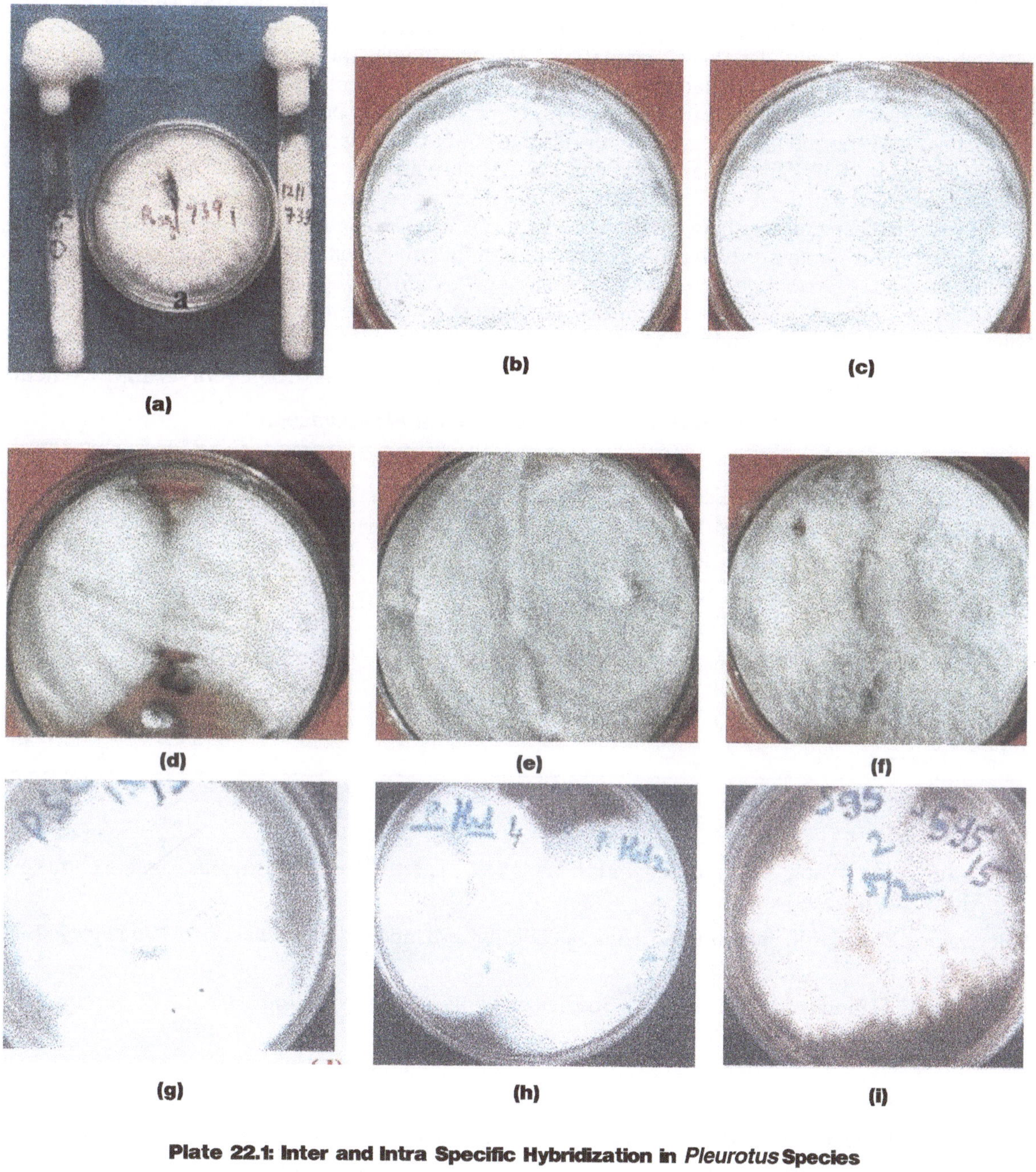

Plate 22.1: Inter and Intra Specific Hybridization in *Pleurotus* Species

(a) Interspecific crossing (*P. djamor* × *P. ostreatus*); (b) Interspecific crossing (*P. ostreatus* × *P. pulmonarius*); (c) Interspecific crossing (*P. pulmonarius* × *P. ostreatus*); (d) Intraspecific crossing (*P. ostreatus*); (e) Intraspecific crossing (*P. pulmonarius*); (f) Intraspecific crossing (*P. djamor*); (g) Intraspecific crossing (*P. sajor caju*); (h) Intraspecific crossing (*P. flabellatus*); (i) Intraspecific crossing (*P. ostreatus*)

nuclear division (Reper, 1966). Roxon and Jong (1997) described 4 mating types of *Pleurotus sajor caju* as $A_1 B_1$, $A_2 B_2$, A_1B_2 and A_2B_1. Koufart (1936) reported 4 types of reactions when he mated monokaryons of *P. ostreatus* and observed mated or compatible, neutral, antagonistic and inhibitory reactions. In the present study the paired monokaryons resulted in three types of colonies which may be termed as mated, non-mated, neutral and antagonostic. This combination may be similar to the result of Koufert (1936). It is also reported that interstock hybrids gave better yield than intrastock hybrids in *Pleurotus* species (Jandaik, 1987, Rajarathnam and Bano 1987, Thakur and Bhandal, 1993). During this study it has been observed that the resultant hybrids produce more or less similar yield as their parents but the hybrid produced by crossing between *P. ostreatus* and *P. djamor* resulted in 25 per cent more yield as compared to their parents, which is in agreement with that of earlier works. In the present work the hybrid obtained between *P. pulmonarius* × *P. ostreatus* (WC-595) having blended characters but higher yield than its parents while other hybrids were almost similar to their parents in terms of yield potential and other morphological features.

Thus, intra and inter-specific hybridization in *Pleurotus* spp. holds promise for its improvement.

Table 22.2: Interspecific Crosses in *Pleurotus* spp.

Sl.No.	Pleurotus Species	No. of Monospores	No. of Crosses	No. of Success Crosses
1.	P. florida × P. ostreatus (WC 739)	42	44	04
2.	P. pulmonarius (WC-537) × P. ostreatus (WC 595)	49	22	02
3.	P. flabellatus × P. ostreatus (WC-595)	48	20	01
4.	P. ostreatus (WC 739) × P. sajor caju	44	18	01
5.	P. pulmonarius (WC 537) × P. columbinus	39	16	–
6.	P. djamor × P. florida	37	15	01
7.	P. sajor caju × P. pulmonarius	38	16	01
8.	P. ostreatus (WC 595) × P. djamor	47	36	02

References

Anderson, N.A., Wang, S.S. and Schwandt, J.W., (1973). The *Pleurotus ostreatus-sapidus* complex. *Mycologia*, 65: 28–35.

Bahakhandi, D. and R.K. Shamo. (2002). Interspecific hybridization in *Pleurotus* spp. *Indian Phytopathol.*, 55: 61–66.

Bahukhandi, D. and R.L. Munjal (1990). Studies on evolving high yielding strains of, *Pleurotus sajor caju* through hybridization. *Indian Phytopath.*, 43(1): 70–73.

Chaudhary, S.K. (2003). Studies on production and management of oyster mushroom. *M.Sc. (Ag.) Thesis*, Department of Plant Pathology, RAU, Bihar.

Eger, G. (1974). Rapid method for breeding *Pleurotus astreatus*. *Mush. Sci.*, 9: 567– 573.

Eger, G. Eden. G and Wissing, E. (1976). *Pleurotus ostreatus* potential of a new cultivated mushroom. *Theor. Appl. Genet.*, 47: 155–163.

Eugenio, C.P. and N.A. Anderson. (1968). The genetics and cultivation of *P. ostreatus*. *Mycologia*, 60: 627.

Ghosh, N. and D.K. Chakravarty. (1991). Studies on evolving new strains or *Pleurotus sojor caju* by selective dicaryotizatan. *Indian Mushroom,* 19: 69–72.

Jandaik, C.L. (1987). Breeding potential of *Pleustus* spp. in India with special reference to *P. sofer caju.* In: *Indian Mush. Sci. II,* (Eds.) T.N. Kaul and B.M. Kapoor. Srinagar, pp. 355–360.

Kaufert, F.H. (1936). The Biology of *Pleurotus corticatus.* Fr. minnesota, *A.E.S. Tech. Bull.,* 114: 3–35.

Liang, Z.R. and S.T. Chang. (1989). A study of intergeneric hybridization between *P. sajor caju* and *Schizophyllum commune* be Protoplast Fusion. *Mush. Sci.,* 12: 125–137.

Rajarathanarn, S. and Bano, Z. (1987). *Pleurotus mushroom* Part I(A): morphology, life cycle, taxonomy, breeding and cultivation. *CRC Critical Reviews in Food Science and Nutrition,* 26: 179–185.

Raper, J.R. (1966). *Genetics and Sexuality of Higher Fungi.* The Ronald Press Company, New York, 283 pp.

Roxon, D. and Jong, S.C. (1977). Sexuality of an edible mushroom, *Pleurotus sajor caju. Mycologia* 69: 203–205.

Roxton, J. and S.C. Jons. (1977). Sexuality of an edible Mushroom, *P. sajor-caju. Mycologia,* 69: 203–205.

Thakur, K. and M.S. Bandal. (1993). Monospores isolates and their intermating in *Pleurotus sapidus* and *P. sajor caju. Mush. Res.,* 2: 41– 44.

Watling, R. (1970). Spore colour. *Bull. Br. Mycal. Soc.,* 4: 48–49.

©R.C. Ram & Asha Sinha (*Editors*)
Microbial Diversity: Modern Trends, **2007**
Daya Publishing House, New Delhi

Chapter 23

Development of New Sterilization Techniques of Substrate for Oyster Mushroom

R.C. Ram and Deepak Gupta
Department of Mycology and Plant Pathology, Institute of Agricultural Sciences, B.H.U., Varanasi – 221 005 (U.P.), India

ABSTRACT

Effect of various substrate sterilization techniques on the growth behaviour and yield potential of *Pleurotus florida* was studied. All the treatments except sunlighting gave significantly fast growth and higher yield over untreated substrate. Higher yield was recorded from formalin and autoclaving treated substrate. However, newly solarization and radiation techniques of substrate sterilization also gave good yield towards the substrate tested.

Keywords: *Pleurotus florida, Sterilization, Substrate, Growth and yield.*

Introduction

Several sterilization methods of substrate preparation, have been developed for cultivation of oyster mushrooms. Steam sterilization, hot water treatment and chemical sterilization are the most effective methods and they are adopted by mushroom growers on large scale in India and abroad too. Some of the important recommendations in this area, use of hot water at 80°C for 1 to 2 hrs. and steaming above 80°C for 1 hr. has been advocated (Junkova, 1971; Zadrazil and Schneidereit, 1972; Zadrazil, 1973; Patil and Jadhav, 1989; Singh and Dwivedi, 1991). These methods of sterilization are both costly and uncomfortable for mushroom growers. Another method of substrate sterilization is soaking of substrate in formalin (500 ppm) and bavistin (75 ppm) for 18 hrs. (Vijay and Sohi, 1987).

This method of substrate sterilization is also costly, unavailable and uncomfortable to farmers, harmful to human health and increasing environmental pollution. The autoclaving is complete sterilization of substrate (Zadrazil and Brunnet, 1980). However this method is not possible for commercial cultivation on large scale and even poor mushroom growers are unable to purchase this autoclave etc.

Oyster mushroom (*Pleurotus florida*) is one of the yield potential mushroom which can be cultivated in polythene bags on wheat straw (Sharma and Jandaik, 1991) after sterilizing the substrate by radiation (ultra voilet-rays) and solarization (sun-light through polythene sheet). The best sterilization method of substrate should be very simple, cheapest, comfortable, free from chemicals and safe to health and environment. Therefore, the present investigation was conducted to study the performance of new sterilization techniques of wheat straw with radiation and solarization on the growth behaviour and yield of *Plelurotus florida.*

Materials and Methods

The culture of *Pleurotus florida* was obtained from NRCM–Solan (H.P.). The culture was sub-cultured and maintained on PDA medium in an B.O.D. incubator at 25 ± 2°C temperature. The boiled and sterilized wheat grains were used for spawn preparation as described method of (Patrabansh and Madan 1995). Well dried and one year old wheat straw was soaked in water for over-night and next morning excess water was drained off. The sterilization of moist wheat straw was done by autoclave, formalin, bavistin, radiation, solarization and sun-lighting.

The substrate was sterilized by autoclave at 10 (lb) pressure for 20 minutes. In chemical sterilization substrate was fumigated with formalin for 12 hrs. in airtight plastic drum. In another chemical treatment substrate was resoaked in 0.2 per cent bavistin solution for 12 hrs. In radiation treatment the substrate was exposed to U.V. light for 45 minutes and again for 45 minutes after turning the substrate. In solarization and sun-lighting, the moist straw was spread on floor in two equal halves in which one half was covered with clean polythene sheet and other half without polythene sheet treated in sun-light from 10 am to 4 pm. Temperatures 55°C and 36°C were recorded from treated substrate with polythene-sheet and without polythene sheet respectively. The soaked wheat straw was kept for control. Prepared substrates were spawned with 3 per cent (moist weight basis) wheat grain spawn and filled into polythene bags (40 × 25 cm). The mouth of bags were tied by cotton thread and 8–9 holes were made for proper aeration. Mushroom beds were incubated in a dark room where temperature ranges between 22–28°C. Polythene bags were teared off and removed after mycelial growth and compact mycelial masses were sprayed with water regularly. The recorded temperature of cropping room was ranged from 18–26°C during harvesting period.

Results and Discussion

Data presented in Table 23.1 show that there was not more difference in the spawn run period ranged between 10–12 days except the sun-lighting treated substrate where it took 18 days. The fruit bodies were initiated after two days which were ready for harvesting of 1st flush in next 4 days. Three flushes were taken during different total cropping periods. These were ranged from 54–60 days however 3rd flush was failure in the untreated substrate.

The number of fruit bodies and their maximum, average and minimum weight in all flushes are given in Table 23.2. Result (Table 23.2) shows that major variabilities were occurred in weight as well as number of fruit bodies among all treated substrates in all three flushes.

Table 23.1: Effect of Different Sterilization Methods of Substrate on Time Taken for Different Growth Stages of *Pleurotus florida*

Treatments	Spawn Run Period (in days)	Initiation for Fruit Body (in days)	Time Taken for Harvesting of Different Flushes (in days)			Total Period of Crop (in days)
			1st Flush	2nd Flush	3rd Flush	
Autoclaving	10	12	16–21	26–38	50–54	54
Formalin (fumigation)	11	13	17–27	33–46	52–60	60
Bavistin (0.2% solution)	12	14	17–25	31–38	51–60	60
Radiation (U.V. light)	12	14	18–23	27–36	47–57	57
Solarization	11	13	17–23	28–38	52–57	57
Sun-lighting	18	20	24–28	38–47	57–60	60
Control (plain water)	20	25	29–34	53–57	–	57

Table 23.2: Effect of Different Sterilization Methods of Substrate on the Weight and Number of Fruit Bodies of *Pleurotus florida*

Treatments	Weight (gms) and Number of Fruit Bodies of Different Flushes															
	1st Flush				2nd Flush				3rd Flush				All Flushes			
	Max. Wt.	Av. Wt.	Min. Wt.	No	Max. Wt.	Av. Wt.	Min. Wt.	No	Max. Wt.	Av. Wt.	Min. Wt.	No	Max. Wt.	Av. Wt.	Min. Wt.	No
Autoclaving	22	8.81	2	16	12	3.66	1	27	6	2.60	1	5	22	5.27	1	48
Formalin (fumigation)	20	7.36	2	11	16	4.80	2	30	9	4.83	2	6	20	5.40	2	47
Bavistin (0.2% solution)	35	11.55	3	9	16	5.58	2	12	7	4.60	2	5	35	7.46	2	26
Radiation (U.V. light)	32	8.53	3	15	10	5.00	2	13	9	4.13	2	8	32	6.27	2	36
Solarization	19	10.2	2	13	14	4.94	2	19	8	2.66	1	6	19	6.39	1	38
Sun-lighting	13	6.16	2	12	9	4.45	2	11	10	2.14	1	7	13	4.60	1	30
Control (plain water)	9	4.50	2	8	11	5.14	2	7	–	–	–	–	9	4.80	2	15

A perusal of data (Table 23.3) indicated that best (254 g) and 2nd best (253 g) yields were obtained from formalin and autoclave treated sustrates respectively and this was followed by solarized straw (243 g). Another good yield (226 g) was obtained from radiated substrate and it was significantly different from bavistin (194 g) and sun-lighting (138 g) treated substrate and plain water treatment gave less 72 g yield.

From the result it was deduced that the 1st flush gave much higher yield than the 2nd and 3rd flushes except formalin treated substrate which yield (144 g) maximum in the 2nd flush than 1st flush (81 g) and 3rd flush (29 g). Suitability of sterilized substrate by means of chemical or autoclaving was recommended by Zadrazil and Brunnert, 1980; Vijay and Sohi, 1987; Bahukhandi, 1990; Singh and Dwivedi, 1991; Jadhav and Jagtop, 1991; Rani and Das, 1998; Khan *et al.*, 2002 and Dogan and Peksen,

2003. Out of six fertilization methods of substrate, the average yield obtained ranged 226–254 g per bag from treated substrate except bavistin and sun-lighting treated substrates. Solarization and radiation techniques of substrate sterilization investigated have shown similar response towards the substrate tested. The fruit bodies raised by these methods are safe for human consumption in comparison to chemically sterilized substrate. These new sterilization techniques are very simple and cheapest for mushroom growers and it is also safe to health hazards and environmental pollution. In future, these new sterilization techniques of substrate will be very successfully adapted by growers in oyster and milky mushroom cultivation.

Table 23.3: Effect of Different Sterilization Methods of Substrate on Yield Potential of *Pleurotus florida*

Treatments	Yield of Sporophores (in gms)			Total Mushroom Yield (in gsm)	Biological Efficiency
	1st Flush	2nd Flush	3rd Flush		
Autoclaving	141	99	13	253	72.28%
Formalin (fumigation)	81	144	29	254	72.57%
Bavistin (0.2% solution)	104	67	23	194	55.43%
Radiation (U.V. light)	128	65	33	226	64.57%
Solarization	133	94	16	243	69.43%
Sun-lighting	74	49	15	138	39.43%
Control (plain water)	36	36	–	72	20.57%
S.D.	1.74	3.46	1.18	(0.12)	2.61
CD: 0.05	4.96	9.86	3.36	(0.34)	7.44
0.01	7.76	15.43	5.26	(0.53)	11.64

Data presented is mean of three replications.

Figures in parenthesis are obtained from transformed data.

The biological efficiency was calculated as per the formula given by Chang et al., 1981

$$BE = \frac{\text{Fresh Weight of Mushroom / Bag}}{\text{Dry Weight of Substrate / Bag}} \times 100$$

References

Bahokhandi, D. (1990). Effect of various treatments on paddy straw on yield of some cultivated spp. of *Pleurotus. Indian Phytopath.*, 43: 471–472.

Chang, S.T., Lao, O.W. and Cho, K.Y. (1981). The cultivation and nutritional value of *Pleurotus sajor-caju. European J. of Applied Microb. and Biotechnol.*, 12: 58–62.

Dogan, H. and Peksen, A. (2003). The effects of substrates prepared by tea waste and disinfection methods on the yield and quality of *Pleurotus sajor-caju. ZiRailt Fakültesi Dergisi*, 18: 39–48.

Jadhav, S.N. and Jagtap, V.D. (1991). Chemical pasteurization of wheat straw used for growing oyster mushroom. *Proceeding of the National Symposium on Mushroom*, pp. 122–124.

Jonkova, A. (1971). Intensivni a extensivni zpusob pestorani hlivy ustricne. *Mykol.*, 8: 53–54.

Khan, S.M., Javed, N., Khan, S.M., Chohan, M.A. (2002). Effect of different sterilization methods on the production of oyster mushroom (*Pleurotus ostreatus*) on different substrates. *Proceedings of 3rd National Conference of Plant Pathology*, NARC, Islamabad, 1–3 October, pp. 159– 160.

Pani, B.K. and Das, S.R. (1998). Effect of pre-treatment of substrate on the yield of oyster mushroom *Pleurotus sajor-caju* (Fr) Singer. *J. Mycopathological Research*, 36: 113–114.

Patil, B.D. and Jadhv, S.W. (1989). Studies on cultivation of *Pleurotus sajor-caju* on different substrates. *Journal of Maharashtra Agriculture University*, 14: 156–158.

Patrabansh, S. and Madan, M. (1995). The microbial conversion of different agricultural residues and its, biological efficiency. *Acta Biotechnol.*, 15: 131–135

Sharma, V.P. and Jandaik, G.L. (1991). Yield response of *Pleurotus ostreatus* and *P. florida* to spent straw and spent compost. *Proceeding of the National Symposium on Mushroom*, pp. 124–127.

Singh, R.P. and. Dwivedi, R.R. (1991). Standardization of substrate for production of *Pleurotus sajor-caju* (Fr) Singer. *Proceeding of the National Symposium on Mushroom*, pp. 102–103.

Vijay, B. and Sohi, H.S. (1987). Cultivation of oyster mushroom *Pleurofus sajor-caju* (Fr) Singer on chemically sterilized wheat straw. *Mush. J. Tropic.*, 7: 67–68.

Zadrazil, F. and Schneidereit, M. (1972). Die grundlagen fur die lnkulturnahone einer nicht Cultivleten *Pleurotus. Art. Champignen*, 12: 25–32.

Zadrazil, F. (1973). Anbauverfahrn fur *Pleurotus florida. Fovose Champignon*, 13: 3–4.

Zadrazil, F. and Bronnert, H. (1980). The influence of ammonium nitrate supplementation on degradation and *in vitro* digestibility of straw colonized by higher fungi. *European Journal of Microbiology and Biotechnology*, 2: 37–44.

©R.C. Ram & Asha Sinha (*Editors*)
***Microbial Diversity: Modern Trends*, 2007**
Daya Publishing House, New Delhi **291–294**

Chapter 24

Comparative Efficiency of Various Substrates and Supplements on Growth Behaviour and Yield Potential of *Pleurotus florida*

K.K. Gautam* and R.C. Ram**

***Department of Botany, **Department of Mycology and Plant Pathology, Institute of Agricultural Sciences, Banaras Hindu University, Varanasi – 221005 (U.P.)**

ABSTRACT

An experiment was conducted to study the comparative efficiency of various substrates *viz.*, wheat straw (*Triticum aestivum* L.), paddy straw (*Oryza sativa* L.), corn stalk (*Zea mays* L.) and fenugreek (*Trigonella foenumgraecum* L.) and different supplements such as wheat bran, rice bran, neem cake and fenugreek bran on growth behaviour and yield potential of oyster mushroom (*Pleurotus florida*). In these investigations, higher mushroom yield with biological efficiency (BE) of 86 per cent was obtained from wheat straw followed by paddy straw (64 per cent biological efficiency). Wheat straw was further used as a substrate for different supplements observation. Wheat bran was recorded significantly superior supplement than other supplements and its. biological efficiency was 96 per cent. The spawn running, premordia initiation and harvesting of first flush were also early in wheat bran supplemented substrate where it completed in the 13 days, 17 days and 20 days respectively.

Keywords: Substrates, Supplements, Yield and Pleurotus florida.

Introduction

Oyster mushroom (*Pleurotus florida*) a delicious edible mushroom is gaining much important due to its short cropping period, lowest cost of cultivation and high yield potential. They can be grown on a wide variety of low value lignocellulosic wastes. The large quantities of which are generated annually from agricultural, forestry and food processing activities. The unexploited crops residues or plant wastes are increasing pressure on environmental pollution and create health hazards and serve as constant sources for survival of pest and pathogen of important crops. Even if one percent of agricultural wastes are used to produce mushroom, India will become a major mushroom producing country in the world (Tiwari and Pandey 2002). Mushroom cultivation therefore represents a prime example of the direct of low-grade residues into a value added commodity. Present experiment was conducted at the Varanasi region to comparative efficiency of various locally available plant wastes and supplements for cultivation of *Pleurotus florida.*

Materials and Methods

The pure mushroom culture of *Pleurotus florida* was obtained from N.R.C.M. Solan (H.P.) and maintained on potato dextrose medium which was used for preparing wheat grain spawn. The present study was carried out with different substrates *viz.*, wheat straw (*Triticum aestivum* L.), paddy straw (*Oryza sativa* L.) corn stalk (*Zea mays* L.) and fenugreek straw (*Trigonella foenumgrecum* L.) and different supplements wheat bran, rice bran, neem cake and fenugreek bran on growth behaviour and yield potential of *Pleurotus florida* which are easily available in Varanasi region of Eastern Uttar Pradesh.

The dried and chopped wheat straw, paddy straw, maize stalk and fenugreek straw were soaked in water for 14 hours and then drain off excess water. These soaked materials are pasteurized by autoclave at 65°C for 45 minutes. Well prepared wheat grain spawn (3 per cent moist weight basis) was mixed with moistened substrates and filled in polythene bags (15″ × 10″). The mouth of each bag was tied with a rubber band and 8–10 holes made in each bag for proper aeration. The polythene bags were removed gently from the substrate after spawn running and these compact masses of substrates were kept in glass house under dark condition at seasonal temperature and humidity ranging between 24–28 °C and 85–90 per cent respectively during spawn running and formation of fruiting bodies.

Another experiment was conducted to comparative efficiency of different supplements such as wheat bran, rice bran, neem cake and fenugreek bran on same mushroom. The sun dried powdered of above supplements were sterilized by autoclave at 15 lb for 30 minutes. Each supplement @ 1.5 per cent dry weight basis was mixed with moistened and pasteurized wheat straw before spawning. The experiment was conducted in a completely randomized block design (CRBD) with five replications for each treatment. Average value of observation with respect to period required for spawn running, premordia initiation, harvesting of first flush and total yield were recorded. Biological efficiency of mushroom on fresh weight basis was calculated as per the formula given by Chang and Miles (1989).

$$\text{Biological Efficiency (B. E.)} = \frac{\text{Fresh Weight of Mushroom per Bag}}{\text{Dry Weight of Substrate per Bag}} \times 100$$

Results and Discussion

Results obtained from the present study are shown in Table 24.1. It is evident that utilized different substrates for growth bahaviour and yield potential of *Pleurotus florida.* Wheat straw was produced significantly highest yield 430 gm/bag than the other substrates and was followed by paddy straw 320 gm/bag. The days required for spawn run period, premordia initiation period and harvesting of first flush were also comparatively less in wheat straw substrate than required by other substrates.

Wheat straw has been found the best substrate for successful cultivation of oyster mushroom. This result was correlated to other workers, Bano *et al.* (1979), Bahukhandi (1990), Savalgi *et al.* (1994) and Hazarika (1998) under different condition.

Table 24.1: Comparative Efficiency of Various Substrates on Growth Bahaviour and Yield of *Pleurotus florida*

Substrates	SRP (Days)	PIP (Days)	FHP (Days)	Total Yield (gms)	Biological Efficiency (%)
Wheat straw	17	21	24	430	86
Paddy straw	18	23	26	320	64
Corn stalk	22	27	30	310	62
Fenugreek stem	24	30	34	301.6	60.32
C.D. (0.05%)				3.86	

* Treatments are means of five replications. SRP: Spawn Run Period; PIP: Premordia Initiation Period; FHP: First Harvesting Period.

The result Table 24.2 indicated that the organic supplements increased the yield significantly. Wheat bran was significantly (B.E. 96 per cent) superior comparatively to other supplements. Wheat bran was also significantly reduced spawn run period by 4 days and hastened the premordia initiation and harvesting of first flush than the control (without supplement). This result is in conformity with the finding of Hazarika (1998). The wheat bran organic supplement has been reported to increase the yield of *Pleurotus florida*. However, wheat bran, a real waste, supplement has been reported earlier by Marimuthu at *al.* (1994). The result of present investigation wheat straw and wheat bran found to be best substrate and supplement respectively for cultivation of oyster mushroom (*Pleurotus florida*).

Table 24.2: Comparative Efficiency of Various Supplements on Growth Behaviour and Yield Potential of *Pleurotus florida*

Substrates	SRP (Days)	PIP (Days)	FHP (Days)	Total Yield (gms)	Biological Efficiency (%)
Wheat bran	13	17	20	480	96
Rice bran	15	19	22	400	80
Neem cake	14	20	23	410	82
Fenugreek bran	15	20	23	350	70
Control	17	21	24	340	68
C.D. (0.05%)				6.50	

* Treatments are means of five replications. SRP: Spawn Run Period; PIP: Premordia Initiation Period; FHP: First Harvesting Period.

References

Bano, Z., Rajarathanam, S. and Nagarajan, N. (1979). *Indian Mushroom Science,* 10: 587–602.

Chang, S.T. and Miles, P.G. (1989). *Edible Mushroom and their Cultivation.* CRC Press, Boca Rattan, pp. 256–074.

Hazarika, D.K. (1998). *Mushroom Res.*, 7(1): 47–48

Marimuthu, T., Krishnamurthy, A.S. and Nalathambi, P. (1994). *Indian J. Mycology and Pl. Pathology*, 24: 103–106.

Savalgi, V. and Savalgi V.K. (1994). *Indian J. Mycology and Pl. Pathology*, 24(3): 190–191.

Tiwari, R.P. and Pandey, M. (2002). *Survey of Indian Agriculture*, 2002: 165–167.

©R.C. Ram & Asha Sinha (*Editors*)
***Microbial Diversity: Modern Trends*, 2007**
Daya Publishing House, New Delhi

295–307

Chapter 25

New Trends in Identification of VA Mycorrhizal Fungi

Sudhir Chandra, H.K. Kehri, Rollie Verma and Sudha Singh
Department of Botany, University of Allahabad, Allahabad - 211 002

Mycorrhiza research is relatively recent. Although, it was in 1885 that A.B. Frank made reference for the first time to the term *"Mycorrhiza"*, it took him nine more years to establish the symbiotic relationship between the fungus and its host plant. It took a much longer time to convince the rest of the scientific community of the validity of Frank's theory. Researches carried out between 1917 to 1950 confirmed, *"Mycorrhizal colonization is a beneficial symbiosis between tree and fungus"*. It is an unique partnership in nature, where the higher plants share their photosynthates with their fungal partners and provide them shelter, the fungal partners in return provide protection to the root system and assist it in performing its functions especially anchorage and absorption of nutrients / water from soil more efficiently.

The endomycorrhizal fungi especially vesicular arbuscular mycorrhizal fungi (VAM fungi) have gained importance as a biological tool in improving the performance of not, only forest trees but also a variety of other plants under a variety of habitats and conditions, only during past four decades.

Credit of identifying the VA mycorrhizal fungi as a biological tool in improving the performance of plants goes to Barbara Mosse (1959) who for the first time experimentally produced VA Mycorrhiza and confirmed its involvement in plant nutrition and water uptake. Since then a lot of information on VAM fungi and its importance as a biological tool in improving the nutrition and water uptake in plants and their resistance to various stresses and root pathogens has accumulated as a result of extensive researches throughout the globe. They have been reported to be associated with all groups of plants growing under all types of habitats. They have been found in, almost all the tropical soils, but land disturbances due to various reasons have been shown to disturb their native population. Attempts are being made at various research centers of the world to explore the possibility of ensuring the cultivation of different crops including legumes, cereals, plantation crops, industrial crops, horticultural crops as well as ornamentals and medicinal plants less expensive thus more profitable employing efficient strains of VA mycorrhizal fungi. They are also being exploited as biocontrol agents against the root pathogens of serious nature. In addition, attempts are being made to gather information about their ultra structure, cytology, genetics, physiology as well as their relationship with the host and their specificity.

At a time when economic and environmental considerations are restricting the use of chemical fertilizers (high energy production costs, and soil and water polluting effects), the interest attached to mycorrhiza as an alternative solution appears legitimate.

VAM Taxonomy and Morphology

Dangeard (1900) was the first to describe a vesicular-arbuscular mycorrhizal fungus *"Rhizophagus populinus"*. Peyronel (1924; 1937) accommodated VAM fungi under *Endogone* Link belonging to family Endogonaceae, (an ancient group which has Devonian fossil representatives) in Mucorales of Zygomycetes. VA Mycorrhizal fungi being obligate biotrophs do not grow on synthetic media and hence are classified according to morphological characteristics of the spores formed in the soil. Schenck has described One Hundred and Twenty species and Perez (1987) and they are grouped in a single family, the Endogonaceae. Until 1990 VA mycorrhizal fungi continued their position under family Endogonaceae in the order Endogonales of the class Zygomycetes and the division Zygomycotina (Benjamin, 1979; Trappe, 1987; Morton, 1988). They were accommodated within six genera of the family *viz., Acaulospora* Gerd. and Trappe em. Berch, *Entrophospora* Ames and Sehn, *Gigaspora* Gerd, and Trappe, *Glomus* Tul. and Tul., *Sclerocysiis* Berk and Br. and *Scutellospora* Walk and Sand. The genus *Endogone* was retained in the family *Endogonaceae* but it included saprophytic members occasionally forming ectomycorrhiza. (Chu-Chou and Grace, 1979). Perozynski and Dalpe (1989) separated *Glomus* and *Sclerocystis* from others and accommodated them in a new family Glomaceae under Endogonales. On the basis of morphological, phylogenetic and physiological considerations, Morton and Benny (1990) proposed that VAM fungi should be termed as AM fungi since many mycorrhizal fungi do not produce intramatrical vesicles. They created a new order Glomales to include all soil fungi, which produce arbuscules and form obligate, mutualistic associations with the roots of terrestrial plants. The order Glomales was divided into two suborders. The suborder Glomineae included members forming chlamydospores, arbuscules as well as vesicles. It accommodated Glomaceae (chlamydospores on a fertile hypha) with *Glomus* (chlamydospores singly) and *Sclerocystis* (chlamydospores in sporocarps) and Acaulosporaceae (chlamydospores from a sporiferous saccule) with *Acaulospora* (lateral chlamydospores) and *Entrophospora* (intercalary chlamydospores). The suborder Gigasporineae included members forming azygospores and only arbuscules. It accommodated only Gigasporaceae (azygospores on sporogenous cells) with *Gigaspora* (lateral azygospores) and *ScuteJlospora* (terminal azygospores). The order Endogonales was retained but it included only *Endogone* under the family Endogonaceae.

The main features of VA Mycorrhizal fungi are summarized by Trappe and Schenck (1982) and Hall (1984). Hall and Abbott (1983) compiled photographic transparencies illustrating diagnostic features of the Endgonaceae. Walker (1983) has suggested a standardized terminology for use in description of endogonaceous spore walls involving a stylized graphic representation of the wall structure ("murograph"). This concept is recommended for future descriptions of new species. Berch (1986) reviewed the taxonomy, specificity, fossil record and phylogeny of Endogonaceous fungi. Schenck and Perez (1987) published a manual for the identification of VA mycorrhizal fungi.

Identification of VAM Fungi

Vesicular-arbuscular mycorrhizal (VAM) fungi are difficult to identify in the absence of knowledge on their complete life cycle. These fungi have also not been cultured on artificial media. The identification of VAM fungi is therefore done on the basis of morphology of azygospores/chlamydospores/sporocarps and hyphae attached to these structures. Precise identification of spores from the field is difficult as morphology may vary. This is major limitation for ecological studies since sporulation depends on environmental conditions and on host and fungal genotype.

The various techniques, which have been employed so far to identify these fungi includes, serological techniques such as fluorescent antibody technique (FAT), radioimmunoassay (RIA),and enzyme-linked immunosorbent assay (ELISA) (Aldwell, 1983; Aldwell *et al.*, 1985), selective enzyme staining following gel electrophoresis, and morphological technique (Walker, 1983).

Morphological Techniques

Some important morphological characteristics used in the identification of taxa are:

Spore

Spores have been useful in identifying taxa within a genus as they are tangible structures with unique morphological characters (Gerdemann and Trappe, 1974; Hall, 1977; Schenck and Smith, 1982; Schenck *et al.*, 1984; Trappe and Schenck, 1982; Morton, 1985; Walker,1983).

Some of the main morphological characteristics of spore which are studied for identification of taxa are:

Colour

Spore colour is a variable character and is influenced by the manner in which the light passes from the object to the observer (Morton, 1988). In most taxa it varies from hyaline to black, with some shades of yellow to brown. Some of the colours generally observed are hyaline, yellow, reddish-brown, orange, brown and black. The young spores in many species may be colourless, but it is not a generalized character.

Size

The size of spore is also a variable character and is influenced by the developmental stage of mycorrhizal fungi. It is generally reported as a range from smallest to largest, with rare extremes noted in parantheses.

Shape

The shape of the spores may be globose, sub-globose, ovoid, pear-shaped, ellipsoid, obovoid, reniform or irregularly elongated. The variations in shape generally depend on the substrate in which the spores are produced.

Content

The cytoplasm within the spores may appear reticulate in a polygonal pattern or may be vacuolated because of the presence of many lipid droplets (oil globules) of variable size. Cytoplasmic content of spores may impart colour to the spores.

Wall Structure

Since 1974, the importance of spore wall characteristics in the taxonomy of Glomaceae has been recognized and these criteria are of major significance in separating new species from the described ones. Studies at ultrastructural level have confirmed that the spore wall is actually a reliable criterion for identifying V AM spores (Bonfonte- Fasolo and Schubert, 1987). Walker (1983), realizing the increasing importance of wall characteristics, standardized the wall terminology and also suggested a system of stylized graphic representation–the MUROGRAPH, to illustrate the position of different wall layers in a composite spore wall.

The different terminologies, which are now being used in describing spore wall characteristics, are:

Wall Group

It is an aggregation of wall layers, which remain in close proximity to each other even after a spore, is crushed

Wall Kinds

The seven kinds of wall layers described so far are mainly distinguished on the basis of their morphological features and their reaction to certain chemicals such as polyvinyl-alcohol-lactphenol, Melzer's reagent, etc. The wall layers described so far are:

Evanescent (E)

It generally a slough off with age and therefore, it is usually absent in mature spores.

Unit (U)

It is a non-laminated rigid layer, which fractures easily, when pressure is applied to a spore.

Laminated (L)

It is a rigid layer made up of fused laminae, which are formed successively as the spore matures.

Membranous (M)

It is a thin, flexible layer that often wrinkles and collapses in spores mounted in hypertonic solutions. It is usually present as the innermost layer in a composite spore wall.

Coriaceous (C)

It is also a flexible layer, but is generally thicker than the membranous layer.

Amorphous (A)

It is a formless (Schenck and Perez, 1988) or elastic (Walker, 1987) layer that differs from other outer wall layers in its differential reaction to different hypertonic solutions.

Expanding (X)

It is a hyaline unit layer that expands markedly (about 20–100 μm away from the inner layer), when exposed to acidic mountants.

Number, Width and Position of Wall Layer

The number, width and position of wall layer differ among species and therefore, they have been increasingly relied upon as taxonomic criterion.

Wall Layer Ornamentation

In certain species, spore wall layers may possess ornamentation on their surface, and this may also form an additional taxonomic criterion for identification of species.

Mode of Spore Germination

Very few germination studies in a relatively smaller number of species have been made. However, in recent years this character has gained an increasing taxonomic importance in separating genera and species.

1. Through the spore wall.
2. Through a "germination shield"
3. Through the old subtending hypha.

Subtending Hypha

The spore bearing hypha (subtending hypha) has been found to be of some taxonomic importance in identification of species. The shape of the subtending hypha, however, forms an important criterion for distinguishing genera belonging to the family Glomaceae and Acaulosporaceae from those of Gigasporaceae.

Occlusion of Spores

The pore connecting spore and hypha may be occluded by different means, which may be a septum formed within 50 µm from the point of attachment to the spore. The occlusion in few spores of certain species may be lacking arid therefore, the spore may remain in open connexion with subtending hypha.

Sporocarp

A sporocarp can be defined as a spore-bearing organ (Hawksworth *et al.*, 1983). Formation of sporocarp in taxa of Glomales is usually hypogeous. However, in certain species, epigeous sporocarps have also been reported. Colour of sporocarps varies considerably among species. Trappe (1982) noted that even the internal and surface colour of sporocarps may vary.

The peridium (a mass of interwoven hyphae) mayor may not be present around sporocarps. Peridium may be in the form of loosely interwoven hyphae compact interwoven hyphae or a patchy covering over the surface of sporocarps.

Auxiliary Cells

Taxa belonging to *Gigaspora* and *Scutellospora* form smooth to ornamented, "sac-like" structures, borne singly or in clusters on a fine coiled hyphae outside the host root. These structures have been variously termed as 'vesicles' (Nicolson and Gerdemann, 1968), 'soil-borne vesicles', 'extramatrical vesicles', ornamented vesicles' and "auxiliary cells". The term auxiliary cell is now a universally accepted terminology.

Anatomy of Mycorrhizal Fungi

Abbot and Robson (1981) suggested that distinct infection patterns in mycorrhiza could be helpful in identifying VAM fungi. On the basis of the assumption, the prominent infection pattern of each mycorrhizal fungus within the host root is independent of host morphology and nutrition. Abbot and Robson (1982) suggested the inclusion of certain key characteristics of mycorrhizal anatomy in the taxonomy of V AM fungi, such as:

1. Mode of branching of hyphae
2. Infection density (proportion of root infected at any point)

3. Projections, constrictions and swellings formed on internal hyphae
4. Pattern, shape and density of vesicles
5. Hyphal diameter (hyphae present in outer and inner cortex), and
6. Staining intensity of external and internal hyphae.

Thus, the morphological features of intraradical structures of VAM fungi may only be used as an additional taxonomic criterion for identification of species.

As early as 1963, Gerdemann and Nicolson distinguished members of Endogonaceae (now segregated into Endogonaceae and Glomaceae) on the basis of size, shape and wall structure of spores. Mosse and Bowen (1968) gave a key to nine spore types on the basis of colour and content. Gerdemann and Trappe (1974) in their monographic study of Endogonaceae described about 30 species of *Endogone, Glomus, Sclerocystis, Acaulospora* and *Gigaspora* and also gave key to the species for each genus. Hall and Fish (1979) employed computer techniques and prepared,a dichotomous key for many named and unnamed species. Trappe (1982) prepared a synoptic key for 78 species of *Glomus, Sclerocystis, Acaulospora* and *Gigaspora.* This key could be expanded and corrected without rewriting the whole key. However, as the number of new taxa is increasing with time, it is becoming more and more difficult to update periodically such keys. The criteria presently used for identification of species are not adequate to fix well-defined boundaries between intergrading taxa. Moreover, with the current status of taxonomic concept in Glomales, where phenotypic variations in mycorrhizal morphology within a. species have been reported, it is difficult to accommodate the range of variation in spore morphology in a key. Therefore, until a stable and a more systematic taxonomic system in Glomales is available, it is not possible to provide a well-equipped key for identification of taxa. Schenck and Perez (1988) prepared a guide to 116 species (known as INV AM species guide) in which they included a synopsis of the morphological characters of spores of 6 genera of VAM and AM fungi to facilitate preliminary identification of the unknown species. The preliminary identification then has to be confirmed with the original species description provided in the literature. They included in their guide spore characteristics such as colour, size, number of wall groups, wall width, wall kinds (in the form of Muronym) and the presence or absence of sporocarps.

Methods that can directly reveal the colonizing fungal taxa within or associated with a root will be of great importance in the elucidation of mycorrhizal biodiversity and ecology. A variety of techniques have therefore been developed to assist in this process. These include the anatomy of infections (Abbott, 1982), selective enzyme staining following gel lectrophoresis (Sen and Hepper, 1986), isozyme techniques, serological techniques, immunological probes, DNA library and fingerprinting and PCR techniques combined with RAPID/RFLP. Out of these, serological and molecular techniques are most potential and have been discussed here.

Serological Techniques

Serological techniques are dependent on the production of highly specific antibodies in animals when they are inoculated with the chemicals contained in or produced by microorganisms (antigens). If the antibody and the antigen that elicited its formation are incubated together they will react. A number of tests have been developed that use this to detect the presence of either the antibody or the antigen. Of these tests, fluorescent antibody technique (FAT), enzyme linked immunosorbent assay (ELISA) and the radioimmunoassay (RIA), are the most sensitive.

Fluorescent Antibody Techniques (FAT)

The FAT has been applied to V AM fungi by Warner and Mosse (1979), Aldwell (1983), Kough *et al.* (1983) and Wilson *et al.* (1983). However, antisera of Warner and Mosse had poor specificity probably because the antigens were prepared from crushed whole spores containing much non-specific chitinaceous spore wall material and were administered by a single intra muscular injection without adjuvant and without boosters (Aldwell, 1983). In contrast Wilson *et al.* (1983) ground, centrifuged and sonicated *Gigaspora decipiens.* Hall and Abbott germ tube hyphae and auxiliary cells gathered from spores germinated between Milipore filters and then produced antisera by subcutaneous injection with Freunds's Complete Adjuvant (FCA) and with intravenous boosters. Using these techniques they were able to distinguish between *G. decipiens* and *Acaulospora laevis* Gerd and Trappe and other non-mycorrhizal mucoraceous fungi.

Aldwell (1983) also deyel oped antisera to mycorrhizal hyphae washed from the surface of roots, later these hyphae also formed the target antigens. Using these techniques he was able to use the FAT to distinguish amongst the genera *Glomus, Gigaspora, Acaulospora* and *Sclerocystis* and to separate these from the non-VAM fungi but was unable to distinguish amongst VAM fungal species within the same genus.

Enzyme-linked Immunosorbent Assay (ELISA)

The ELISA is now well established as a practical method for the qualitative and quantitative detection of hormones, drugs (Pal, 1978) and plant pathogens (Clark, 1981 a,b). The most significant characteristic of ELISA is their ability to detect antigens occurring in concentrations much lower than those that can be detected by the FAT and have sensitivities comparable with RIA. However, unlike the RIA, the ELISA employs stable reagents and requires less expensive equipment. Aldwell (1983) and Aldwell *et al.* (1983, 1985) therefore chose to apply the ELISA to VAM rather than RIA following Aldwell's failure to distinguish between species within a genus using the FAT.

With the ELISA and antisera prepared using techniques essentially similar to those used to produce antisera for the FAT, Aldwell *et al.* (1983, 1985). Obtained these results:

1. VAM fungal species within a genus could be distinguished.
2. There was little difference in the antigenic composition of hyphae washed from surface of roots when the host plant was changed.
3. Although hyphae washed from the surface of roots were some times contaminated with small quantities of root material, *Pseuqomonas* sp. and *Bacillus* sp., these did not elicit the production of detectable quantities of antibodies.
4. Although the antigenic composition of hyphae was affected by the way the hyphae were raised, *i.e.* normal pot culture, hydroponic sand pot cultures, nutrient film cultures, or germ tube hyphae, these differences were not as great as the differences between species within a genus.
5. There was little difference in the antigenic composition of hyphae from cultures of a species originally from the same source but held and subcultured in different geographical locations for ten years.

Serological Techniques for Identification of Mycorrhizal Fungi

Serological techniques, in particular the ELISA show considerable promise for use as taxonomic tools for the VAM fungi. Unfortunately Aldwell *et al.* (1983, 1985) only prepared antisera to six VAM fungal species and this would be insufficient to identify all VAM fungi. Consequently, considerably more antisera to a wider range of species would need to be raised before the routine use of the ELISA for identifying VAM fungi could be envisaged.

Molecular Techniques

As mycorrhizal fungi are usually identified on the basis of morphological characters shown by the fruit bodies, spores, vegetative mycelia *of* symbiotic structures, it is proposed by Lanfranco *et al.* (1995), that the development of molecular techniques provides a valuable alternative approach to identify mycorrhizal fungi, particularly when it is difficult to gather a sufficient amount of data on morphological features. Short arbitrary oligonucleotides were used as primers for the amplification of genomic DNA extracted from the spores of arbuscular fungi. The random amplified polymorphic DNA (RAPD) fingerprints showed banding patterns, which allowed differentiation between species or even isolates within Glomales. In order to identify mycorrhizal fungi during their symbiotic phase, a non-polymorphic RAPD band, identified as a marker for some isolates of *Glomus mosseae,* was purified from agarose gels and cloned in a blue script vector. The fragment was sequenced and specific primers (PO-M3) were designated for the mycorrhizal fungus, these primers specifically amplified the DNA le not only from *G. mosseae* spores but also from the roots of pea, clover, onion plants when they were colonized by *G. mosseae* spores.

A prerequisite for these techniques is the recovery of DNA from biological samples. DNA can be extracted from both mycorrhizal spores and mycorrhizal roots. The number of spores necessary to obtain amplifiable DNA has reduced from several thousand to single spores. Justin and Simon have been frequently successful in amplifying from single spores using a Chelax resin DNA extraction method based on Walsh *et al.* (1991).

Various buffers are used to extract DNA from mycorrhizal roots. They tend to contain a combination of hexadecyltrimethylammonium bromide (CTAB), sodium dodecyl sulphate (SDS), and B-mercaptoethanol or Proteins K and involve incubation at 50–65 °C for 30–60 minutes (Gardes *et al.*, 1991; Henrion *et al.*, 1992; Lanfranco *et al.*, 1994). Homogenization with liquid nitrogen prior to extraction is only really necessary if very small quantities of root material are used or quantification of DNA is to be performed (as other methods of homogenization are less standardized). Otherwise, homogenization within the extraction buffer will suffice. The root homogenate is usually incubated with phenol to remove protein. Nucleic acids are then retrieved by alcohol precipitation.

Genus-specific Primers and PCR

By coupling the sensitivity of the polymerase chain reaction and the specificity afforded-taxon-specific primers, a variety of samples can be analyzed including small amounts of colonized roots. Family-specific primers as well as generic primers are described and can be used to amplify small sub-unit rRNA fragments from endomycorrhizal fungi by polymerize chain reaction. The amplified products are then subjected to single-strand conformation polymorphism analysis to detect sequence differences. Among the advantages of this approach are the possibilities of directly identifying the fungi inside field-collected roots without having to rely on the fortuitous presence of spores. Primers specific to the order Glomales (VANSI) and to the three genera *Gigasporal Scutellospora* (VAGIGA), *Acaulospora* (VAACAU), and *Glomus* (VAGLO) have been used in conjunction with PCR (Simon *et al.*, 1992).

RAPD-DNA Analysis

Random Amplified Polymorphic (RAPD)-DNA analysis is a variation of standard PCR. It uses single, short, arbitrary primers, which anneal more or less frequently throughout the genome. Where the primers anneal in inverted orientation on opposite stands within approximately 4 KB of each other, amplifiable products can be produced. Usually, several products of varying size are formed, which can be visualized by agarose gel electrophoresis and ethidium bromide staining. DNA differences that disrupt or create primer-annealing sites will determine the pattern of products visualized. In this manner DNA polymorphism can be detected without specific nucleotide sequence information. Wyss and Bonfante (1993) have used RAPD analysis to show polymorphism between and within VAM fungal species. Lanfranco et al. (1994) isolated species-specific primers for use in standard PCR. In this case a RAPD band, specific for *Glomus mosseae* was identified and sequenced. This information was then used to design primers, which amplified the *G. mosseae* RAPD fragment alone. It was found that, using these primers, a fragment was only amplified from *G. mosseae.*

PCR-RFLP Technique

A possibility of identifying individual VAMF spores from the field on the basis of a molecular marker, namely the nuclear ribosomal DNA encoding the highly conserved 5.8 S rRNA with the two flanking internal transcribed spacers (ITS region) known to vary between species was explored by Sanders *et al.* (1995). A technique involving polymerase chain reaction followed by restriction fragment length polymorphism analysis (PCR-RFLP) was developed to amplify and characterize the ITS region from single VAMF spore. PCR reactions with extracts from single spores of three VAMF species raised under glasshouse conditions yielded reproducibly a single amplification product of the ITS region in sufficient amounts to allow cleavage with several restrictions enzymes. The size of the ITS region C.600 base pairs varied only slightly between species. Digestion of the PCR products with the restriction enzymes Hinfl and TaqIresulted in banding patterns that were reproducible, for different individual spores of a given.

Species but showed clear differences between the three species tested. The sum of the fragment sizes was sometimes greater than the size of the original PCR product *e.g.*, in *Glomus mosseae.* Clones of the amplification product from a single spore of this fungus were obtained and sequenced. This yielded two closely related but different sequences indicating that two different ITS regions co-existed in the spore. The RFLP pattern of the amplification product of the spore was a result of an amalgamation of these two sequences. Sanders *et al.* (1995) Applied this technique to the AMF spores collected from a species rich grassland. They sorted out the spores on the basis of morphological characters *e.g.* their colour, size and shape and then subjected to PCR-RFLP analysis. A group of *Glomus* spores yielded amplification products in the majority of cases was further investigated. PCR-RFLP analysis on 10 individual spores from the field produced 10 different patterns. Similar results were obtained with other groups of spores. The results suggest that the diversity in natural VAM communities and the genetic diversity within individual spores might be much greater than previously thought.

SSCP

The SSCP technique (single-stranded conformational polymorphism) has not been greatly exploited in the study of Mycorrhizae. It is most useful for rapidly screening amplified sequences from large number of isolates for variation (Simon *et al.,* 1003). As such, it is particularly suitable for assessing intra-specific variation. SSCP has been reported to be capable of detecting single-base substitution (Hayashi, 1991) in moderately-sized PCR products (up to 450 base pairs). A full description of this technique, as applied to AM fungi, is given by Simen (1994).

Epilogue

Molecular techniques are a major impetus in mycorrhizal research. They are tremendously powerful and offer huge advantages over conventional methods. While it has been impossible to describe all the techniques in detail, it is important that one sees that a number of them can be applied to several areas. The new information obtained by molecular techniques, especially in mycorrhizal biodiversity, must be founded in current taxonomy. However, it is likely that as the weight of molecular data rises, the current classification of many species and genera will have to be reviewed. It is also becoming apparent that researchers in this field must become conversant with molecular methods. It is only by doing so that invaluable information being made accessible by these methods will be accepted into current opinion.

References

Abbott, L.K. (1982). Comparative anatomy of vesicular-arbuscular mycorrhizas formed on subterranean clover. *Aust. J. Bot.*, 30: 485–500.

Abbott, L.K. and Robson, A.D. (1981). Infectivity and effectiveness of five endomycorrhizal fungi: competition with indigenous fungi in field soils. *Aust. J. Agric. Res.*, 32: 621–630.

Abbott, L.K. (1982). Comparative anatomy of vesicular-arbuscular mycorrhizas formed on subterranean clover. *Aust. J. Bot.*, 30: 485–500.

Abbott, L.K. and Robson, A.D. (1982). The role of vesicular-arbuscular mycorrhizal fungi in agriculture and the selection of fungi for inoculation. *Aust. J. Agric. Res.*, 33: 389–408.

Abbott, L.K. and Robson, A.D. (1984). Colonization of the root system of subterranean clover by three species of vesicular-arbuscular mycorrhizal fungi. *New Plytologist*, 96: 275–281.

Aldwell, F.E.B. (1983). Immunological techniques to identify VAM fungi. *Ph.D. Thesis*, Otago University.

Aldwell, F.E.B., Hall, I.R., and Smith, J.M.B. (1983). *Soil Biol. Biochem.*, 15: 377–378. c.f. A review of serological techniques for the identification of mycorrhizal fungi. (Eds.) F.E.B. Aldwell and I.R. Hall). In: *Mycorrhiza News* 3 (3) (1991), TERI, New Delhi.

Aldwell, F.E.B., Hall, I.R. and Smith, J.M.B. (1985). Enzyme-linked immunosorbent assay as an aid to taxonomy of the Endogonaceae. *Trans. Br. Mycol. Soc.*, 84: 399–402.

Benjamin, R.K. and Mehrotra, B.S. (1963). Obligate azygospore formation in two species of *Mucor* (Mucorales). *Aliso*, 5: 235–245.

Benjamin, R.K. (1979). Zygomycetes and their spores. In: *The Wholefungus*, (Ed.) B. Kendrick. National Museums of Canada, Ottawa, Chapter 23.

Berch, S.M., Gamiet, S. and Deom, E. (1988). Mycorrhizal status of some plants of South Western British Columbia. *Can J. Bot.*, 68(10): 1924–1928.

Bonfnte-Fasolo, P. and Schubert, A. (1987). Spore wall architecture of *Glomus* spp. *Can. J. Bot.*, 65: 539–546.

Chu-Chou, M. and Grace, L.G. (1979). Endogone Flammicorona as a mycorrhiza symbiont of Douglas fir in New Zealand. *N.Z. J. For. Sci.*, 9: 344–347.

Clark, M.F. (1981). In: *Immunoassays for the Eighties*, (Ed.) A. Voller, Lancaster, M.Y.P. Dangeard, P.A. (1900). Le "Rhizophagus populinus" Dangeard, *Botaniste*, 7: 285.

Gardes, M., Mueller, G.M., Fortin, J.A. and Kropp, B.R. (1991). Mitochondrial DNA polymorphism in *Laccaria bicolor, L. laccata, L. proxima* and *L. amythystina. Mycological Research*, 95(2): 206–216.

Gerdemann, J.W. and Trappe, J.M. (1974). Endogonaceae in the Pacific Northwest. *Mycologia Mem.*, 5: 1–76.

Hall, I.R. (1977). Species of mycorrhizal infections of New Zealand (Endogonaceae). *Trans. Br. Mycol. Soc.*, 68: 341–356.

Hall, I.R. and Fish, B.J. (1979). A key to the Endogonaceae. *Trans. Br. Mycol. Soc.*, 73: 261–270.

Hall, I.R. and Abbott, L.K. (1983). Some Endogonaceae from southwestern Australia. *Trans. Br. Mycol. Soc.*, 83: 203–208.

Hall, I.R. (1984). Taxonomy of VA mycorrhizal fungi. In: *VA Mycorrhiza*, (Eds.) C.L. Powell and Bagyaraj. CRC Press, Boca Raton, pp. 57–94.

Hawksworth, D.L., Sutton, B.C. and Ainsworth, G.C. (1983). *Ainsworth and Bisby's Dictionary of the Fungi*, 7th Ed. Commonwealth Agricultural Bureaux, England, pp. 445.

Hayashi, K. (1991). PCR-SSCP: A simple and sensitive method for detection of mutations in the genomic DNA. *PCR Methods and Applications*, 1: 34–38.

Henrion, B., Le, Tacon F. and Martin, F.(1992). Rapid identification of genetic variation of ectomycorrhizal fungi by amplification of ribosomal RNA genes. *New Phytologist*, 122: 289–298.

Kough, J., Malajczuk, N. and Linderman, R.C. (1983). *New Phytol.*, 94: 57–62.

Koske, R.E., Miller, D.D. and Walker, C. (1983). *Gigaspora reticulata:* A newly described endomycorrhizal fungus from New England. *Mycotaxon.*, 16: 429–435.

Koske, R.E. (1984). Spores of VAM fungi inside spores of VAM fungi. *Mycologia*, 76: 853–862.

Koske, R.E. (1985). *Glomus aggregatum* emended: A distinct taxon in the *Glomus fasciculatum* complex. *Mycologia*, 77: 619–630.

Lanfranco, L., Wyss, P., Marzachi, C. and Bonfante, P. (1994). Generation of RAPD–PCR primers for the identification of isolates of *Glomus mosseae*, an arbuscular mycorrhizal fungus. *Molecular Ecology.*

Mortan, J.B. (1988). Taxonomy of VA mycorrhizal fungi: Classification, nomenclature and identification. *Mycotaxon.*, 32: 267–327.

Mortan, J.B. and Benny, G.L. (1990). Revised classification of arbuscular mycorrhizal fungi (Zygomycetes): A new order, Glomales, two new sub order, Glominae and Gigasporinae and two new families, Acaulosporaceae and Gigasporaceae with an emendation of Glomaceae. *Mycotaxon.*, 37: 471–491.

Morton, J.B., Bentivenga, S.P. and Bever, J.D. (1994). Discovery, measurement and interpretation of diversity in symbiotic endomycorrhizal fungi (Glomales, Zygomycetes). *Can. J. Bot.*

Mosse, B. and Bowen, G.D. (1968). A key to the recognition of some *Endogone* spore types. *Trans. Br. Mycol. Soc.*, 51: 469–483.

Nicolson, T.H. and Gerdemann, J.W. (1968). Mycorrhizal *Endogone* species. *Mycologia*, 60: 313–325.

Pal, S.B. (1978) (Ed). *Enzyme-Labelled Immunoassay of Hormones and Drugs.* Walter de Guyter. Berlin.

Peyronel, B. (1924). Specifie di Endogone productrici micorrizae endotrofische. *Ball. Staz. Patol. Veg. Rome*, 5: 73–75.

Peyronel, B. (1937). Le Endogone qualiproductrici di micorrizae endotrofische nelle fanerogame alpestri. *Nuova G. Bot. Ital.*, 44: 584–586.

Pirozynski, K.A and Dalpe, Y. (1989). Geological history of the Glomaceae with particular reference to mycrorrhizal symbiosis. *Symbiosis*, 7: 1–36.

Sanders, R.I., Alt, M., Groppe, K., Holler, T. and Wiemken, A. (1995). Identification of ribosomal DNA polymorphism among and within spores of the Glomales: Application to studies on the genetic diversity of arbuscular mycorrhizal fungal communities. *New Phytologist*, 130(3): 419–427.

Schenck, N.C. and Smith, G.S. (1982). Additional new and unreported species of mycorrhizal fungi (Endogonaceae) from *Florida. Mycologia*, 74: 77–92.

Schenck, N.C., Spain, J.L., Sieverding, E. and Howeler, R.H. (1984). Several new and unreported vesicular–arbuscular mycorrhizal fungi (Endogonaceae) from Columbia. *Mycologia*, pp. 685–699.

Schenck, N.C. and Perez, Y. (1988). *Manual for the Identification of VA Mycorrhizal Fungi.* 2nd Edn. INVAM, Gainesvile, Florida, USA, pp. 245.

Schenck, N.C. and Perez, Y. (1990). *Manual for the Identification of VA Mycorrhizal Fungi*, 3rd Edn. INVAM, Gainesville, Florida, University of Florida.

Sen, R. and Hepper, C.M. (1986). *Soil. Biol. Biochem.*, 18: 29–34. c.f. A review of serological techniques for the identification of mycorrhizal fungi. (Eds.) F.E.H. Aldwell and I.R. Hall. *Mycorrhiza News*, 3(3), 1991.

Simon, L., Lalonde, M., and Hruns, T.D. (1992). Specific amplification of 18S fungal ribosomal genes from vesicular-arbuscular endomycorrhizal fungi colonizing roots. *Applied and Environmental Microbiology*, 58(1): 291–295.

Simon, L., Levesque, R. C. and Lalonde, M. (1992). Rapid quantitation by PCR of endomycorrhizal fungi colonizing roots. *PCR Methods and Applications*, 2. 76–80.

Simon, L., Levesque, R.C., and Lalonde, M. (1993). Identification of endo-mycorrhizal fungi colonizing roots by fluorescent single-strand conformational polymorphism-polymerase chain reaction. *Applied and Environmental Microbiology*, 59(12): 4211–4215.

Simon, L. (1994). Single stranded conformational polymorphism (SSCP) for the identification of endomycorrhizal fungi. In: *Methods in Molecular Biology*, (Ed.) J.P. Clapp. Humana Press, New Jersey, USA.

Trappe, J.M. (1982). Synoptic keys to the genera and species of Zygomycetous mycorrhizal fungi. *Phytopath.*, 72: 1102–1108.

Trappe, J.M. and Schenck, N.C. (1982). Taxonomy of the fungi forming endomycorrhizae. In: *Methods and Principles of Mycorrhizal Research*, (Ed.) N.C. Schenck). *Am. Phytopath. Soc.*, St. Paul, pp. 1–9.

Trappe, J.M. (1987). Phylogenetic and ecological aspects of mycotrophy in the Magnoliophyta (Angiosperms) from an evolutionary stand point. In: *Ecophysiology of Vesicular Arbuscular Mycorrhizal Plants*, (Ed.) G. Safir. C.R.C. Press, Boca Raton, Fl., pp. 5–25.

Walker, C. (1983). Taxonomic concepts in the Endogonaceae: Spore wall characteristics in species descriptions. *Mycotaxon.*, 18: 443–445.

Walsh, P.S., Metzger, D.R., Higuchi, R. (1991). Chelex 100 as a medium for simple extraction of DNA for PCR-based typing from forensic material. *Biotechniques*, 10(4): 506–513.

Warner, A. and Mosse, B. (1978). In: Rothamsted Report for 1977. Harpenden, Rothamsted Experimental Station, pp. 239.

Wilson, J.M., Trinick, M.J. and Parker, C.A. (1983). *Soil Biol. Biochem.*, 15: 439–445. c.f. A review of serological techniques for the identification of mycorrhizal fungi, (Eds.) F.E.B. Aldwell and I.R. Hall. *Mycorrhiza News*, 3: 1991.

Wyss, P. and Bonfante, P. (1993). Amplification of genomic DNA of arbuscular-mycorrhizal (AM) fungi by PCR using short arbitrary primers. *Mycological Research*, 95(11): 1351–1357.

Subject Index

Author Index

9 789359 191577

Printed at Repro India Ltd.